Separation Techniques
in
Nuclear Waste Management

Edited by

Thomas E. Carleson, Ph.D.
Department of Chemical Engineering
University of Idaho
Moscow, Idaho

Nathan A. Chipman, Ph.D.
Lockheed Martin Idaho Technologies
Idaho National Engineering Laboratory
Idaho Falls, Idaho

Chien M. Wai, Ph.D.
Department of Chemistry
University of Idaho
Moscow, Idaho

CRC Press
Taylor & Francis Group
Boca Raton London New York

CRC Press is an imprint of the
Taylor & Francis Group, an **informa** business

First published 1996 by CRC Press
Taylor & Francis Group
6000 Broken Sound Parkway NW, Suite 300
Boca Raton, FL 33487-2742

Reissued 2018 by CRC Press

A Library of Congress record exists under LC control number: 95010771

Publisher's Note
The publisher has gone to great lengths to ensure the quality of this reprint but points out that some imperfections in the original copies may be apparent.

Disclaimer
The publisher has made every effort to trace copyright holders and welcomes correspondence from those they have been unable to contact.

ISBN 13: 978-1-138-50567-4 (hbk)
ISBN 13: 978-1-138-56182-3 (pbk)
ISBN 13: 978-0-203-71038-8 (ebk)

Visit the Taylor & Francis Web site at http://www.taylorandfrancis.com and the CRC Press Web site at http://www.crcpress.com

PREFACE

This book is intended to be a reference work for scientists and engineers interested in a single source containing summaries and examples of separation techniques for nuclear waste management. The book is divided into four sections treating liquid wastes, solid wastes, gases and waste disposal, and summaries of current separation technologies in the world. Section I is subdivided into individual articles dealing with various techniques, such as liquid extraction, adsorption and ion exchange, electrochemical processes, and other processes. Emphasis has been put on liquid extraction (three chapters) since that is the technique of choice for transuranic separations. There is also a chapter on case studies for wastes at West Valley, New York and the Idaho National Engineering Laboratory, Idaho Falls, Idaho. Section II concerns solid wastes and contains articles on pyrochemical techniques of salts, pyrochemical techniques of oxides and metals, surface treatments, and physical techniques. Section III consists of one chapter on gaseous separations and process waste disposal. Often separation processes result in off-gases that must be treated. In this chapter calcination and various vitrification methods are also discussed. Section IV concerns current separation technologies and research in the United States, Western Europe, Eastern Europe and the Commonwealth of Independent States, and Asia.

An attempt has been made for the book to be comprehensive, but not all possible separation techniques have been covered. Authors from the various laboratories have written about techniques of interest to them and their laboratories. Most of the government laboratories in the United States dealing with significant amounts of radiological wastes were solicited for contributions. Articles by individuals from the Pacific Northwest Laboratory in Richland, Washington; the Idaho National Engineering Laboratory in Idaho Falls, Idaho; Los Alamos National Laboratory in Los Alamos, New Mexico; Lawrence Livermore National Laboratory in Livermore, California; and Argonne National Laboratory, in Argonne, Illinois were received and are included in the book. Other contributors in the United States include scientists and engineers from Florida State University and the University of Idaho. In addition, contributions were received from international scientists and engineers at British Nuclear Fuels Limited in England; the V.I. Vernadsky Institute of Geochemistry and Analytical Chemistry of the Russian Academy of Science, the Japan Atomic Energy Research Institute, Tsinghua University of the Peoples Republic of China, and the Korean Advanced Institute of Science and Technology. These last contributors certainly make this an international work.

The future of nuclear power in the United States remains uncertain. While other countries, such as France, China, and Japan, actively develop nuclear energy, it looks like only two new reactors will come on line in the U.S. before the end of this century. Certainly waste problems and safety issues are causes for concern. Analysis and evaluation of methods to handle the former problems were reasons for the writing and compilation of articles in this book. Over the last three decades there has been a considerable effort spent on resolving the problems of nuclear waste in this country. Some of the nations best scientists and engineers have developed nuclear power technology and continue now to address the stored wastes. Good engineering solutions to the waste problem are presented in this book. It is hoped that these methods will be further developed and applied to the current stored wastes. With this problem resolved perhaps the United States can then join other nations in the use of nuclear power as a significant energy source for its citizens.

This book is dedicated to the scientists and engineers working at the national laboratories.

Thomas E. Carleson
Nathan A. Chipman
Chen M. Wai

EDITORS

Thomas E. Carleson, Ph.D., P.E. (Idaho), is Professor of Chemical Engineering at the University of Idaho. Professor Carleson has also worked as a consultant for contractors at the Idaho National Engineering Laboratory (INEL) and the Pacific Northwest Laboratory over the last twenty years.

Professor Carleson obtained his Ph.D. degree in Chemical Engineering from the University of Washington in 1982. He also has degrees from the University of Idaho (M.S. Chemical Engineering, 1977) and Oregon State University (B.S. Chemistry, 1966). Professor Carleson has taught and conducted research at three universities in the Peoples Republic of China (1989-1990, 1992) and is currently an affiliate faculty member at the Chemical Engineering Institute of South China University of China in Guangzhou, China. He has also served on two process review boards for operations at the Idaho Chemical Processing Plant (at INEL).

Professor Carleson is a member of the American Institute of Chemical Engineers, the American Chemical Society, the Air and Waste Management Association, and the Idaho Academy of Science. He has presented papers at more than 20 national meetings and has given seminars on separations research at ten different universities in two countries. He has published approximately 20 research papers, six technical reports, and one book chapter. Professor Carleson has been the recipient of two Associated Western University fellowships in which he performed engineering assessments at the Idaho National Engineering Laboratory. Professor Carleson's current research interests include evaluation of methods for removal of heavy metals from acidic wastes, numerical remodeling of transport processes in droplets, supercritical fluid leaching and extraction, and measurements and modeling of emissions from combustion of alternate fuels.

Nathan A. Chipman, Ph.D., is Mission Acquisition Manager for Lockheed Martin Idaho Technologies in Idaho Falls, Idaho. He is a member of the Graduate Faculty of the University of Idaho where he also is an Adjunct Professor of Chemical Engineering. Dr. Chipman is the University of Idaho Chemical Engineering Curriculum Coordinator at the Idaho Falls Center for Higher Education where he has taught evening classes over the last fifteen years.

Dr. Chipman earned his Ph.D. degree in chemical engineering in 1993 from the University of Idaho. He also holds B.S. and M.S. degrees in chemical engineering and a Master of Arts in Teaching (College of Education) granted from that university.

He is a member of the American Instutute of Chemical Engineers and has served as the Chairman of its National Career Guidance Committee. He is also a member of the Idaho Academy of Science, the American Society for Engineering Education, and the American Nuclear Society. Dr. Chipman has worked for Department of Energy contractors at the Idaho National Engineering Laboratory for twenty-one years in the general areas of nuclear fuel reprocessing, nuclear waste management, and spent-nuclear fuel management and dispositioning. In positions both technical and managerial in nature he has authored sixty-five publications, presentations, and technical reports and has chaired seven technical sessions at national technical society meetings.

Chien M. Wai, Ph.D., is Professor of Chemistry at the University of Idaho. He received his Ph.D. degree from the University of California, Irvine in 1967, followed by two years postdoctoral work at the University of California, Los Angeles. He has been teaching and conducting research at the University of Idaho since 1969. Professor Wai was also a visiting scientist at Argonne National Laboratory in 1982, Battelle Pacific Northwest Laboratory in 1990, and at the Japan Atomic Energy Research Institute, Tokai-Mura, in 1991.

Professor Wai is a member of the American Chemical Society and the Idaho Academy of Science. He has presented more than fifty papers at the national and international meetings over the last thirty years and has published over 100 scientific papers in refereed journals. Professor Wai's current research interests include trace metal analysis, chemical speciation studies, separation of lanthanides and actinides with macrocyclic compounds, and supercritical fluid extraction of metal chelates.

CONTRIBUTORS

Mikhail K. Beklemishev
Department of Chemistry
University of Idaho
Moscow, Idaho

Richard D. Boardman
Lockheed Martin Idaho Technologies
Idaho Falls, Idaho

Lane A. Bray
Pacific Northwest Laboratory
Battelle Memorial Institute
Richland, Washington

Mark C. Bronson
Lawrence Livermore National Laboratory
Livermore, California

Mark F. Buehler
Pacific Northwest Laboratory
Battelle Memorial Institute
Richland, Washington

Thomas E. Carleson
Buchanan Engineering Laboratory
Department of Chemical Engineering
University of Idaho
Moscow, Idaho

Renato Chiarizia
Chemistry Division
Argonne National Laboratory
Argonne, Illinois

Nathan A. Chipman
Lockheed Martin Idaho Technologies
Idaho Falls, Idaho

Gregory R. Choppin
Department of Chemistry
Florida State University
Tallahassee, Florida

Jerry D. Christian
Lockheed Martin Idaho Technologies
Idaho Falls, Idaho

Niel Christiansen
Lockheed Martin Idaho Technologies
Idaho Falls, Idaho

Karen E. Dodson
Lawrence Livermore National Laboratory
Livermore, California

Harry Eccles
British National Fuels Limited
Springfields Works, Salwick
Preston, United Kingdom

John R. FitzPatrick
Los Alamos National Laboratories
Los Alamos, New Mexico

Robert G.G. Holmes
British Nuclear Fuels Limited
Springfields Works, Salwick
Preston, United Kingdom

E. Philip Horwitz
Chemistry Division
Argonne National Laboratory
Argonne, Illinois

Joon-Hyung Kim
Nuclear Environment Management
  Center
Korea Atomic Energy Research Institute
Taejeon, Korea

Dean E. Kurath
Process Technology and Engineering
  Systems
Pacific Northwest Laboratory
Battelle Memorial Institute
Richland, Washington

Jack Douglas Law
Lockheed Martin Idaho Technologies
Idaho Falls, Idaho

Igor A. Lebedev
V.J. Vernadsky Institute of Geochemistry
  and Analytical Chemistry
Russian Academy of Sciences
Moscow, Russia

Yadong Li
Department of Environmental
  Engineering
Tsinghua University
Beijing, People's Republic of China

Charles W. McPheeters
Argonne National Laboratory
Argonne, Illinois

James A. Murphy
Lockheed Martin Idaho Technologies
Idaho Falls, Idaho

Boris F. Myasoedov
V.J. Vernadsky Institute of Geochemistry
  and Analytical Chemistry
Russian Academy of Sciences
Moscow, Russia

Arlin L. Olson
Lockheed Martin Idaho Technologies
Idaho Falls, Idaho

Layne F. Pincock
Lockheed Martin Idaho Technologies
Idaho Falls, Idaho

Linfeng Rao
Pacific Northwest Laboratory
Battelle Memorial Institute
Richland, Washington

Jeffrey E. Surma
Pacific Northwest Laboratory
Battelle Memorial Institute
Richland, Washington

Thomas R. Thomas
Applied Engineering Development
  Laboratory
Lockheed Martin Idaho Technologies
Idaho Falls, Idaho

Chien M. Wai
Department of Chemistry
University of Idaho
Moscow, Idaho

Laura A. Worl
Los Alamos National Laboratory
Los Alamos, New Mexico

Zenko Yoshida
Advanced Science Research Center
Japan Atomic Energy Research Institute
Tokai-Mura, Japan

CONTENTS

Introduction

Historically, separation processes in the nuclear industry were centered on liquid–liquid extraction for the recovery of weapons-usable materials and highly enriched uranium for nuclear fuel, distillation for the recovery of heavy water and radioactive gases, gaseous and thermal diffusion for the isotopic separation of uranium and other materials, and chemisorption and filtration for off-gas and water cleanup and treatment.

With the end of the Cold War, the decreased demand for weapons-usable plutonium and uranium enrichment, and the cessation of nuclear fuel reprocessing for the recovery of highly enriched uranium, many of the facilities in the United States nuclear complex became no longer needed for their original missions. These facilities include those located at Richland, Washington (at the Hanford Reservation); Idaho Falls, Idaho (at the Idaho National Engineering Laboratory); Aiken, South Carolina (at the Savannah River site); Oak Ridge, Tennessee; West Valley, New York, and other sites across the U.S. Where alternative missions are not identified, cleanup activities have and will be initiated at these sites and others to prepare facilities and equipment for reuse, disposal, or other disposition. Many of these efforts will result in waste streams that need separation processes for cost-efficient and environmentally sound management. Some of the separation processes may be small when compared to the major processes of just a few years ago.

Although the United States has chosen to reduce its Department of Energy (DOE) nuclear complex, other nations continue to expand their nuclear facilities and capabilities. Thus, traditional separation processes remain important in these expanding areas, and traditional and nontraditional processes will also remain important as the United States utilizes this type of technology to close out many of its facilities and treat by-product effluents. Separations of wastes are often required to simplify the hazard/risk issues to a level that can be resolved.

The wastes considered for separation in this book are generated by the fabrication of nuclear fuel elements and weapons and the processing of spent nuclear fuel and other nuclear materials. Wastes generated by the processing of weapons materials and spent nuclear fuel constitute the majority of radioactive wastes. This book is mainly directed toward separation processes for these wastes.

In order to select a separation process for a waste, one needs to develop a menu of options that reflects the impact of selecting which government regulations will dictate how (and where) the waste will be stored and/or disposed. The regulations are based upon the waste classification. The waste is classified based upon how the waste was generated and/or its constituents. More hazardous wastes require more elaborate and expensive disposal methods and sites. Wastes may be classified as high-level radioactive waste, low level radioactive waste hazardous wastes, mixed wastes (low or high level with hazardous components), and transuranic wastes. Low-level wastes are separated into Classes A, B, and C. Radioactive waste definitions for these classifications are set by various government agencies such as the DOE, the Nuclear Regulatory Commission (NRC), the Environmental Protection Agency (EPA), and some state environmental agencies.[1] The DOE in Order 5820.2A specifies high-level waste (HLW) as the "highly radioactive waste material that results from the reprocessing of spent nuclear fuel, including liquid waste produced directly in reprocessing and any solid waste derived from the liquid, that contains a combination of transuranic waste and fission products in concentrations

Table 1 Maximum Allowable Activities for Low-Level Wastes

Radionuclide	Activity (Ci/m³)		
	Class A	Class B	Class C
^{3}H	40	—	—
^{63}Ni	3.5	70	700
^{90}Sr	0.04	150	7000
^{137}Cs	1	44	4600

requiring permanent isolation".[2] A similar definition appears in 10 CFR 60 where irradiated reactor fuel is included and in 40 CFR 191 and 10 CFR 50, Appendix F.[2-5] As noted above, the definition of high-level waste is based upon the waste source. An example of high-level waste is the calcine stored at the Idaho Chemical Processing Plant, which is considered high-level waste because it was produced by the processing of spent nuclear fuel.[1]

Other waste categories that may include high-level wastes are (1) transuranics (TRU), which are wastes that contain alpha-emitting transuranium nuclides with half-lives greater than 20 years, having an activity greater than 100 nCi/g, and not including ^{233}U and (2) greater than Class C (GTCC) classified wastes, which are more contaminated than low-level wastes (LLW), but are not classified as either transuranic or high-level wastes.[2-5] The DOE also classifies low-level waste as radioactive waste not classified as HLW, TRU, or spent nuclear fuel. The EPA exempts uranium/thorium mill tailings from this waste category.[6]

The EPA, DOE, and NRC have various regulations classifying low-level wastes. In addition, the Low-Level Radioactive Waste Policy Act of 1980 also classifies low-level wastes. The three classes of low-level wastes are defined by the NRC and appear in the Code of Federal Regulations.[7] These are Classes A, B, and C. (See Table 1 for maximum allowable activities for each class. These classifications do not include nuclear fuels, uranium and mill tailings, or high-level waste.)

Wastes containing concentrations of the specified radionuclides above the following activities (in Ci/m³) may also be considered low-level wastes: ^{14}C, 8; ^{99}Tc, 3; ^{129}I, 0.08; ^{241}Pu, 3500; and alpha-emitting radionuclides with half-lives greater than 5 years, 100 nC/g.

Examples of low-level waste resulting from the processing of spent nuclear fuel include "incidental" wastes generated in fuel reprocessing. These include hulls or other fuel hardware left over from reprocessing, salt residues from separation processes such as those stored at the West Valley and Savannah River sites, ion-exchange resins, some sludges, etc.[8]

Hazardous wastes are defined by EPA Resource Conservation and Recovery Act (RCRA) standards based upon their source or constituents (compounds that appear on an EPA list) or characteristics (ignitability, corrosivity, reactivity, or toxicity).[9] Listed wastes retain their hazardous nature throughout their existence, while a characteristic waste is no longer considered a hazardous waste if the characteristic causing the waste to be classified as a hazardous waste is removed.

Mixed wastes contain both radioactivity (at levels specified by the NRC, DOE, or EPA) and are hazardous (based upon EPA RCRA regulations). Many fuel-reprocessing wastes are categorized as mixed waste due to the presence of nuclear poisons such as cadmium.

Waste categorization is generally conducted on a case-by-case basis from discussions between the source contractor and the pertinent government agencies. Both waste catego-

Table 2 Estimated Inventories of Spent Fuel and Wastes at U.S. Sites

Site	Spent fuel (metric ton)	HLW (m³)	TRU (m³)	Mixed, LLW (m³)	LLW (m³)
Hanford	2,133	277,000		20,990	563,700
INEL	289	6,110	102,000	5,680	145,000
Savannah	202	128,800	9,900	1,700	
Oak Ridge	3				
West Valley		2,500			
Other DOE	28				
Total	2,655				

Compiled from DOE EIS (1994)[10] and WINCO report (1994).[11]

rization as well as approved disposal methods and locations are mutually agreed upon. In many cases, however, this process is quite lengthy. A case in point is the siting of a final HLW repository in the United States after discussion and studies that have been going on for more than 10 years. Continuous testing and delays may force the selection of a disposal program that requires lifetime monitoring and control via man-made facilities. This approach could prove to be more cost-effective and acceptable to the general public if success is not realized for geologic repository construction.

Current estimated inventories of spent fuel and wastes at several different national laboratories are listed in Table 2.

In Table 2 the HLW at INEL is composed of 4970 m³ of calcined waste and 300,000 gal (1140 m³) of liquid waste. TRU and LLW wastes at INEL are mostly solid process and refuse waste stored at the Radioactive Waste Management Complex in drums. There are about 1.5×10^6 gal (5680 m³) of sodium salt solution, low-level, mixed waste called sodium-bearing waste. About 99% of the waste stored in tanks at Hanford is mixed HLW waste (60 million gal, about 227,000 m³). Most of Savannah River HLW is liquid waste (34×10^6 gallons, about 128,800 m³). West Valley HLW is liquid waste (660,000 gal, about 2500 m³).

The spent fuel stored at the various locations will have to be prepared for interim storage and dispositioning. This may involve processing to remove actinides and fission products as a waste management technique. Approximately 0.6 m³ (after evaporation) of aqueous acidic (4 to 7 M nitric acid) waste is produced per metric ton of metal in the fuel according to Benedict.[12] Based upon the same reference, for fuel cooled 150 days and 6 years after discharge, the actinide concentration is about 10 g/l and the fission product concentration about 50 g/l. The activity of the waste will depend upon the waste age. After 6 years, the actinide activity is about 4 Ci/l and the fission product activity is about 600 Ci/l. Most of the actinide activity is due to curium-244, americium-241, and plutonium-241 and -238. Most of the fission product activity is due to strontium-90, yttrium-90, cesium-134 and -137, barium-137, and promethium-147.

The stored wastes vary in terms of source (processing wastes, contaminated clothing and equipment, fuel or weapon fabrication residues, etc.), physical state (solid or liquid), and nature (calcine oxides, contaminated refuse or metal parts, fuel element metals, sludges or salt cakes, water solutions, spent ion-exchange resin, etc.). The liquid wastes in turn differ in pH (predominantly acid liquid wastes are stored at INEL while other sites have alkaline wastes) and constituents (concentrations of actinides, fission products, and RCRA-classified hazardous materials). Waste characteristics for specific wastes are described in the chapters in this text. For example, in Chapter 8 the waste characteristics of the West Valley waste are discussed. This stored liquid/sludge waste consists of an alkaline (pH 10 to 11) supernatant over a metal hydroxide sludge (about 10% of tank volume). The sludge contains most of the metals, including strontium and the transuranics. The supernatant consists of a 6 M sodium nitrate and nitrite solution and contains most of the cesium. The cesium-137 activity is 3.4 Ci/l. This waste is somewhat typical of the alkaline wastes stored at the Hanford, Oak Ridge, and Savannah River plants. Based upon a Hanford report, the major cations in the sludge are sodium (about 10% of sludge) and zirconium (about 6% of the sludge).[13] The Hanford tank sludge contains an average of about 650 nCi/g of plutonium activity and 450 nCi/g of americium activity.

Chapter 7 of this volume describes the sodium-bearing liquid waste stored at the Idaho Chemical Processing Plant. The average acid concentration of the waste is about 1.5 M. The waste contains about 1.2 M sodium and 4.4 M nitrate ions. The cesium-137 activity is about 94 mCi/l and the strontium-90 concentration is about 81 mCi/l. Uranium is the major actinide and is present at a concentration of about 78 mg/l (activity of about 1.6 μCi/l based upon ^{235}U). The other actinides (plutonium, americium-241, and neptunium-237) are present in a concentration of about 2.5 mg/l (activity of about 15 mCi/l based upon ^{238}Pu and ^{241}Am).

Many processes that will be used to prepare radioactive liquid and solid wastes for interim storage and disposal will generate off-gas streams. These gaseous streams may contain entrained particulates, volatilized constituents, and gases. A chapter is devoted to the separation techniques available for off-gas stream cleanup. In the same chapter there are short descriptions of vitrification technologies to prepare the treated wastes for final disposal.

Current proposed waste processing scenarios at the different DOE sites involve separation of the stored radioactive wastes based upon activity into high- and low-level fractions (INEL, Hanford, West Valley, Savannah River, and Oak Ridge plants). The high-level fractions will be solidified and stored temporarily, pending construction of a geologic repository. The low-level wastes will be grouted and generally disposed in vaults on site in accordance with RCRA regulations. The separation processes described in this text have been evaluated at the various locations.

Many of the studies presented in this book document work conducted in the United States. However, Chapters 15 to 17 are devoted to present work conducted in several foreign nations. The studies and references presented in those chapters afford an opportunity to review worldwide applications of separation processes.

The intent of the authors and editors of this book is to create a reference book for separation processes used in the nuclear complex. Clearly, because of the wide spectrum of planned and ongoing applications of separation processes in the nuclear complex, this book cannot be an exhaustive study. Instead, representative examples are presented, sometimes as case studies, to allow the reader to obtain a broad scope of separation techniques and potential applications. The references prepared for each chapter also provide an excellent resource for investigating separation processes.

REFERENCES

1. **Boardman, R.D., K.N. Brewer, P.T. Grahovac, G.F. Kessinger, R.L. Nebeker, A.L. Olson, E.C. Thiel,** "Final Report of the Waste Definition and Disposal Committee Criteria Committee," WINCO Report WIN-360, Westinghouse Idaho Nuclear Company, Idaho Falls, ID, July (1993).
2. Department of Energy, "Radioactive Waste Management," DOE Order 5820.2A, September (1988).

3. Nuclear Regulatory Commission, "Disposal of High-Level Wastes in Geologic Repositories," Code of Federal Regulations, 10 CFR 60, December (1982).
4. Nuclear Regulatory Commission, "Domestic Licensing of Production and Utilization Facilities," Code of Federal Regulations, 10 CFR 50, May (1975).
5. Environmental Protection Agency, "Environmental Radiation Protection Standards for Management and Disposal of Spent Nuclear Fuel, High-Level and Transuranic Radioactive Waste," Code of Federal Regulations, 40 CFR 191, September (1985).
6. Environmental Protection Agency, "Environmental Protection Standards for Low-Level Radioactive Wastes," Code of Federal Regulations, 40 CFR 193, draft (1993).
7. Nuclear Regulatory Commission, "Licensing Requirements for Land Disposal of Radioactive Waste," Code of Federal Regulations, 10 CFR 61, December (1982).
8. Nuclear Regulatory Commission, "States of Washington and Oregon: Denial of Petition for Rulemaking," 58 Federal Register, 12342, March 4, 1993.
9. Environmental Protection Agency, "Identification and Listing of Hazardous Wastes," Code of Federal Regulations, 40 CFR 261, May (1980).
10. Department of Energy, "Department of Energy Progress of Spent Nuclear Fuel Management and Idaho National Engineering Laboratory Environmental Restoration and Waste Management Programs Draft Environmental Impact Statement," DOE/EIS-0203-D, June (1994).
11. Westinghouse Idaho Nuclear Company (WINCO), "High-Level Waste Issues and Resolutions Document," WINCO-1210, May (1994).
12. **Benedict, M., T.H. Pigford, H.W. Levi,** *Nuclear Chemical Engineering,* McGraw-Hill, New York (1981).
13. **Swanson, J.L.,** "Initial Studies of Pretreatment Methods for Neutralized Cladding Removal Waste (NCRW) Sludge," Battelle Pacific Northwest Laboratory report, PNL-7716, June (1991).

Section I

Treatment of Aqueous Streams

Aqueous waste streams in the nuclear industry have been the focus of substantial attention since the late 1980s. This interest is fueled by the health hazards associated with radionuclides and toxic metals contained in the streams and the large volumes that have been, and will continue to be produced. These streams originate from various activities, including reactor operations, spent fuel storage, previous spent fuel-reprocessing procedures, and decontamination and decommissioning of old operational facilities. Frequently, these streams are monitored and regulated by both the Environmental Protection Agency (EPA) through the Resource Conservation and Recovery Act (RCRA) and the Nuclear Regulatory Commission (NRC). This regulatory environment significantly complicates the treatment of these streams.

The technical community is in general agreement that radionuclides and toxic metals should be separated and consolidated such that the final volume of material, which must be disposed of, is minimized to the extent practical. This practice would minimize the cost and the quantity of land required for final disposal.

Many old and emerging technologies are available or are being developed to provide the separations capability required for radionuclides and toxic metals in aqueous streams. It is sometimes bewildering as to which technologies are most efficient and suitable for processing certain streams. The purpose of this section is to first provide the reader with a suite of separation methods, including experimental and modeling techniques, for various components. Rigorous technical details are not presented; however, references that are furnished in the individual chapters do provide that detail. Second, some individual studies are presented that provide templates for selection of individual technologies to process these aqueous streams. For the casual reader this section should provide an excellent overview of available separation technologies; and for the more serious researcher it will provide the foundation for more serious examination.

Arlin Olson

1

Chapter 1

Liquid Extraction, the TRUEX Process — Experimental Studies

E. Philip Horwitz and Renato Chiarizia

I. INTRODUCTION - GENERAL PROCESS SCHEME

Nuclear waste comes in a variety of forms. Perhaps the most troublesome is liquid waste generated during the processing of nuclear material for the extraction and recovery of plutonium and uranium. In this process called PUREX, a considerable quantity of highly radioactive transuranium elements — mostly americium-241 — remains dissolved in the nitric acid waste solution. In the near future the Department of Energy will probably require that all liquid waste containing more than 100 nCi/g of transuranium elements (TRU) be converted to glass and buried in a geological repository.

At Argonne National Laboratory a new solvent extraction process, called TRUEX — for transuranium extraction — has been developed, with the specific objective of reducing the transuranic concentration in acidic waste solutions to less than 100 nCi/l. Once TRU concentration has been reduced below this level, waste solutions can be much more safely and cheaply converted to grout or glass for near-surface storage. Only the recovered transuranics, which will be reduced in volume by a factor of 100 to 1000, will require deep burial. The result will be savings of hundreds of millions of dollars in vitrification and burial costs, plus the value of recovered plutonium, although no final decision has been reached yet on its possible use.

During the last two decades a number of new processes for the removal of TRUs from nitric acid waste streams have been reported. A few of these processes are based on the use of novel extractants. With the exception of TRUEX, all these new processes suffer from some serious drawback, being either inefficient or cumbersome, or not adaptable for waste processing in existing fuel-processing facilities. The major problem in waste processing is the removal or trivalent transuranium elements, which are very difficult to extract. This is an especially serious problem because the major source of alpha radiation in nuclear waste is the significant concentration of ^{241}Am, which occurs in waste solutions only in the trivalent oxidation state.

The key ingredient in the TRUEX process is the extractant octyl(phenyl)-N,N-diisobutylcarbamoylmethylphosphine oxide, abbreviated as OΦD(iB)CMPO, or, more simply, CMPO. Its chemical structure is shown in Figure 1. This molecule is capable of selectively extracting the transuranic elements from most of the fission products and inert constituents present in nuclear waste. CMPO was designed, synthesized, and characterized at Argonne National Laboratory by members of the Chemical Separations Group in the Chemistry Division. It contains a unique chemical structure that enables it to efficiently extract transuranics in the trivalent, tetravalent, and hexavalent oxidation states from both concentrated nitric and hydrochloric acids. Because CMPO also has good chemical stability in the acidic and radioactive environment encountered in waste processing, it possesses all of the properties needed for an extractant in processing nuclear waste streams.

3

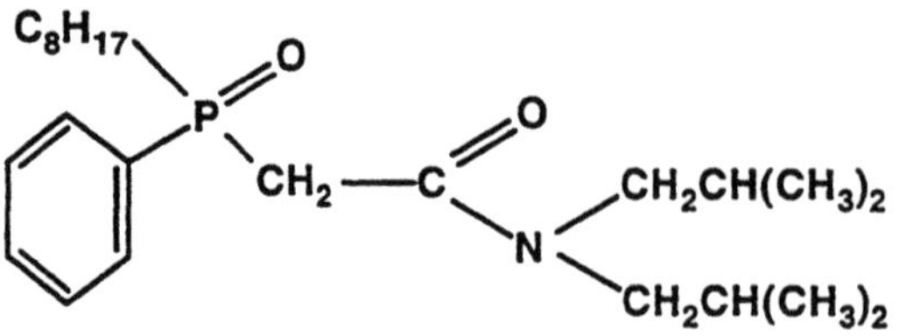

Figure 1 Chemical structure of OΦD(iB)CMPO.

The TRUEX process solvent contains not only CMPO, but also a large excess of tributylphosphate (TBP). The CMPO–TBP mixture is dissolved in a diluent, which, depending on the specific application, is a normal paraffinic or a chlorinated hydrocarbon. TBP in paraffinic hydrocarbons is the standard solvent in the PUREX process for isolating and separating uranium and plutonium from spent nuclear fuel. In other words, the extractant mixture used in the TRUEX process is simply the extractant used in the PUREX process combined with a relatively small concentration of CMPO. This means that the TRUEX process is compatible with existing nuclear-processing facilities.

The basic TRUEX process separation scheme is reported in Figure 2. Overall, the operation of the basic TRUEX flow sheet is as follows. The TRU waste is fed into a centrifugal contactor bank, where the TRU elements are extracted by the TRUEX process solvent; as a result, the aqueous raffinate containing the bulk of the waste exits as a non-TRU waste. The loaded solvent is scrubbed to effect a cleaner separation ; the solvent is then stripped with dilute acid to recover some TRU elements, e.g., americium and curium, as well as the rare-earth fission products. In a second strip, the plutonium is recovered by adding a small amount of complexing or reducing agents in the aqueous phase. Any uranium in the feed is removed from the solvent in the solvent cleanup operation, as are acidic impurities and solvent degradation products.

Some aspects of the chemistry of metal extraction by CMPO and by other carbamoylphosphoryl extractants are discussed in the following sections, together with the role of TBP in the process solvent, the diluent effects that characterize the chemistry of CMPO, the hydrolytic and radiolytic degradation of CMPO, and the solvent cleanup procedures. Finally, a more detailed description of some of the possible TRUEX flow sheets will be presented. Because of the limited length of this chapter, the discussion of the above aspects of CMPO chemistry cannot be exhaustive. Readers interested in a deeper understanding of the subjects covered in this chapter should refer to the original literature.

II. CHEMISTRY OF METAL EXTRACTION BY CARBAMOYLPHOSPHORYL EXTRACTANTS

A. EFFECT OF SUBSTITUENTS ON THE EXTRACTIVE PROPERTIES OF CMPO

Octyl(phenyl)-*N,N*-diisobutylcarbamoylmethylphosphine oxide (OΦD(iB)CMPO), or more simply CMPO, is a neutral bifunctional organophosphorus extractant with a unique combination of substituent groups that give it very favorable properties as an actinide extractant. CMPO was the result of several years of basic studies performed at Argonne National Laboratory on the extraction of Am(III) and other metal species from nitrate and chloride media by extractants belonging to the carbamoylphosphoryl class. The most important literature contributions from the Argonne groups, eventually leading to the development of CMPO and of the TRUEX process, are reported in the references list.[1-43]

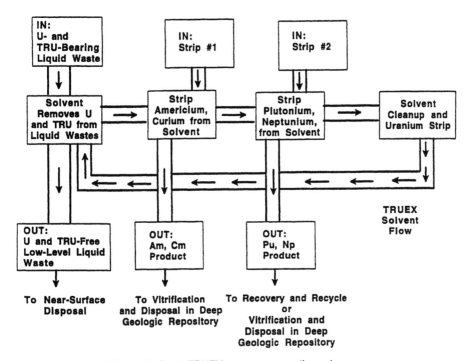

Figure 2 Basic TRUEX process separation scheme.

The extraction reaction of actinide ions in different oxidation states by CMPO from low-acidity solutions can be written as:[2,3,22]

$$UO_2^{2+} + 2NO_3^- + \overline{2CMPO} \Leftrightarrow \overline{UO_2(NO_3)_2 \cdot 2CMPO} \tag{1}$$

$$Pu^{4+} + 4NO_3^- + \overline{2CMPO} \Leftrightarrow \overline{Pu(NO_3)_4 \cdot 2CMPO} \tag{2}$$

$$Am^{+3} + 3NO_3^- + \overline{3CMPO} \Leftrightarrow \overline{Am(NO_3)_3 \cdot 3CMPO} \tag{3}$$

in nitrate solutions, and

$$UO_2^{2+} + 2Cl^- + \overline{2CMPO} \Leftrightarrow \overline{UO_2Cl_2 \cdot 2CMPO} \tag{4}$$

$$Pu^{4+} + 4Cl^- + \overline{3CMPO} \Leftrightarrow \overline{PuCl_4 \cdot 3CMPO} \tag{5}$$

$$Am^{3+} + 3Cl^- + \overline{3CMPO} \Leftrightarrow \overline{AmCl_3 \cdot 3CMPO} \tag{6}$$

in chloride solutions (the bar denotes organic phase species).

Because of the bifunctional nature of CMPO and of the other extractants belonging to the carbamoylphosphoryl class, it was initially thought that these compounds could efficiently extract Am(III) through chelation by the primary and secondary donor groups. However, it was early recognized that the carbamoyl–phosphoryl extractants behave essentially as monodentate ligands and their ability to extract Am(III) from highly acidic solutions was rather due to other effects.[2,3] To better understand the basic extraction chemistry of CMPO and related compounds, the extraction of Am(III) was studied as a function of a series of parameters, such as:

- Substituents attached to P=O
- Substituents attached to the amide nitrogen
- Number of carbon atoms bridging P=O and C=O

The results can be summarized by saying that, in general, the extraction chemistry of these compounds can be satisfactorily explained in terms of basicity of the primary donor group (P=O), intramolecular buffering effect of the amidic part of the molecule, and inductive and steric effects of the substituents.

Figure 3, for example, shows a comparison of the distribution ratio of Am(III), D_{Am}, as a function of nitric acid concentration for CMPO and a series of monofunctional extractants of increasing basicity, that is, dibutyl butylphosphonate (DB[BP]), butyl dibutylphosphinate (B[DBP]), and trioctylphosphine oxide (TOPO). The effect of increasing basicity of P=O, obtained by increasing the number of alkyl groups directly attached to the phosphorus atom, is quite evident. The monofunctional extractants exhibit a maximum in the Am acid dependency curves, which is shifted to lower acidity for higher P=O basicity, indicating that the decline of Am(III) extraction is due to competition by HNO_3 for the donor group. The CMPO data, on the other hand, show that the americium extraction is dramatically less sensitive to acidity. To explain this behavior, an intramolecular buffering effect was hypothesized. This hypothesis states that nitric acid bonds to the carbonyl oxygen in the metal ion nitrato–extractant complex at high acidity, thereby reducing the direct attack by hydrogen ions on the metal ion–phosphoryl oxygen bond.[3] The protonated carbonyl group probably acts as a shield against incoming acid through electrostatic interactions. The existence of the intramolecular buffering effect has been demonstrated by the difference of extraction behavior from solutions where a high concentration of nitrate ions was introduced as nitric acid vs. lithium nitrate.[2,3] Also, the effect does not exist when a monofunctional phosphine oxide is mixed with an aliphatic amide. It is also important to mention that the effect is almost absent when carbamoylphosphoryl extractants contain two, instead of one, carbon atoms bridging the P=O and C=O groups,[16] and is therefore characteristic of extractants containing the carbamoyl-*methyl*phosphoryl moiety. The different roles of the two donor groups in metal extraction from highly acidic solutions has been confirmed by infrared, NMR, and loading studies.[36,37] The comparison between CMPO and TOPO also shows the effect on D_{Am} of the reduced basicity of CMPO relative to TOPO due to the inductive effect of the phenyl group. It is thus evident why CMPO is a much more effective americium extractant than TOPO: at low acidity it exhibits lower distribution ratios (easier to strip) because of the phenyl group inductive effect, while at high acidity the distribution ratios are much higher and less sensitive to HNO_3 competition, because of the presence of the amidic part of the molecule.

A comparison among carbamoylphosphoryl extractants exhibiting P=O groups with different basicities is shown in Figure 4, where D_{Am} acid dependencies are shown for the

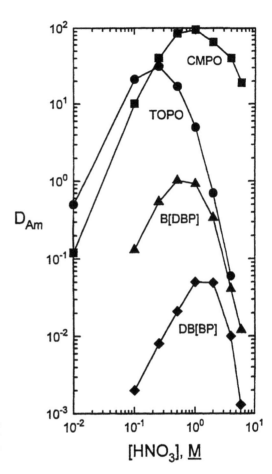

Figure 3 Acid dependency of D_{Am} with OΦD(iB)CMPO and some monofunctional extractants; 0.5 M extractants in diisopropylbenzene, 25°C.

series dihexyl-N,N-diethylcarbamoylmethylphosphonate (DHDECMP), hexyl hexyl-N,N-diethylcarbamoylmethylphosphinate, (HHDECMP), and dihexyl-N,N-diethyl-carbamoylmethylphosphine oxide (DHDECMPO). Again, D_{Am} correlates well at low acidity with the P=O basicity of the three extractants. However, as the aqueous acidity increases, the difference in D_{Am} for the three derivatives reduces drastically, as a consequence of the increasing competition between Am(III) and nitric acid for the phosphoryl oxygen. From the standpoint of a possible process application the data of Figure 4 indicate that both the phosphonate and phosphinate would be too inefficient as Am(III) extractants, although they show a better selectivity over fission products and Fe(III) than the phosphine oxide.[16] The latter, on the other hand, exhibits values of D_{Am} at low acidity that are too high for easy Am(III) stripping.

A more favorable Am(III) extraction behavior is obtained when at least one of the alkyl groups attached to the P=O is replaced by phenyl groups, to reduce the excessive basicity of the dialkyl-CMPO phosphoryl. The pronounced effect that phenyl substituents have on the acid dependency of D_{Am} is reported in Figure 5, comparing the behavior of DO-, OΦ-, and DΦD(iB)CMPO. Again, a reversal in extractant strength among the three CMPOs at high and low acidity is observed. Since the dioctyl derivative has the highest

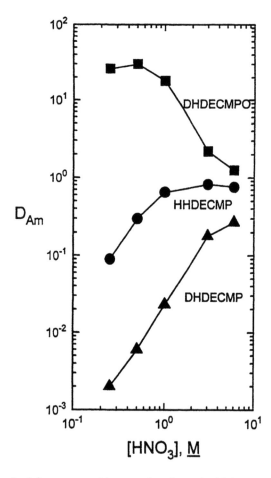

Figure 4 Acid dependency of D_{Am} with 0.25 M DHDECMP, HHDECMP, and DHDECMPO in tetrachloroethylene at 25°C.

basicity, one would expect it to have the highest values of D_{Am} at low HNO_3 concentrations. However, at high HNO_3 concentrations the competition between the nitric acid and Am(III) for the donor oxygen becomes significant. Under these conditions the extractant strength is reversed, because the nitric acid interacts less strongly with the least basic CMPO, the diphenyl derivative.

Based on the data of Figure 5, it appears that the OΦ- and DΦD(iB)CMPO compounds would be the most effective Am(III) extractants. They also exhibit a good selectivity for Am(III) over fission products.[16] However, the DΦ derivative suffers from very poor solubility in normal paraffinic hydrocarbon diluents, especially with carbon chains in the C_{12} to C_{14} range, even in the presence of phase modifiers. Therefore, OΦD(iB)CMPO has the best combination of high D_{Am} values at high acidity, good selectivity, and good phase compatibility.

Regarding the amidic part of the CMPO molecule, studies on the influence of lengthening and branching the amide substituents have shown only a modest effect on D_{Am} values. The fact that D_{Am} is much more suppressed by steric hindrance in the phosphoryl substituents than in the amidic substituents is further evidence that the phosphoryl oxygen is the donor group primarily involved in metal bonding. Isobutyl groups attached to the

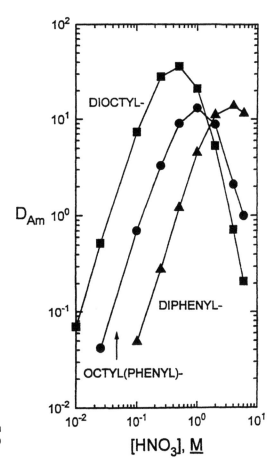

Figure 5 Acid dependency of D_{Am} with 0.25 M DOD(iB)CMPO, OΦD(iB)CMPO and DΦD(iB)CMPO in CCl$_4$ at 25°C.

amide nitrogen are preferred to *n*-butyl substituents because they impart to the CMPO molecule substantially improved solubility and phase compatibility.

The better shielding of the effective positive charge on the amidic nitrogen by branched chain substituents is believed to be responsible for the improved phase compatibility of isobutyl and *sec*-butyl CMPOs.[17]

Figure 6 summarizes the properties affected by substituents on the carbamoylphosphoryl moiety. For a more detailed discussion of the substituents effects on extraction strength, selectivity, and phase compatibility, the original literature should be consulted.[16]

The efficiency of CMPO solutions for extracting tri-, tetra-, and hexavalent actinides from a wide range of HNO$_3$ concentrations is reported in Figure 7. Similar data for extraction from hydrochloric solutions are also available.[22] The differences in the distribution ratios of actinides in the various oxidation states are typical of their behavior with less complex monofunctional extractants, but the D values are greatly elevated with CMPO. The data of Figure 7 show clearly that Am(III) can be easily stripped from a loaded CMPO solution simply by reducing the acidity of the aqueous phase. For tetra- and hexavalent actinides, on the contrary, more drastic stripping conditions are required and aqueous complexing or reducing agents are needed.

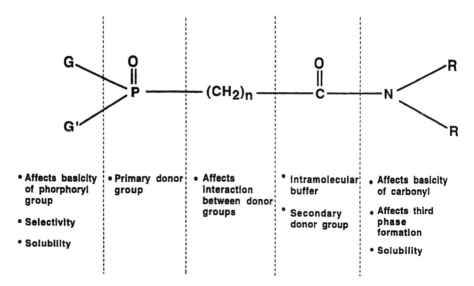

- **Affects basicity of phorphoryl group**
- **Selectivity**
- **Solubility**

- **Primary donor group**

- **Affects interaction between donor groups**

- **Intramolecular buffer**
- **Secondary donor group**

- **Affects basicity of carbonyl**
- **Affects third phase formation**
- **Solubility**

Figure 6 Effect of substituents on the properties of the carbamoylphosphoryl moiety.

B. EXTRACTION OF ACIDS BY CMPO

In the previous section the possibility of interaction between the donor groups of CMPO and the aqueous HNO_3 was mentioned on several occasions. This interaction is of course expected, given the basicity of the donor groups, especially of the P=O. The equilibrium constant of the nitric acid extraction reaction at low acidity (<1 M)

$$H^+ + NO_3^- + \overline{E} \Leftrightarrow \overline{E \cdot HNO_3} \tag{7}$$

where E is a generic carbamoylphosphoryl extractant, has been used as a quantitative measure of the P=O basicity for comparisons among selected carbamoylphosphoryl extractants.[16] The extraction of HNO_3 by TRUEX process solvent from solutions up to 10 M in nitric acid has been investigated and modeled for flow sheet design and optimization.[25]

A more complete investigation of the extraction of nitric and hydrochloric acid by CMPO solutions to understand the effect of acid extraction on the D_{Am} acid dependencies has been reported.[21] Figure 8 shows the striking difference in Am(III) extraction between the CMPO/chloride and CMPO/nitrate systems and also provides a comparison of the behavior of CMPO and of the monofunctional extractant TOPO. The data show that CMPO is a very effective extractant for Am(III) only at high HCl concentrations. The extremely high acid dependency (approximately sixth power) is in marked contrast to the corresponding nitric acid dependency of third power or less. The very rapid increase in the activity of HCl and the third power dependency of D_{Am} on the HCl activity are responsible for the steep rise in D_{Am} shown in Figure 8. Since the activity of HCl increases much more rapidly with concentration than for HNO_3, D_{Am} has a much higher dependency on HCl than on HNO_3 concentration. The pronounced leveling of D_{Am} at high HNO_3 concentration has been attributed previously to the intramolecular buffering effect of the carbamoyl group. An analogous effect does not seem to occur at high HCl concentrations. To understand the difference in behavior of Am at high HNO_3 and HCl concentrations,

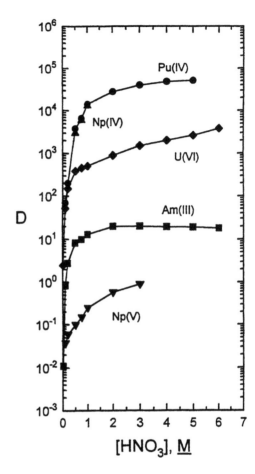

Figure 7 Distribution ratio of selected actinide ions as a function of the aqueous nitric acid concentration with 0.2 M CMPO–1.2 M TBP in dodecane at 25°C.

it is necessary to compare the extractions of HNO_3 and HCl by CMPO. The acid extraction data are reported in Figure 9, which shows that significantly less hydrochloric than nitric acid is extracted by CMPO. With HCl, the mole ratio of extracted acid to CMPO never exceeds one, indicating that only the phosphoryl oxygen is readily protonated by HCl in the extractant, while in the case of HNO_3 the involvement of both donor groups in acid extraction is demonstrated by a limiting value of two for the mole ratio of HNO_3 to CMPO. The comparison between CMPO and TOPO shown in Figure 8 is particularly interesting. As already discussed, the difference between the two extractants in the nitric acid system and the very evident presence of the intramolecular buffering effect suggest no evidence of chelation between CMPO and the Am(III) ion. The behavior of Am(III) extracted from HCl by the same two extractants is in marked contrast. Under no condition is TOPO a better extractant than CMPO for Am(III) from HCl. The data of Figure 8 suggest that the involvement of the carbonyl oxygen in the CMPO is greater in the Am–chloro complexes than in the Am–nitrato complexes, leading to a probable increased bidentate behavior of CMPO in the chloro complexes.

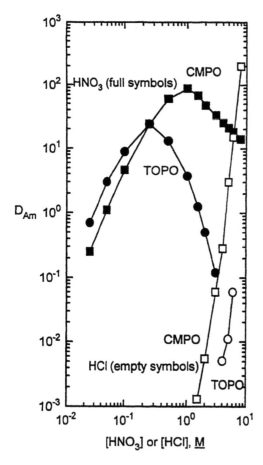

Figure 8 Comparison of the distribution ratio of Am(III) vs. concentration of HCl and HNO₃ for 0.5 M OΦD(iB)CMPO and 0.5 M TOPO in tetrachloroethylene at 25°C.

Because of the extraction of HNO_3 and HCl by CMPO, the Am(III) extraction is better represented by the following equations, which apply to the whole range of acid concentrations:

$$Am^{+3} + 3NO_3^- + \overline{3E \cdot (HNO_3)_n} \Leftrightarrow \overline{Am(NO_3)_3 \cdot E_3 \cdot (HNO_3)_m} + (3n - m)HNO_3$$

$$n = 0 \text{ to } 2, \quad m = 0 \text{ to } 3 \tag{8}$$

and

$$Am^{+3} + 3Cl^- + \overline{3E \cdot (HCl)_n} \Leftrightarrow \overline{AmCl_3 \cdot E_3 \cdot (HCl)_m} + (3n - m)HCl$$

$$n = 0 \text{ to } 1, \quad m = 0 \text{ to } 1 \tag{9}$$

The extraction of mineral acids such as HNO_3 and HCl by CMPO has important implications for the TRUEX process. The concentration of acid present in the organic

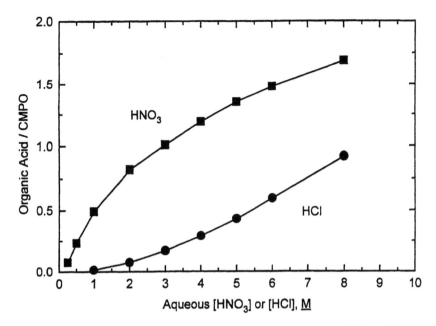

Figure 9 Molar ratio of extracted nitric and hydrochloric acids to OΦD(iB)CMPO in tetrachloroethylene as a function of the aqueous phase acidity at 25°C.

phase must be reduced to a very low level before stripping of Am(III) is attempted by contacting the organic phase with a diluted HNO_3 concentration. The removal of excess acid from the solvent is accomplished through a suitable scrubbing operation and is discussed in more details in the section where the TRUEX flow sheets are described.

The extraction of other acids by CMPO has also been investigated.[9,19,23] Of particular relevance for the TRUEX process is the extraction of pertechnetic ($HTcO_4$) and oxalic ($H_2C_2O_4$) acids. Pertechnetate and oxalate ions are generally present in the feed solution of the TRUEX process for different reasons. Technetium is an important long-lived fission product. Oxalic acid is added to the feed solution to increase the selectivity of CMPO for actinides by inhibiting the extraction of troublesome fission products (i.e., Pd, Zr, and Mo) and iron by forming inextractable aqueous phase oxalato-complexes. In solutions containing HNO_3 at a concentration below 1 M, both $HTcO_4$ and $H_2C_2O_4$ are quite well extracted by CMPO (distribution ratio between one and three) and report mostly in the organic phases. If oxalic acid is not removed from the organic phase, the risk exists of lanthanide and TRU oxalate salts precipitation during stripping. The fate of technetium in the TRUEX process and the oxalic acid problem is discussed in more detail in a later section.

C. SYNTHESIS AND PURIFICATION OF CMPO

Several synthetic reactions have been utilized for the preparation of CMPO and related compounds.[24] The methods were chosen among the Arbuzov reaction, Michaelis–Becker reaction, Grignard reaction, and various combinations. A commonly used reaction scheme is reported in Figure 10. The starting step is the esterification of phenylphosphinic acid (I) using triethylphosphite (II). The resultant ethylphenylphosphinate (III) is reacted with

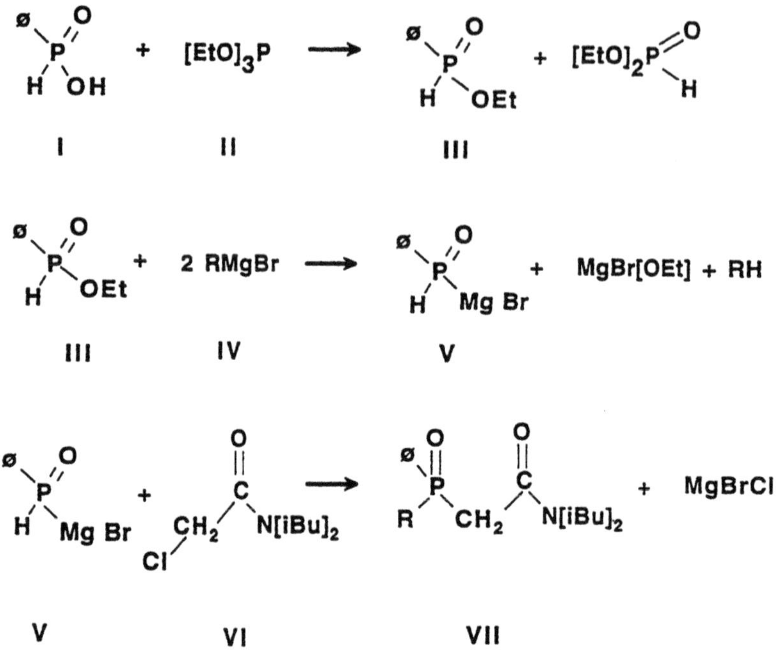

Figure 10 Synthesis route for OΦD(iB)CMPO.

the Grignard reagent (IV) to give the Grignard intermediate (V), which, through reaction with the *N,N*-diisobutylchloroacetamide (VI), leads to formation of CMPO (VII). The experimental details of this and several other synthetic routes can be found in Reference 24.

The synthesis of CMPO must be conducted with great care to minimize the formation of some acidic impurities that can strongly affect the stripping of Am(III) with dilute acid. Among these impurities dioctylphosphinic acid and octyl(phenyl)phosphinic acid have been reported. The former arises if the starting material (III) contains unremoved diethylphosphite, and the latter forms from the incomplete reaction of the intermediate (V) with the chloroacetamide (VI), followed by air oxidation of the secondary phosphine oxide.[24] Fortunately, both of these acids are effectively removed from a CMPO solution by repeatedly washing the organic phase with an aqueous base solution or by treatment with a macroporous anion-exchange resin on a basic cycle.[24]

A much more deleterious acidic impurity, resistant to removal by the above methods, is (*p*-octylphenyl)phosphinic acid (POPPA). This very lipophilic compound is a powerful acidic extractant. Its presence in CMPO samples prepared as shown in Figure 10 has been demonstrated and its formation during the synthesis sequence has been explained.[24]

The investigations on POPPA and its effect on Am(III) stripping from CMPO, led to the development of a relatively simple and effective CMPO purification procedure called MIX, for mixed ion exchange.[43] The procedure consists of treating a solution of CMPO in heptane with a macroporous cation-exchange resin on the H+ cycle at 50°C, followed by addition, without removal of the first resin, of an equivalent amount a macroporous anion-exchange resin on the OH– cycle. Although the mixed ion-exchange procedure is not completely understood, a probable explanation of its chemistry has been formulated.[24] The MIX procedure provides CMPO samples in which the presence of the derivatized acidic impurity POPPA is undetectable by gas chromatography. CMPO of this quality

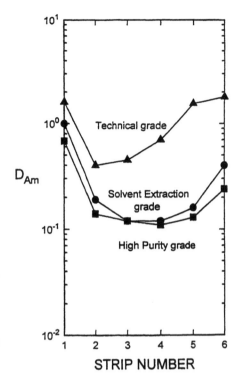

Figure 11 D_{Am} vs. successive strip number for different grades of CMPO. Organic phase: 0.25 M CMPO–1 M TBP in tetrachloroethylene; strip conditions: 0.04 M HNO$_3$, O/A = 1, T = 25°C. Each organic phase had been initially made 3×10^{-5} M in Am(III) from a 1 M HNO$_3$ solution and scrubbed three times with 0.2 M HNO$_3$ at O/A = 3.

represents material acceptable for use in a solvent extraction system and is referred to as SX grade.

The availability of an effective and simple method, such as MIX, for purifying CMPO up to a SX grade is of paramount importance if large-scale application of the TRUEX process is implemented. Existing traditional methods, in fact, although very effective in providing high-purity material (HP grade), are either too slow, or too expensive, or environmentally unacceptable. A critical discussion of these methods, namely, crystallization, distillation, and mercury salt precipitation, can be found in Reference 24.

The acceptable performance of SX-grade CMPO in the TRUEX process conditions has been demonstrated in Reference 29. A comparison of D_{Am} obtained by successive stripping of Am(III) from three different grades (technical, solvent extraction, and high purity) of TRUEX process solvent in tetrachloroethylene (TCE) with 0.04 M HNO$_3$ is shown in Figure 11. In this system 0.04 M HNO$_3$ was chosen because it is the acidity in the stripping stages of the TRUEX process. The Am(III) was extracted into the process solvent from 1 M HNO$_3$. The pattern of the distribution coefficients for HP and SX grades is similar, but the Am(III) distribution coefficients are much higher (poorer strip) for the tech-grade solvent. The initially high values of D_{Am} obtained in the first and second contact are due to HNO$_3$ in the aqueous phase coming from stripping the HNO$_3$ remaining in the organic phase after the scrubbing step. The D_{Am} attained on the third and fourth contact of the organic phase corresponds closely to the reference value measured extracting Am(III) from a 0.04 M HNO$_3$. It is interesting to note, however, that D_{Am} increases slightly after the fourth contact, even for the HP grade. This increase in D_{Am} indicates that small amounts of acidic constituents are even present in the HP grade of solvent. In the

TRUEX process five Am(III) strip stages are planned. While D_{Am} is somewhat higher on the fifth stripping contact, it is still quite low for the HP and SX grades and so should not significantly reduce the total Am(III) removed from the organic phase for these two solvent grades. The Am(III) concentration in the organic phase after the sixth strip is 4×10^{-10}, 1×10^{-9}, and 2×10^{-7} M for the HP-, SX-, and tech-grade solvents, respectively.

III. TBP AS A PHASE MODIFIER

Extractants belonging to the CMPO class suffer from a lack of high loading capacity and, more seriously, from an incompatibility with aliphatic diluents, which can cause third-phase formation.[7,4,6] Attempts to overcome this problem led to the addition of a second polar compound to the organic phase. Among the several phase modifiers tested, tributylphosphate (TBP) offered the best compromise of chemical and physical properties.[8-10,44,45]

A dramatic improvement in phase compatibility and loading capacity was brought about by the modification of the solvent properties upon addition of TBP. Loading and third-phase formation studies are discussed in the section on diluent effects. It was also found that, following TBP addition, the Am(III) extraction from nitric acid solutions by CMPO–TBP mixtures was profoundly altered.[12,20,38,44,45] Acid dependencies of D_{Am} obtained in decalin or tetrachloroethylene (TCE) for progressively increasing amounts of TBP added to 0.25 M CMPO are reported in Figures 12 and 13.

The presence of TBP has a somewhat similar effect in all diluents, but to varying degrees. In the low-acidity region, a reduction of D_{Am} values is always observed (antagonism), while in the high-acidity region an enhancement of distribution ratios is obtained (synergism). The acidity where the D_{Am} acid dependencies with or without TBP intersect is always in the 1 to 2 M range. Above this nitric acid concentration, the P=O of TBP interacts with HNO_3, reducing its competition for the phosphoryl group of CMPO. It is well known that the P=O of TBP, though less basic than that of CMPO, is still basic enough to react with nitric acid.[46] The lowering of D_{Am} at low acidity can be explained in a different way. Although a continuous variation study showed no evidence of mixed Am–CMPO–TBP complexes,[10] it is likely that TBP hydrogen bonds through water molecules and hydrogen ions to uncomplexed CMPO, or interacts through dipole-induced dipole bonds with the P=O and C=O of the CMPO molecule. This interaction effectively lowers the basicity of the donor groups of CMPO, reducing their availability for Am(III) extraction. An association phenomenon can also be utilized to further explain the D_{Am} enhancement at high acidity brought about by the presence of TBP. It is likely that the high concentration of TBP and the increased polarity of the TBP-diluent mixture enhance the stability and solubility in the organic phase of the complex $Am(NO_3)_3(CMPO)_3(HNO_3)_m$ by solvation of the HNO_3 present in the complex. In this case, the TBP molecules would be present in the outer coordination sphere and the CMPO molecules in the inner coordination sphere of the above complex. A more complete explanation of the effect of TBP on D_{Am} may be found in the section on diluent effects.

The presence of TBP in mixtures with CMPO in a TRUEX process solvent has also a remarkable effect in minimizing the solvent degradation due to hydrolysis and radiolysis.[20,38] The addition of TBP (0.73 M) to a solution of CMPO in CCl_4 (0.25 M) heavily degraded by previous exposure to an absorbed gamma dose of 97 watt hour/liter (Wh/l), caused a very pronounced decline of the Am(III) distribution ratios at 0.01 M HNO_3.[20,38] This indicates that TBP suppresses the influence of acidic degradation products, probably through hydrogen bonding, preventing them from extracting Am(III). The overall effect

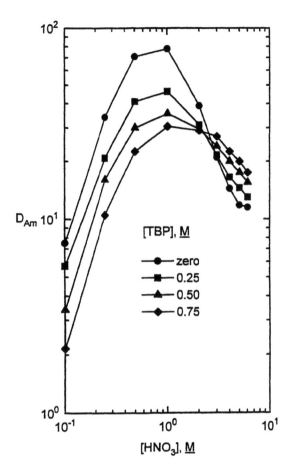

Figure 12 Acid dependency of D_{Am} with 0.25 M OΦD(iB)CMPO in decalin as a function of added TBP at 25°C.

of the presence of TBP is a remarkable (although only apparent) increase in the CMPO stability toward hydrolytic and radiolytic degradation, as measured by the Am(III) distribution ratio at low acidity. The addition of TBP to CMPO affects positively also the distribution ratio of CMPO between organic and aqueous phases.[30] A more detailed discussion of this effect will be given in the next section.

In conclusion, the addition of TBP to CMPO in the formulation of the TRUEX process solvent was a major breakthrough for the development of a solvent showing good "solvent extraction" characteristics. TBP was initially added to CMPO as a phase modifier to reduce third-phase formation (especially when a normal paraffinic hydrocarbon was used) and to increase the loading capacity of the solvent. Serendipitously, a number of other beneficial effects were found, all brought about by the TBP addition. The role of TBP in the TRUEX process solvent can be summarized as follows:

- It enhances D_{Am} above 1 M HNO$_3$ (more efficient extraction).
- It decreases D_{Am} below 1M HNO$_3$ (easier strip).
- It reduces the sensitivity of D_{Am} to acidity variations above 1 M (buffer effect).
- It improves phase compatibility (higher loading and less third-phase formation).
- It raises D_{CMPO}, lowering extractant losses to the aqueous phase.
- It minimizes the effect of acidic degradation products.

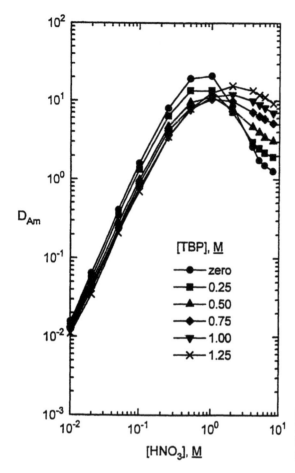

Figure 13 Acid dependency of D_{Am} with 0.25 M OΦD(iB)CMPO in tetrachloroethylene as a function of added TBP at 25°C.

IV. DILUENT EFFECTS ON CMPO EXTRACTION CHEMISTRY

A. DISTRIBUTION OF CMPO

Molecules of the carbamoylmethylphosphine oxide (CMPO) class contain two polar groups and are capable of interacting with the diluent through dipole–dipole and dipole-induced dipole interactions. For solubility to occur the extractant must be able to fit into the network of diluent molecules and the interaction between the extractant and the diluent must be stronger than the interaction between the extractant molecules with each other. One means of studying these interactions is through the measurement of the distribution of the extractant between the organic and aqueous phases. When CMPO interacts strongly with the diluent, less extractant reports to the aqueous phase, which results in an increase of its distribution ratio. The greater the interaction with the diluent, the lower the equilibrium concentration of CMPO in the aqueous phase.

The values of D_{CMPO} for 0.20 to 0.25 M CMPO solutions in several diluents and for different formulations of the TRUEX process solvent are reported in Reference 30. The data show that, in general, D_{CMPO} is lowest in aliphatic diluents and highest in aromatic

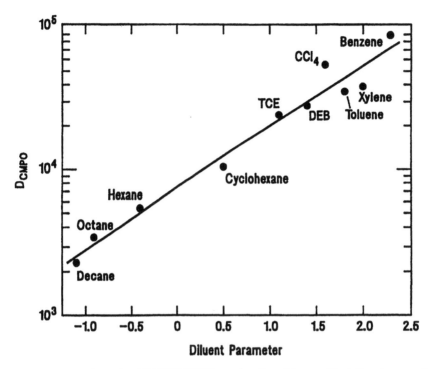

Figure 14 Distribution ratio of OΦD(iB)CMPO as a function of the empirical diluent parameter of Schmidt. Conditions: 0.2 M CMPO in diluent–0.25 M HNO$_3$ at 25°C; DEB = diethylbenzene; TCE = tetrachloroethylene. (Reprinted from Reference 30 [Figure 3], by courtesy of Marcel Dekker, Inc.)

and chlorocarbon diluents. All values of D_{CMPO} are quite high (1.1×10^3 to 8.3×10^4), indicating that CMPO is a very lipophilic molecule. Thus, very little loss of extractant to the aqueous phase occurs even after numerous contacts.

D_{CMPO} in a wide variety of diluents correlates approximately with the solubility parameter of the diluent (δ) using equations applicable to regular solutions.[47] However, δ_{H_2O} must be adjusted by numerical analysis to obtain the optimum fit. A much better correlation of the D_{CMPO} data is obtained using the empirical diluent parameter (DP) derived by Schmidt,[48] which characterizes the ability of a liquid to effect nonspecific solvation of molecules being extracted. The correlation is shown in Figure 14. The DP parameter is quite useful because it enables one to predict approximately (probably within 20%) the distribution ratio of CMPO for a number of diluents in which measurements of D_{CMPO} are difficult because of the solubility of the extractant, the volatility of the diluent, or because the partition constant is too high for reliable measurements (10^5).

Particularly interesting, because of its relevance on the TRUEX process, is the effect of TBP addition on the distribution ratio of CMPO: in all the diluents investigated in Reference 30, D_{CMPO} increases upon TBP addition. For example, D_{CMPO} values show a five- to sixfold increase when TBP is added to normal paraffinic hydrocarbons. This effect has been quantitated[30] by calculating the change in solubility parameter with addition of TBP. Overall, the influence of TBP on the distribution ratio of CMPO can be explained on the basis of solvent effects.

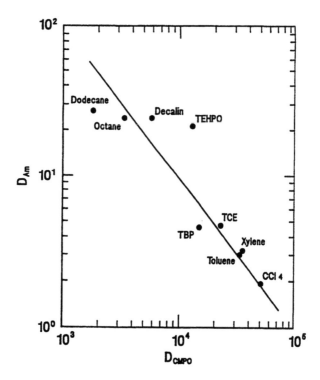

Figure 15 Distribution ratio of Am(III) as a function of the distribution ratio of OΦD(iB)CMPO. Conditions: 0.2 M CMPO in diluent–0.25 M HNO₃ at 25°C. TEHPO = tris(2-ethylhexyl)phosphine oxide; TCE = tetrachloroethylene. (Reprinted from Reference 30 [Figure 8], by courtesy of Marcel Dekker, Inc.)

B. CORRELATION BETWEEN D_{CMPO} AND D_{AM}

Since the value of D_{CMPO} is, in general, a manifestation of the degree of interaction between the diluent and CMPO, one would expect that the distribution ratio of a given metal ion, e.g., Am(III), in the same diluents would be inversely affected, i.e., the higher D_{CMPO}, the lower D_{Am}. The reason for the expected correlation is that the same phosphoryl and carbonyl groups that interact with the diluent through dipole– dipole and dipole-induced dipole interactions also coordinate to the metal ion. Although the metal ion–extractant complexes also interact with diluents, the strength of this interaction should be much reduced, relative to the uncomplexed extractants, because the major polar groups of the extractant are involved in much stronger metal ion–dipole interactions.

Figure 15 shows an approximate inverse first power relationship between log D_{Am} and log D_{CMPO} for selected diluents. The large positive deviation in the case of tris(2-ethylhexyl)-phosphine oxide (TEHPO) may be explained by its substantial steric hindrance, which creates sufficiently large holes for the Am(NO₃)₃·3CMPO complex as well as for ucomplexed CMPO to occupy. Conversely, dodecane shows a negative deviation, which may be explained by its significantly larger molar volume and greater cohesiveness than octane, which make accommodation of the larger Am(NO₃)₃·3CMPO species relatively more difficult than accomodation of the smaller uncomplexed CMPO. The above results fully expain why differences in D_{Am} are obtained with TRUEX process solvent prepared with different diluents.[30]

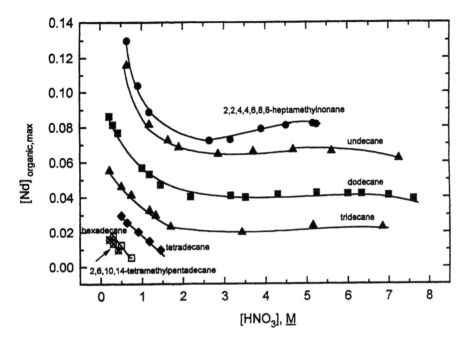

Figure 16 Solubility of Nd(III) nitrate in 0.2 M OΦD(iB)CMPO–1.2 M TBP in alkane diluents as a function of the aqueous HNO$_3$ concentration at 25°C.

C. EFFECT OF DILUENTS ON THE ORGANIC PHASE LOADING

At the recommended composition of the TRUEX-dodecane process solvent, namely 0.2 M CMPO + 1.2 to 1.4 M TBP, the solvent loading capacity is often determined by the limited solubility, rather than by the stoichiometry, of the extracted complex. This is because a second heavy organic phase is sometimes formed at high solvent loading. A detailed study of the loading capacity of solvents containing CMPO was performed in systems with both limited and unlimited solubility of the extracted complexes from nitric acid solutions.[26] Nd(III) was used as a stand-in for Am(III), but some data were also obtained with U(VI) and Pu(IV). The metal-loading and third-phase formation study was extended to chloride media in Reference 28, where the extraction of Pu(IV) and Th(IV) was investigated.

The results confirmed a strong dependence of the maximum attainable concentration (referred to as solubility) of extracted metal nitrates in the organic phase on the equilibrium concentration of nitric acid. Figure 16 shows this quite complex relationship. The figure also shows the increase of the solubility obtained with decreasing size and increasing branching of the diluent molecule. The effect of branching on solubility is particularly noteworthy. When a highly branched diluent with an average chain length of 12, such as Isopar-L, was used, no solubility limit was found for Nd(III) nitrate. The dependence of the solubility on the nitric acid concentration varies to some extent with the nature of the extracted metal, following in general, the order M(III) > M(IV) > M(VI), where M is a generic actinide or lanthanide ion. The solubility also markedly increases with temperature.

The primary importance of the phase modifier for the solubilization of extracted metal nitrates in the organic phase is illustrated by a sharp increase of the maximum attainable Nd(III) concentration in the solvent upon TBP addition. The solubilizing efficiency of the phosphoryl modifier increases very strongly with the alkyl chain length of trialkylphosphates and with elimination of an ester oxygen atom from the TBP modifier molecule, that is, in the order TOP > TBP and DB[BP] > TBP (TOP and DB[BP] are trioctylphosphate and dibutyl butylphosphonate, respectively).

Another interesting aspect of the loading behavior of the CMPO-TBP mixtures was the discovery that CMPO can be loaded to a CMPO/Nd(III) ratio of about two instead of the expected value of three. Because trisolvates appear to be typical species of lanthanide(III) nitrates extracted by neutral phosphoryl extractants, possible formation of mixed complexes like $Nd(NO_3)_3 \cdot 2CMPO \cdot TBP$, or aggregate formation associated with the bifunctional character of CMPO is indicated.[31,32]

V. HYDROLYTIC AND RADIOLYTIC DEGRADATION OF CMPO

The hydrolytic and radiolytic degradation of the TRUEX process solvent in its various formulations and of its major component, CMPO, in various diluents, have been reported in a number of papers.[15,38-40]

In the first investigation,[15] the Am(III) extraction at various HNO_3 concentrations was measured using 0.25 M OΦD(iB)CMPO + 0.75 M TBP in CCl_4 that had been subjected to vigorous stirring with HNO_3 up to 5 M at temperatures up to 70°C in absence of radiation. Analogous experiments were also performed while irradiating the biphasic system with up to an absorbed dose of about 100 Wh/l. After hydrolysis or radiolysis, the extraction of Am(III) was examined at the following nitric acid concentrations:

- 2 M, where the D_{Am} is at its maximum
- 0.05 M, which is the condition chosen for Am(III) stripping
- 0.01 M, where acidic impurities generated by degradation would most effectively alter D_{Am}

In the hydrolysis experiments, it was observed that only under the most severe conditions did significant variations of D_{Am} take place. This led to the conclusion that the hydrolytic damage would be of no consequence for periods equivalent to several years of continuous operaton of the process. It was also concluded that the observed effects on the extraction behavior of the solvent could be largely explained by the degradation of TBP to dibutylphosphoric acid (HDBP).[49] Similar results were obtained in the radiolysis experiments. Only after high radiation doses (>10 Wh/l) were significant changes in extraction behavior observed. Of particular interest was the experimental evidence that a reduction of D_{Am} at 2 M HNO_3 (considered indicative of loss of CMPO) was found only after doses above 70 Wh/l. Also, in this case, it was concluded that the radiolysis of TBP should be of more concern than that of CMPO. Since both hydrolysis and radiolysis of the TRUEX–CCl_4 solvent appeared to be essentially dominated by the TBP decomposition, no information was obtained on the behavior of CMPO itself, with the exception that it appeared to have a rather high apparent stability. Therefore, a more specific investigation was performed where CMPO and related compounds were studied in the absence of TBP, using again CCl_4 as the diluent.[38]

The hydrolysis of a 0.25 M CMPO solution, in the presence, under stirring, of 5 M HNO_3 at 50°C, was followed for 110 d and its D_{Am} values determined for different hydrolysis periods. The results showed very small changes in the extraction behavior of the CMPO solution over the entire hydrolysis period that is, a slight decline of D_{Am} at 2 M HNO_3 and D_{Am} values at 0.01 M HNO_3 only from two to three times higher than the pristine value. D_{Am} at 0.05 M HNO_3 was practically unchanged. This finding agrees with the previous observation that in the TRUEX–CCl_4 solvent the hydrolysis of TBP is the major factor affecting the extraction characteristics of the solvent. However, the gas chromatographic analysis of the organic solution revealed that only 78% of the initial CMPO was still present at the end of the experiment. This was taken as evidence that hydrolytic degradation of CMPO did indeed take place and the resulting products should contain at least one neutral extractant that is capable of extracting Am(III) not very differently from CMPO itself, to justify the almost constant value of D_{Am} at 2 M HNO3.

The data collected during radiolysis experiments conducted under the same conditions as the hydrolysis (5 M HNO$_3$ at 50°C under stirring) showed an extremely steep increase of D_{Am} at low acidity. D_{Am} at 0.01 M HNO$_3$, for example, increased by almost six orders of magnitude when given an absorbed dose of 97 Wh/l. Correspondingly, D_{Am} at 2 M HNO$_3$ showed a substantial drop already at 20 Wh/l and was about 20 times lower at about 100 Wh/l. The concentration of residual CMPO was determined by gas chromatographic analysis at different absorbed doses. The concentrations calculated from the D_{Am} values, based on a third-power extractant dependency, were systematically higher than those obtained gas chromatographically. This was taken as indication that the irradiation of CMPO has two major effects:

1. Formation of acidic products that extract Am(III) at low acidity
2. Formation of neutral compounds that extract or participate in the extraction of Am(III) at high acidity

From a comparative study of the radiolysis of substituted CMPOs (DO-, OΦ-, and DΦD(iB)CMPO) it was concluded that the octyl(phenyl)phosphinic acid (H[O(Φ)P]), is responsible for the increase of D_{Am} at low acidity after extensive radiolysis. However, even for long hydrolysis times and for absorbed doses of up to 10 Wh/l, the effect of H[O(Φ)P] on D_{Am} at low acidity is only minor. The presence of neutral extractants, e.g., CMPO and TBP, in large excess over H[O(Φ)P] significantly suppresses its ability to extract Am(III) due to hydrogen bonding between the acidic hydrogen of the phosphinic acid and the phosphoryl oxygen of the neutral extractants.

Methyl(octyl)(phenyl)phosphine oxide (MOΦPO) was tentatively identified as a neutral degradation product of CMPO. Studies of Am(III) extraction by mixtures of CMPO and MOΦPO showed that the latter is capable of replacing CMPO in the Am nitrato–CMPO complex, preventing the expected decline in D_{Am} from taking place as the concentration of CMPO decreases. The lower than expected decline of D_{Am} at 2 M HNO$_3$ was, therefore, tentatively attributed to the formation of MOΦPO.

A further investigation of the hydrolytic and radiolytic degradation of both CMPO and CMPO–TBP mixtures was performed in Reference 39, where a more specific quantitation of CMPO, TBP, and their degradation products by gas chromatographic and mass spectrometric analysis was attempted. In addition, CCl$_4$ was replaced by tetrachloroethylene (TCE) because of the excessive vaporing of the former as a process diluent. The conditions of the investigation were similar to those of the previous study.[38]

The same degradation products were observed in both radiolysis and hydrolysis experiments, but the relative yields of each were different. The only identified products from TBP were dibutylphosphoric acid, (HDBP) and small amounts of monobutylphosphoric acid (H$_2$MBP). A greater variety of degradation products was observed from CMPO. Neutral products included MOΦPO and octyl(phenyl)monoisobutyl-CMPO (OΦM(iB)CMPO). The acidic degradation products were H[O(Φ)P] and octyl(phenyl)phosphorylacetic acid (H[OΦPOAc]). The probable correlation of these degradation products is shown in Figure 17. Separate experiments in which the two degradation products MOΦPO and H[O(Φ)P] were hydrolyzed for 6 weeks in the same

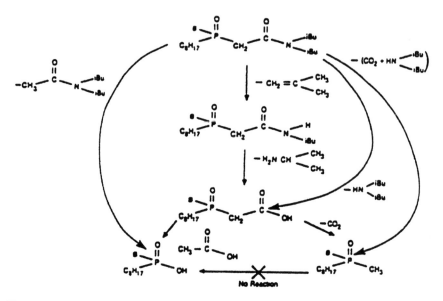

Figure 17 OΦD(iB)CMPO degradation pathways. (Reprinted from Reference 39 [Figure 2], by courtesy of Marcel Dekker, Inc.)

conditions as for CMPO showed complete recovery of the starting materials, suggesting that they are terminal products and are not subject to further degradation.

The rate of disappearance of CMPO in TCE was found to be lower than in CCl₄ (5.2 vs. 7.4),[38] thus indicating a greater radiation stability of CMPO in TCE. It has also to be mentioned that with highly irradiated solutions (68 Wh/l) the mass balance for the radiolysis of CMPO in either TCE or TBP–TCE mixtures was only 70%. The chromatograms of these solutions showed the presence of numerous unidentifiable components. The missing material might have been polymeric or volatile species impossible to detect under the analytical conditions used. In particular, the neutral compound MOΦPO was found in the degraded solutions at a concentration always below 0.01 M. Because MOΦPO is not an important degradation product, the explanation given previously[38] of partial replacement of CMPO by MOΦPO in the metal complexes could not be confirmed in this case. Yet, despite the destruction of nearly 40% of the original CMPO in solution after an absorbed dose of 68 Wh/l, little decline was observed in D_{Am} at 2 M HNO₃, implying that another unidentified product or some more subtle alteration of the medium was acting to maintain D_{Am} approximately constant.

The degradation studies were also extended to the TRUEX–dodecane process solvent.[40] In this case the concentrations of CMPO and TBP were 0.2 M and 1.2 M, respectively. The results of the previous investigations were essentially confirmed in terms of species formed upon degradation. It was also confirmed that the most important acidic extractant produced in the degradation of TRUEX process solvent is HDBP deriving from the phase modifier TBP. The concentration profiles for the degradation products determined by gas chromatography were successfully fit assuming pseudo-first-order kinetics and a dealkylation mechanism was established for HDBP production. The G value for the disappearance of CMPO in the TRUEX–dodecane solvent was found equal to 1.2. This very low value shows that CMPO is definitely more stable toward radiolysis in normal paraffinic hydrocarbon diluents than it is in chlorinated solvents.

VI. TRUEX PROCESS SOLVENT CLEANUP

In the previous section it was shown that both CMPO and TBP are subject to degradation under the conditions met during the processing of nuclear waste, for example, high acidity, elevated temperature, presence of radiation. The loss of CMPO due to degradation is not, however, a major concern, because little decline of the distribution ratio at 2 M HNO_3 is observed, although the reason for such behavior is not completely understood. Much more important is the effect of acidic degradation products of both TBP and CMPO on the distribution ratio of actinides at low acidity, i.e., under the stripping conditions of the TRUEX process. The major organic soluble acidic compounds generated by hydrolysis and/or radiolysis of TBP is dibutylphosphoric acid (HDBP). In the case of CMPO a complete identification of all the degradation products brought about by hydrolysis and radiolysis has not been achieved. Gas chromatographic investigations on degraded CMPO have allowed, however, the identification of n-octyl(phenyl)phosphinic acid H[O(Φ)P] and n-octyl-(phenyl)phosphinylacetic acid (H[O(Φ)POAc]) as major acidic degradation products.[39,40] The presence of HDBP, H[O(Φ)P], H[O(Φ)POAc], and other unidentified acidic compounds in degraded TRUEX process solvent is responsible for the drastic elevation (up to two to three orders of magnitude) of the Am(III) distribution ratio at 0.01 M HNO_3. A higher than expected D_{Am} in the stripping stages of the TRUEX process can complicate the back-extraction of Am(III). It is therefore important that a solvent cleanup procedure capable of removing the acidic impurities from the TRUEX process solvent be developed.

A sodium carbonate scrub has been generally considered to be an effective primary solvent cleanup operation, in analogy with the practice in PUREX plants.[50-52] An extensive Na_2CO_3 scrub is indeed capable of restoring the pristine Am(III) extraction characteristics of the TRUEX process solvent when the diluent is CCl_4[38] or tetrachloroethylene.[39] With aliphatic diluents the situation is different. The carbonate scrub procedure is effective in restoration of the D_{Am} at 0.01 M HNO_3 in the hydrolyzed TRUEX process solvent, but it is increasingly less effective at higher absorbed doses in radiolyzed solutions.[40] It follows that, at least when a normal paraffinic hydrocarbon is used as the diluent, a secondary cleanup operation is needed for the TRUEX process solvent if restoration to pristine condition is desired.

Based on literature information regarding PUREX process solvent,[51,52] a detailed investigation has been performed[42] on the use of a number of solid sorbents for a secondary cleanup operation, following the carbonate scrub. The effect of each cleanup operation was quantitatively assessed using gas chromatographic analysis, Am(III) distribution ratio determinations at different acidities, and measurement of D_{Am} during successive contacts of a loaded organic phase using fresh portions of 0.04 M HNO_3. The last criterion was chosen because, as shown previously, the successive strip experiment is a powerful tool in determining the purity of a TRUEX process solvent.

The results of the investigation showed that a secondary cleanup with an excess of strong base macroporous anion-exchange resin is capable of quantitatively removing the acidic compounds left behind by the carbonate scrub. The behavior of the purified solution in the stripping conditions of the TRUEX process is fully restored. Figure 18 shows the effect of the treatment of samples degraded at progressive absorbed doses with different amounts of strong base macroporous resins. The data are presented as D_{Am} at 0.04 M HNO_3 measured for successive strips, after the resin treatment. The curves of Figure 18 clearly show that treatment with an excess of resin is capable of restoring the pristine D_{Am} values.

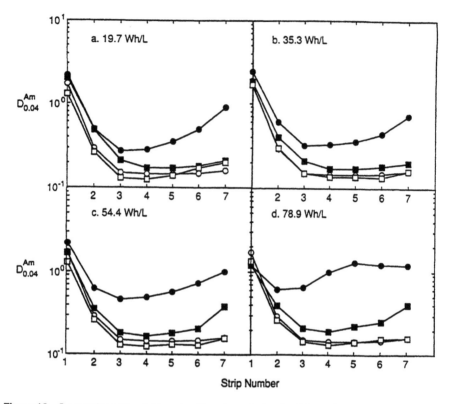

Figure 18 Successive strip experiments with undegraded and radiolyzed TRUEX process solvent at different absorbed doses: ○ reference (undegraded solution); ● after carbonate scrub; ■ after strong base resin treatment, following carbonate scrub (2 mg/ml for a and b; 5 mg/ml for c and d); □ after excess strong base resin treatment, following carbonate scrub (100 mg/ml for a and b; 150 mg/ml for c and d). (Reprinted from Reference 42 [Figure 5], by courtesy of Marcel Dekker, Inc.)

Acid-washed activated carbon is also very effective in restoring the original distribution behavior of the degraded process solvent. Among the inorganic oxides tested as adsorbents of the acidic degradation compounds, alumina gave the best results. The acid-washed alumina, in particular, performs well enough to be considered for a process application, even if the removal of acidic degradation compounds does not seem to be quantitative.

Based on the above results, a secondary cleanup with either one of the above systems seems therefore possible. Each one of them, however, offers advantages and disadvantages. The strong base resin is very effective but presents operational difficulties such as cost, availability, and waste disposal. Also, the regeneration of the strong base resin is possible but this operation generates more waste than simply discarding the used resin. The activated carbon offers simplicity of use because the technology of adsorption with granular and powdered activated carbon is well developed. One drawback of the activated carbon is its high flammability, which might make its use unacceptable in nuclear waste-processing facilities. The use of activated alumina is simpler, if its slightly reduced effectiveness is acceptable. The technology of adsorption with alumina is also well developed and the spent adsorbent could be simply converted into grout for shallow burial.

Table 1 Extractabilities of Elements from
1 to 3 M HNO$_3$ Using 0.2 M CMPO–1.4 M
TBP in C$_{12}$ to C$_{14}$-NPH

Distribution ranges			
<5 × 10^{-2}		10^{-2} to 1	1 to 20	>20
Na–Cs	Ni	Zr[a]	La–Eu	Th
Be–Ba	Cu	Mo(VI)[a]	Y	U(VI)
Al	Zn	Ru	Tc	Np(IV)
Cr(III)	Rh	Pd[a]		Pu(IV)
Mn(II)	Ag		Am	
Fe(III)	Cd		Cm	
Co(II)	In			
	Sb			
	Tc			

[a] Ds lowered by oxalate ion.

Table 2 Extractabilities of Elements from
6 M HCl Using 0.5 M CMPO in TCE

Distribution ranges			
<10^{-2}	10^{-2} to 1	1 to 10^2	>10^2
Li–Cs	Mn(II)	V	Fe
Be–Ba	Co(II)	Zn	Ga
Al	Ni	Cd	Zr
Cr(III)	Pb	Sn	Mo
		Am	Th
		Cm	U(VI)
			Np(IV)
			Pu(IV)

VII. TRUEX PROCESS FLOW SHEETS

A. INTRODUCTION

TRUEX process flow sheets and related development studies have been described in References 1, 8, 9, 14, 15, 18, 22, 25, 33, and 34. More recent TRUEX test runs are described in References 53 to 55. An extraction chromatographic resin that utilizes the CMPO–TBP system is described in Reference 56. All TRUEX flow sheets involve extraction from moderate to high acidity and stripping at low acidity. Differences in individual flow sheets arise from the different types of waste solutions, the scrubbing and stripping cycles, and the level of decontamination required. TRUEX process flow sheets designed to treat high-level liquid wastes are the most complicated because of the large number of constituents. Tables 1 and 2 summarize the distribution ranges using TRUEX process solvents for a number of fission products and inert constituents that are frequently found in waste solutions.

B. SCRUBBING

The objective of the scrubbing cycle is twofold. First, scrubbing the loaded organic phase with nitric acid or nitric acid containing oxalic acid provides additional decontamination of the TRU fraction from fission products and inert constituents. The use of oxalic acid is particularly important if Zr, Mo, and Pd are present in the waste stream. The oxalato complexes of the above three fission products are water soluble, thus providing an effective means of facilitating their removal from the organic phase. By maintaining the nitric acid concentration in the scrub above 1 M, the TRUs will remain in the organic phase.[1,11]

The second purpose of the scrub is to reduce the nitric acid concentration in the organic phase. This is necessary to facilitate stripping, which requires low acidity to effectively lower the distribution ratios of the TRUs. Because the two different goals of scrubbing require different nitric concentrations, a split scrub is frequently required in TRUEX process.[1,11]

C. STRIPPING

The purpose of stripping is to recover the TRUs and U from the organic phase, thus allowing the TRUEX process solvent to be recycled. Several stripping options are possible with TRUEX because of the vast differences in the complexing strengths of tri-, tetra-, and hexavalent actinides. Table 3 summarizes the stripping reagents and partitioning for TRUEX process solvent–nitrate media. The commercially available 1-hydroxyethane-1,1-diphosphonic acid (HEDPA) is an extremely effective stripping agent even when the process solvent is extensively degraded both hydrolytically and radiolytically.[57] However, recovery of actinides from HEDPA generates phosphoric acid, and is somewhat cumbersome. In addition, phosphorus is an undesirable constituent in the vitrification of the actinide fraction. Ammonium oxalate is also a good overall stripping agent for actinides and does not have the degradation and disposal problems of HEDPA.[54] The high oxalate concentration helps to solubilize the tri- and tetravalent TRUs.

The most useful partitioning step in TRUEX is the selective stripping of trivalent (e.g., Am and Cm) from tetra- and hexavalent actinides, as shown in Figure 2. This partitioning is performed using dilute (e.g., 0.04 M) nitric acid. As is discussed below, very good separations of Am from Pu, and vice versa, have been achieved using a dilute nitric acid strip. After the dilute HNO_3 strip, tetravalent actinides are usually back-extracted using HNO_3–HF followed by a Na_2CO_3 wash to remove U(VI) and process solvent degradation products. A more recent innovation involves the use of tetrahydrofuran-2,3,4,5-tetracarboxylic acid (THFTCA) to partition TRUs (both tri- and tetravalent) from U(VI). The separation of TRUs from U is important because uranium is frequently a major constituent in high-level waste but does not require vitrification for storage in a deep geologic repository.

D. HOT TEST RUNS OF TRUEX

Batch and continuous countercurrent test runs have been performed on a variety of acidic nitrate waste streams. The first countercurrent test runs were performed at Argonne National Laboratory using a synthetic waste stream that was formulated to simulate a dissolved sludge waste from the Hanford storage tanks.[9,11,14] The concentrations of Am, Pu, and Np in the waste were reduced by 5×10^4, 1×10^5, and 6, respectively, with a once-through pass of the solvent. All extraction, scrub, and strip stages were performed in a batch mode.

Table 3 Summary of Stripping Reagents and Partitioning for TRUEX
Process Solvent–Nitrate Media.

	Actinides (macro concentrations)			
	Np(V)[a]	Am(III), Cm(III)	Pu(IV),[b] Np(IV)	U(VI)
0.25 M HEDPA	+	+	+	+
0.25 M HEDPA–1 M HNO$_3$	–	–	+	+
0.1 M H$_2$C$_2$O$_4$	+	+ '	+ '	+
0.3 M (NH$_4$)$_2$C$_2$O$_4$	+	+	+	+
0.25 M THFTCA	+	+	+	–
0.2 M HNO$_3$	+	–	–	–
0.04 M HNO$_3$	+	+	–	–
0.04 M HNO$_3$–0.04 M HF	+	+	+ '	–
0.25 M Na$_2$CO$_3$	+	+ '	+ '	+

[a] H$_2$O$_2$ will reduce Np(VI) to Np(V) with higher oxidation state initially in the organic phase.
[b] HAN will reduce Pu(IV) to Pu(III) with higher oxidation state initially in the organic phase.

Note: HEDPA, 1-hydroxyethane-1,1-diphosphonic acid; THFTCA, tetrahydrofuran tetracarboxylic acid; H$_2$C$_2$O$_4$, oxalic acid; +, effectively stripped; –, not effectively stripped; ', precipitate formed.

Based on the above successful test run, a continuous countercurrent test run of the TRUEX process was carried out at the Westinghouse Hanford Plutonium Finishing Plant (PFP) using the ANL 4-cm centrifugal contactors.[33] The feed solution consisted of a clarified PFP waste, 3 to 4 M in HNO$_3$, containing 603 μCi/l of Pu plus [241]Am. Decontamination of TRUs in the raffinate was 10[4]. Additional batch test runs have been carried out at Pacific Northwest Laboratory (PNL) on an acidified waste solution containing chelating agents, a waste solution generated during cladding removal operations, and a Purex raffinate generated from a fuel reprocessing operation. The results are summarized in Reference 1. In each case, the TRU concentration in the effluent was reduced to 1 to 2 nCi/g of disposed form. The TRU decontamination factor with the high-level liquid waste from the full reprocessing stream was 5 × 10[5].

More recent continuous countercurrent test runs have utilized both mixer-settlers[53] and centrifugal contactors.[54] The latter studies are particularly noteworthy because of the large TRU decontamination factors achieved as well as the excellent partitioning of Am from Pu, and Pu from Am using 0.04 M HNO$_3$ to strip Am and then 0.3 M (NH$_4$)$_2$C$_2$O$_4$ to strip Pu. For example, the Am fraction contained 99.1% of the Am and only 0.2% of the Pu while the Pu product stream contained 99.8% of the Pu and 0.9% of the Am. Recent TRUEX experiments at PNL have focused on acidic waste streams generated by the dissolution of sludges from the Hanford storage tanks.[55] The TRUEX process has performed satisfactorily in all cases. Although practically all studies on TRUEX have involved acidic nitrate media, the process can be applied to highly acidic chloride waste streams.[1,21,22] Applications of the TRUEX–chloride system have been conducted at Los Alamos National Laboratory and have utilized primarily the extraction chromatographic mode.[58]

E. INTEGRATED TRUEX PROCESS FLOWSHEETS

Studies at ANL have attempted to combine TRUEX with the strontium extraction process (SREX) to simultaneously extract TRUs, Tc, and [90]Sr.[59-62] Such integrated processes have the advantage of simplifying the pretreatment of high-level waste streams where both TRUs and ^{90}Sr must be removed. However, process solvent formulations become more complex and flow sheets become more involved because of the difficulty of reducing the HNO_3 concentration in the process solvent without removing ^{90}Sr from the organic phase. Studies involving integrated extraction processes that incorporate TRUEX are very preliminary and, thus, the merits of this approach have yet to be validated.

ACKNOWLEDGMENT

Work was performed under the auspices of the Office of Basic Energy Sciences, Division of Chemical Sciences, U.S. Department of Energy, under contract number W-31-109-ENG-38.

REFERENCES

1. **Horwitz, E. P. and Schulz, W. W.,** The TRUEX Process: a vital tool for disposal of U.S. defense nuclear waste, in *New Separation Chemistry Techniques for Radioactive Waste and Other Specific Applications,* Cecille, L., Casarci, M., and Pietrelli, L., Eds., Elsevier Applied Science, New York, 1991, 21.
2. **Horwitz, E. P., Kalina, D. G., and Muscatello, A. C.,** The extraction of Th(IV) by dihexyl-*N,N*-diethylcarbamoylmethylphosphonate from aqueous nitrate media, *Sep. Sci. Technol.,* 16, 403, 1981.
3. **Horwitz, E. P., Muscatello, A. C., Kalina, D. G., and Kaplan, L.,** The extraction of selected transplutonium(III) and lanthanide(III) ions by dihexyl-*N,N*-diethyl-carbamoylmethylphosphonate from aqueous nitrate media, *Sep. Sci. Technol.,* 16, 417, 1981.
4. **Kalina, D. G., Horwitz, E. P., Kaplan, L., and Muscatello, A. C.,** The extraction of Am(III) and Fe(III) by selected dihexyl-*N,N*-dialklycarbamoylmethylphosphonates, -phosphinates and -phosphine oxides from nitrate media, *Sep. Sci. Technol.,* 16, 1127, 1981.
5. **Muscatello, A. C., Horwitz, E. P., Kalina, D. G., and Kaplan, L.,** The extraction of Am(III) and Eu(III) from aqueous ammonium thiocyanate by dihexyl-*N,N*-diethylcarbamoylmethylphosphonate and related compounds, *Sep. Sci. Technol.,* 17, 859, 1982.
6. **Horwitz, E. P., Kalina, D. G., Kaplan, L., Mason, G. W., and Diamond, H.,** Selected alkyl(phenyl)-*N,N*-dialkylcarbamoylmethylphosphine oxides as extractants for Am(III) from nitric acid media, *Sep. Sci. Technol.,* 17, 1261, 1982.
7. **Horwitz, E. P., Diamond, H., and Kalina, D. G.,** Carbamoylmethylphosphoryl derivatives as actinide extractants. Their significance in the processing and recovery of plutonium and other actinides, in *Plutonium Chemistry,* Carnall, W. T. and Choppin, G. R., Eds., ACS Symposium Series 216, American Chemical Society, Washington, D.C., 1983, chap. 27.
8. **Horwitz, E. P., Diamond, H., Kalina, D. G., Kaplan, L., and Mason, G. W.,** Octyl(phenyl)-*N,N*-diisobutylcarbamoylmethylphosphine oxide as an extractant for actinides from nitric acid waste, in Proc. Intern. Solvent Extraction Conference ISEC '83, Denver, 1983, 451.
9. **Vandegrift, G. F., Leonard, R. A., Steindler, M. J., Horwitz, E. P., Basile, L. J., Diamond, H., Kalina, D. G., and Kaplan, L.,** Transuranic Decontamination of Nitric Acid Solutions by the TRUEX Solvent Extraction Process — Preliminary Development Studies, Report ANL-84-45, Argonne National Laboratory, Argonne, IL, 1984.
10. **Horwitz, E. P. and Kalina, D. G.,** The extraction of Am(III) from nitric acid by octyl-(phenyl)-*N,N*-diisobutylcarbamoylmethylphosphine oxide–tri-*n*-butyl phosphate mixtures, *Solvent Extr. Ion Exch.,* 2, 179, 1984.
11. **Horwitz, E. P., Kalina, D. G., Diamond, H., Vandegrift, G. F., and Schulz, W. W.,** The TRUEX process — a process for the extraction of the transuranic elements from nitric acid wastes utilizing modified PUREX solvent, *Solvent Extr. Ion Exch.,* 3, 75, 1985.

12. **Kalina, D. G. and Horwitz, E. P.**, Variations in the solvent extraction behavior of bifunctional phosphorus-based compounds modified with TBP, *Solvent Extr. Ion Exch.*, 3, 235, 1985.

13. **Horwitz, E. P. and Schulz, W. W.**, Solvent extraction and recovery of the transuranic elements from waste solutions using the TRUEX process, in *Solvent Extraction and Ion Exchange in the Nuclear Fuel Cycle*, Logsdail, D. H. and Mills, A. L., Eds., Ellis Horwood Limited, Chichester, England, 1985, 137.

14. **Horwitz, E. P., Kalina, D. G., Diamond, H., Kaplan, L., Vandegrift, G. F., Leonard, R. A., Steindler, M. J., and Schulz, W. W.**, TRU decontamination of high-level purex waste by solvent extraction utilizing a mixed octyl(phenyl)-*N,N*-diisobutyl-carbamoylmethylphosphine oxide/TBP/NPH (TRUEX) solvent, in *Actinide/Lanthanide Separations*, Choppin, G. R., Navratil, J. D., and Schulz, W. W., Eds., World Scientific Publication, Republic of Singapore, 1985, 43.

15. **Leonard, R. A., Vandegrift, G. F., Kalina, D. G., Fisher, D. F., Bane, R. W., Burris, L., Horwitz, E. P., Chiarizia, R., and Diamond, H.**, The Extraction and Recovery of Plutonium and Americium from Nitric Acid Waste Solutions by the TRUEX Process. Continuing Development Studies, Report ANL-85-45, Argonne National Laboratory, Argonne, IL, 1985.

16. **Horwitz, E. P., Martin, K. A., Diamond, H., and Kaplan, L.**, Extraction of Am(III) from nitric acid by carbamoyl-phosphoryl extractants: the influence of substituents on the selectivity of Am over Fe and selected fission products, *Solvent Extr. Ion Exch.*, 4, 449, 1986.

17. **Gatrone, R. C., Dietz, M. L., and Horwitz, E. P.**, The effect of steric hindrance on the amidic substituents of the carbamoylmethylphosphine oxides on third phase formation, *Solvent Extr. Ion Exch.*, 11, 411, 1993.

18. **Horwitz, E. P. and Schulz, W. W.**, Application of the TRUEX process to the decontamination of nuclear waste streams, in Proc. Int. Solvent Extraction Conference ISEC '86, Munich, 1986, Vol. 1, 81.

19. **Chiarizia, R. and Horwitz, E. P.**, The solvent extraction of selected organic complexants by TRUEX process solvent, *Solvent Extr. Ion Exch.*, 5, 175, 1987.

20. **Chiarizia, R. and Horwitz, E. P.**, The influence of TBP on americium extraction by octyl(phenyl)-*N,N*-diisobutyl-carbamoylmethylphosphine oxide, *Inorg. Chim. Acta*, 140, 261, 1987.

21. **Horwitz, E. P., Diamond, H., Martin, K. A., and Chiarizia, R.**, Extraction of Am(III) from chloride media by octyl(phenyl)-*N,N*-diisobutyl-carbamoylmethylphosphine oxide, *Solvent Extr. Ion Exch.*, 5, 419, 1987.

22. **Horwitz, E. P., Diamond, H., and Martin, K. A.**, The extraction of selected actinides in the (III), (IV), and (VI) oxidation states from hydrochloric acid by OΦD(iB)CMPO: the TRUEX–chloride process, *Solvent Extr. Ion Exch.*, 5, 447, 1987.

23. **Clark, G. A., Gatrone, R. C., and Horwitz, E. P.**, The extraction of carboxylic acids by OΦD(iB)CMPO, *Solvent Extr. Ion Exch.*, 5, 471, 1987.

24. **Gatrone, R. C., Kaplan, L., and Horwitz, E. P.**, Synthesis of the carbamoylmethylphosphine oxides, *Solvent Extr. Ion Exch.*, 5, 1075, 1987.

25. **Leonard, R. A., Vandegrift, G. F., Gay, E. C., Fredrickson, D. R., Sabau, C. S., Chajko, D. J., Burris, L., Horwitz, E. P., Chiarizia, R., Diamond, H., Gatrone, R. C., Clark, G. A., and Martin, K. A.**, The TRUEX Process for Recovery of Plutonium and Americium from Nitric Acid Waste Solutions — Continuing Development Studies, FY 1986, Report ANL-87-3, Argonne National Laboratory, Argonne, IL, 1987.

26. **Kolarik, Z. J. and Horwitz, E. P.**, Extraction of metal nitrates with octyl(phenyl)-*N,N*-diisobutylcarbamoylmethylphosphine oxides in alkane diluents at high solvent loading, *Solvent Extr. Ion Exch.*, 6, 61, 1988.

27. **Kolarik, Z. J. and Horwitz, E. P.**, Extraction of neptunium and plutonium nitrates with octyl(phenyl)-*N,N*-diisobutylcarbamoylmethylphosphine oxide, *Solvent Extr. Ion Exch.*, 6, 247, 1988.

28. **Kolarik, Z. J. and Horwitz, E. P.**, Extraction of Pu(IV) and Th(IV) chlorides with *n*-octyl(phenyl)-*N,N*-diisobutylcarbamoylmethylphosphine oxide in tetrachloroethylene, *Solvent Extr. Ion Exch.*, 6, 649, 1988.

29. **Horwitz, E. P., Chiarizia, R., and Gatrone, R. C.**, The behavior of americium in the strip stages of the TRUEX process, *Solvent Extr. Ion Exch.*, 6, 93, 1988.

30. **Horwitz, E. P., Martin, K. A., and Diamond, H.,** Influence of diluent on the distribution behavior of octyl(phenyl)-N,N-diisobutylcarbamoylmethylphosphine oxide, *Solvent Extr. Ion Exch.,* 6, 859, 1988.

31. **Thiyagarajan, P., Diamond, H., and Horwitz, E. P.,** Small-angle neutron scattering studies of the aggregation of $Pr(NO_3)_3$-CMPO and $PrCl_3$-CMPO complexes in organic solvents, *J. Appl. Cryst.,* 21, 848, 1988.

32. **Diamond, H., Thiyagarajan, P., and Horwitz, E. P.,** Small angle neutron scattering studies of praseodymium–CMPO polymerization, *Solvent Extr. Ion Exch.,* 8, 503, 1990.

33. **Schulz, W. W. and Horwitz, E. P.,** The TRUEX process and the management of liquid TRU waste, *Sep. Sci. Technol.,* 23, 1191, 1988.

34. **Leonard, R. A., Vandegrift, G. F., Chamberlain, D. B., Chajko, D. J., Fredrickson, D. R., Stangel, J. E., Burris, L., Horwitz, E. P., Gatrone, R. C., Nash, K. L., and Rickert, P. G.,** The TRUEX Process for Recovery of Plutonium and Americium from Nitric Acid Waste Solutions — Final Report, Report ANL-90-6, Argonne National Laboratory, Argonne, IL, 1990.

35. **Schulz, W. W. and Horwitz, E. P.,** Multifunctional organophosphorus extractants: a status report on development and applications, in Proc. Int. Solvent Extraction Conference ISEC '88, Moscow, 1988, vol. 1, 25.

36. **Martin, K. A., Horwitz, E. P., and Ferraro, J. R.,** Infrared studies of bifunctional extractants, *Solvent Extr. Ion Exch.,* 1149, 1986.

37. **Kalina, D. G.,** Application of lanthanide induced shifts for the determination of solution structures of metal ion–extractant complexes, *Solvent Extr. Ion Exch.,* 2, 381, 1984.

38. **Chiarizia, R. and Horwitz, E. P.,** Hydrolytic and radiolytic degradation of octyl(phenyl)-N,N-diisobutylcarbamoylmethylphosphine oxide and related compounds, *Solvent Extr. Ion Exch.,* 4, 677, 1986.

39. **Nash, K. L., Gatrone, R. C., Clark, G. A., Rickert, P. G., and Horwitz, E. P.,** Hydrolytic and radiolytic degradation of OD(iB)CMPO: continuing studies, *Sep. Sci. Technol.,* 23, 1355, 1988.

40. **Nash, K. L., Rickert, P. G., and Horwitz, E. P.,** Degradation of TRUEX–dodecane process solvent, *Solvent Extr. Ion Exch.,* 7, 655, 1989.

41. **Chiarizia, R. and Horwitz, E. P.,** Diluent effects in the extraction of Am(III) from nitric acid solutions by selected carbamoyl-phosphoryl extractants and related monofunctional compounds, *Solvent Extr. Ion Exch.,* 10, 101, 1992.

42. **Chiarizia, R. and Horwitz, E. P.,** Secondary cleanup of TRUEX process solvent, *Solvent Extr. Ion Exch.,* 8, 907, 1990.

43. **Horwitz, E. P., Gatrone, R. C., and Chiarizia, R.,** U.S. Patent 4,741,857, A Method of Purifying Neutral Organophosphorus Extractants, 1988.

44. **Navratil, J. D. and Martella, L. L.,** Extraction behavior of americium and plutonium with mixed solvent extractants, *Sep. Sci. Technol.,* 16, 1147, 1981.

45. **Muscatello, A. C., Navratil, J. D., and Killion, M. L.,** The extraction of americium(III) by mixtures of tributylphosphate with dibutyl-N,N-diethylcarbamoyl-phosphonates and methylphosphonate (DBDECP and DBDECMP), *Solvent Extr. Ion Exch.,* 1, 127, 1983.

46. **Marcus, Y. and Kertes, S.,** *Ion Exchange and Solvent Extraction of Metal Complexes,* Wiley Interscience, New York, 1969, chap. 9.

47. **Barton, A. F. M.,** *CRC Handbook of Solubility Parameters and Other Cohesion Parameters,* CRC Press, Boca Raton, 1983.

48. **Schmidt, V. S., Rybakov, K. A., Shemenkov, S. A., and Rubisov, V. N.,** Physical distribution of extractants and extractives between aqueous solutions and various diluents. I. Distribution of trialkylamines and trialkylphosphates, *Sov. Radiochem.,* 23, 272, 1981.

49. **Davis, Jr., W.,** Radiolytic behavior, in *Science and Technology of Tributylphosphate,* Vol. 1, Schulz, W. W., and Navratil J. D., Eds., CRC Press, Boca Raton, 1984, 221.

50. **Bond, W. D.,** Purex solvent extraction chemistry, in *Light Water Reactor Nuclear Fuel Cycle,* Wymer, R. G. and Vondra, B. L., Eds., CRC Press, Boca Raton, 1981, 139.

51. **Vandegrift, G. F.,** Diluents for TBP extraction systems, in *Science and Technology of Tributylphosphate,* Vol. 1, Schulz, W. W. and Navratil, J. D., Eds., CRC Press, Boca Raton, 1984, 69.

52. **Mailen, J. C. and Tallent, O. K.,** Solvent cleanup and degradation: a survey and recent ORNL results, in Fuel Reprocessing and Waste Management, Proceedings, American Nuclear Society International Topical Meeting, Jackson, WY, August 1984, 431.

53. **Ozawa, M., Nemoto, S., Togashi, A., Kawata, T., and Onishi K.,** Partitioning of actinides and fission products in highly active raffinate from PUREX process by mixer-settlers, *Solvent Extr. Ion Exch.,* 10, 829, 1992.

54. **Vandegrift, G. F., Betts, S., Chamberlain, D. B., Copple, J. M., Dow, J. A., Everson. L., Farley, S. E., Hutter, J. C., Jaskot, R. J., Leonard, R. A., Nunez, L., Redfield, D., Regalbuto, M. C., Sedlet, J., Srinivasan, B., Tranovich, M. A., Van Deventer, E., Weber, S., and Wygmans, D. G.,** Separation Science and Technology Semiannual Progress Report, Report ANL-93/38, Argonne National Laboratory, Argonne, IL, 1993, 43.

55. **Lumetta, G. J.,** Pretreatment of Neutralized Cladding Removal Waste Sludge: Results of the Second Design Basis Experiment, Battelle Pacific Northwest Laboratory Report, PNL-9747, 1994.

56. **Horwitz, E. P., Chiarizia, R., Dietz, M. L., Diamond, H., and Nelson, D. M.,** Separation and preconcentration of actinides from acidic media by extraction chromatography, *Anal. Chim. Acta,* 281, 361, 1993.

57. **Horwitz, E. P., Diamond, H., Gatrone, R. C., Nash, K. L., and Rickert, P. G.,** TUCS: a new class of aqueous complexing agents for use in solvent extraction processes, in *Solvent Extraction,* Sekine, T., Ed., Elsevier, New York, 1990, 357.

58. **Schulte, L. D.,** Private communication.

59. **Horwitz, E. P.,** Combining extractant systems for the simultaneous extraction of transuranic elements and selected fission products, Proc. 1st Hanford Separation Science Workshop, Battelle Pacific Northwest Laboratory Report, PNL-SA-21775, 1993, Vol. II, 29.

60. **Horwitz, E. P., Dietz, M. L., Diamond, H., Rogers, R. D., and Leonard R. A.,** Combining TRU•SREX Extraction/Recovery Process, in *Proc. Int. Solvent Extraction Conf. (ISEC'93),* Vol. 3, Elsevier, New York, 1993, 1805.

61. **Horwitz, E. P., Dietz, M. L., Diamond, H., Rogers, R. D., and Leonard R. A.,** Advanced chemical separations in support of the clean option strategy, Global 93: Int. Conference and Technology Exhibition, in *Future Nuclear System: Emerging Fuel Cycles and Waste Disposal Options,* Vol. 1, American Nuclear Society, LaGrange Park, IL, 1993, 39.

62. **Horwitz, E. P., Dietz, M. L., Diamond, H., Rogers, R. D., and Leonard R. A.,** Combined TRUEX–SREX extraction/recovery process, in *Chemical Pretreatment of Waste for Disposal,* Schulz, W. W. and Horwitz, E. P., Eds., Plenum, New York, NY, 1994, 81.

Chapter 2

Liquid Extraction, the TRUEX Process — Mathematical Modeling

Jack D. Law and Thomas E. Carleson

I. INTRODUCTION

The plutonium–uranium extraction (PUREX) process has been universally used for the reprocessing of spent nuclear fuel. Many computer codes have been developed for the mathematical modeling of the PUREX process including the SEPHIS,[1] PULSTX,[2] SOLVEX,[3] and MIXSET[4] computer codes. A detailed description of computational techniques available for TBP extraction processes has been published.[5]

The transuranic extraction (TRUEX) process was developed by E. P. Horwitz and co-workers at Argonne National Laboratory (ANL) for the separation of transuranic elements from acidic nuclear waste solutions.[6] This solvent extraction process can be used to reduce the quantity of transuranics in a waste solution to a level where the waste can be classified as non-TRU. Non-TRU is defined as containing less than 100 nCi of alpha-emitting nuclides per gram of material. The generic TRUEX model (GTM) was developed by Vandegrift and co-workers at Argonne National Laboratory.[7] The GTM is a computer modeling tool used for development and design of TRUEX processes for the treatment of site-specific waste streams. A large database of experimental distribution coefficients was generated to support the model.[7]

The GTM has been used to develop flow sheets for the separation of actinides from Idaho Chemical Processing Plant (ICPP) sodium-bearing waste (SBW) and from dissolved Hanford neutralized cladding removal waste (NCRW) sludge. A comparison of the predicted GTM results with experimental results has been made to determine the applicability of the GTM for these specific waste streams.

II. GENERIC TRUEX MODEL DESCRIPTION

The GTM contains four main sections: SASSE, SASPE, SPACE, and INPUT/OUTPUT.[7] The GTM is executed using Microsoft Excel software on Macintosh and IBM PC or compatible computers. The spreadsheet algorithm for stagewise solvent extraction (SASSE) section calculates user specified multistage flow sheets using distribution coefficients calculated in the spreadsheet algorithm for speciation and partitioning equilibria (SASPE) section. For a specific feed solution input and a set of process goals and constraints, the GTM will calculate the compositions of all streams exiting each stage of a solvent extraction flow sheet at steady-state conditions. The size of plant and cost estimation (SPACE) section estimates the space and cost requirements for installation of a user-specified TRUEX process. Calculations can be performed based on processing with centrifugal contactor, mixer-settler, or pulsed-column equipment in a glove box, shielded cell, or canyon facility. The INPUT/OUTPUT section is a menu-driven interface for input of information required by the GTM for the particular option chosen. A list of the ten options available with the GTM are listed in Table 1.

0-8493-4876-5/96/$0.00+$.50
© 1996 by CRC Press, Inc.

Table 1 Generic TRUEX Model Options[7]

Option No.	Description
1	Calculate complete generic TRUEX model for a specific feed solution
2	Calculate the following:
	charge balance check of complex aqueous solution
	density of complex aqueous solution
	ionic strength of complex aqueous solution
	activities of H^+, NO_3^-, and water
3	Calculate oxalic-acid additions to fission-product-containing waste
4	Calculate D values for user-specified aqueous phase (org. phase assumed equilibrated)
5	Calculate D values for equilibration of user-specified aqueous and organic phases
6	Flow sheet analysis with user-specified distribution ratios
7	Generate a TRUEX flow sheet for a user-specified feed
8	Estimate space and costs for user-specified flow sheet
9	Estimate solvent degradation for specific TRUEX process
10	Generate reports from existing TRUEX flow sheets or space and cost calculations

III. APPLICATION OF THE GTM FOR TRUEX PROCESSING OF ICPP SBW

The GTM has been used at the ICPP to develop flow sheets for the separation of actinides from acidic SBW solution. Reduction in the activity of actinides to <10 nCi/g, along with processes for the separation of strontium and cesium, would allow most of the SBW to be disposed of as low-level waste.

A. EXPERIMENTAL DESCRIPTION

Experimental data generated for the separation of actinides from SBW were compared to results predicted by the GTM in order to determine the applicability of the GTM to ICPP SBW. These data were generated from TRUEX tests performed on simulated ICPP SBW with radioactive tracers.[8] The composition of the SBW simulant is given in Table 2. Extraction, scrub, strip, and solvent wash distribution coefficients, D, were determined experimentally for americium, neptunium, plutonium, uranium, and technetium. The distribution coefficient is defined as the activity of the isotope in the organic phase divided by the activity of the isotope in the aqueous phase.

Individual tracers were spiked into simulated SBW. The isotopes used as tracers for the actinides and technetium were ^{241}Am, ^{237}Np, ^{238}Pu, ^{233}U, and ^{95m}Tc. All contacts were performed in test tubes for 1 min at an O/A = 1 and room temperature ($27 \pm 2°C$). Phases were separated by centrifuging at 5000 rpm for 1 min. The plutonium is believed to have been in the (IV) valence state. An experimental schematic of the batch contacts is given in Figure 1.

The TRUEX solvent (0.2 M CMPO and 1.4 M TBP in n-dodecane) was preequilibrated with nonspiked waste simulant by contacting the solvent three times, fresh simulant being used each time, at an O/A = 1. The preequilibrated TRUEX solvent was contacted with the spiked simulant to determine the extraction distribution coefficient. This was defined as contact E1. The aqueous raffinate from the first extraction was used as a feed in the second extraction, where it was contacted with fresh preequilibrated TRUEX solvent and a second extraction distribution coefficient was obtained (contact E2). This process was then repeated to obtain a third extraction distribution coefficient (contact E3).

Table 2 Composition of Sodium-Bearing
Waste Simulant

Component	M	Component	M
Acid (H⁺)	1.35	K	1.43E–1
Al	6.48E–1	Mn	1.42E–2
B	1.71E–2	Mo	1.49E–3
Cd	2.29E–3	Na	1.26
Ca	3.91E–2	NO₃	4.46
Ce	3.88E–4	Ni	2.20E–3
Cl	3.52E–2	Pb	1.08E–3
Cs	7.52E–5	PO₄	<9.18E–3
Cr	6.56E–3	Sr	1.70E–4
F	9.66E–2	SO₄	3.86E–2
Fe	3.10E–2	Zr	1.00E–3
Hg	2.88E–3		

Three consecutive scrubs were performed, contacts Sc1 through Sc3. The scrub solution used was 0.2 M HNO₃. The scrubbed organic was then contacted with a stripping agent three consecutive times, contacts St1 through St3. The stripping agents utilized included 0.01 M HNO₃, 0.1 M H₂C₂O₄, and a thermally unstable complexant (TUCS). The TUCS compound used was 1-hydroxyethane-1,1-diphosphonic acid. The strip solution used consisted of 0.25 M of this TUCS compound in 0.05 M HNO₃, which is identical to the composition utilized in the GTM for a complexant wash. Previous studies indicate this to be an effective stripping agent for the TRUEX process.[9,10] The organic raffinate from the 0.01 M HNO₃ strip contacts was then washed with 0.25 M Na₂CO₃ one or more times.

B. COMPARISON OF GTM TO EXPERIMENTAL DATA WITH ICPP WASTE SOLUTION

The experimental distribution coefficients and the distribution coefficients predicted by the GTM, Version 2.5, are presented in Table 3. The GTM results were generated by modeling the experimental flow sheets in as exact a manner as possible. This included using the GTM to simulate batch contacts with no entrainment, specifying fresh organic which has been preequilibrated with simulant, and using solution compositions identical to experimental compositions.

As can be seen from Table 3, the experimental extraction distribution coefficients obtained for uranium are lower than predicted by the GTM. However, the second and third forward extractions (E2 and E3) were approaching the detection limits of the analytical instrumentation and cannot be accurately compared to the GTM results.[8] The GTM and the experimental data both indicate that uranium is readily extracted from the SBW. For the 0.2 M HNO₃ scrub the experimental uranium distribution coefficient data compare very well with the GTM. Both the GTM and experimental data indicate that a 0.01 M HNO₃ strip solution is not effective as a stripping reagent for uranium. The GTM predicts better stripping results with 0.1 M H₂C₂O₄ than were experimentally obtained. The GTM and experimental data indicate very low distribution coefficients for uranium with a TUCS strip or 0.25 M Na₂CO₃ solvent wash.

The experimental extraction distribution coefficients obtained for americium were also lower than predicted by the GTM. However, both indicate that americium is readily

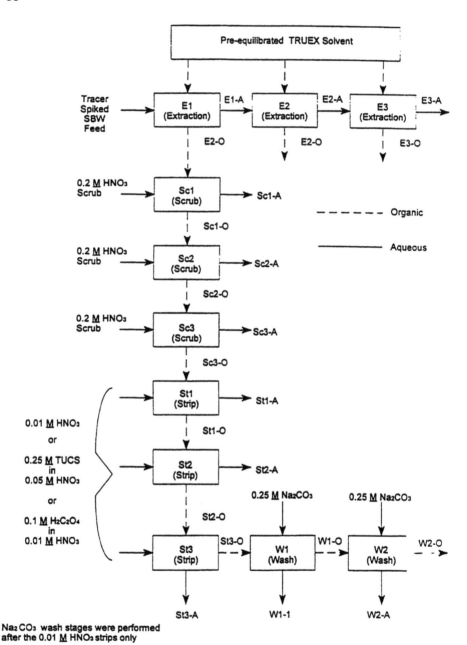

Figure 1 Experimental schematic drawing of batch contacts with ICPP SBW.

extractable from the SBW. The experimental scrub distribution coefficient data for americium compare very well with the GTM. For the 0.01 M HNO$_3$ strip, the GTM predicts better stripping than was obtained experimentally. Both the GTM and the experimental data indicate very low americium distribution coefficients for a 0.1 M H$_2$C$_2$O$_4$ strip, a TUCS strip, and a 0.25 M Na$_2$CO$_3$ solvent wash.

Table 3 Experimental vs. Estimated (GTM) Distribution Coefficients for TRUEX Extractions from Simulated SBW[a]

Sample	Description	D_{U-233}		D_{Am-241}		D_{Pu-238}		D_{Np-239}		D_{Tc-95m}	
		Exptl.	GTM	Exptl.	GTM	Exptl.	GTM	Exptl.	GTM	Exptl.	GTM
Forward	E1	1710	3700	33	170	486	3400	2.28	2.40	4.01	0.80
extraction	E2	170	3800	30	200	7.1	3500	1.90	2.30	4.57	0.98
	E3	BDL	3900	27	230	BDL	3500	1.40	2.10	4.10	1.15
0.2 M HNO₃	Sc1	460	470	19.5	21.6	94.5	1100	—	0.217	10.7	7.6
scrubs	Sc2	240	258	9.3	10.3	44.5	430	—	0.116	12.6	10.4
	Sc3	160	154	4.9	5.55	20.5	220	—	0.083	11.7	10.3
0.01 M HNO₃	St1–N	33	17.2	0.59	0.421	6.0	32.1	2.60	0.185	7.8	5.45
strips	St2–N	7.6	11.7	0.18	0.013	1.6	2.52	1.83	0.074	1.5	1.25
	St3–N	3.3	0.522	0.91	4.2E-03	22	1.15	0.366	0.022	0.65	0.81
0.1 M H₂C₂O₄	St1–O	2.38	0.041	0.09	2.4E-02	0.24	1.0E-03	—	0.18	1.94	4.51
strips	St2–O	0.81	1.1E-03	0.02	1.0E-03	—	1.0E-03	—	6.8E-02	—	2.85
	St3–O	0.59	1.1E-03	BDL	1.0E-03	—	1.0E-03	—	1.4E-02	—	2.92
TUCS	St1–T	2.5E-04	1.2E-03	5.0E-04	7.0E-05	2.1E-03	3.5E-02	—	1.0E-03	—	1.9
strips	St2–T	BDL	1.2E-03	BDL	7.0E-05	2.8	3.5E-02	—	1.0E-03	—	1.9
	St3–T	BDL	1.2E-03	BDL	7.0E-05	2.1	3.5E-02	—	1.0E-03	—	1.9
0.25 M Na₂CO₃	St1–C	3.1E-03	8.8E-04	0.003	7.0E-05	2.6E-03	1.2E-3	0.1555	0.0015	—	0.079
Solvent Wash	St2–C	BDL	8.8E-04	BDL	7.0E-05	0.8	1.2E-3	BDL	—	—	0.079

†BDL = below detection limits: activity in either the organic or aqueous phase was below detection limits of the analytical instrumentation.

The experimental extraction distribution coefficients obtained for plutonium were also lower than predicted by the GTM. However, the second and third forward extractions (E2 and E3) were approaching the detection limits of the analytical instrumentation and cannot be accurately compared to the GTM results.[8] The GTM and the experimental data both indicate that plutonium is readily extractable from SBW. The experimental scrub distributions were also lower than predicted by the GTM. The GTM and experimental data both indicate that plutonium is not stripped with 0.01 M HNO_3. The GTM predicts better stripping of plutonium with 0.1 M $H_2C_2O_4$ than was obtained experimentally. Both the GTM and experimental data indicate very low plutonium distribution coefficients with a TUCS strip or 0.25 M Na_2CO_3 solvent wash.

The experimental extraction distribution coefficients for neptunium compare very well with the GTM. The GTM predicts much better stripping of neptunium with 0.01 M HNO_3 than was obtained experimentally. The GTM also predicts a lower distribution coefficient with the 0.25 M Na_2CO_3 solvent wash than was obtained experimentally. Experimental scrub, oxalic acid strip, and TUCS strip data were not obtained.

Experimental results indicate that technetium is extracted from the SBW (D ≈ 4). The GTM predicts technetium to be largely inextractable (D < 1). Scrub, 0.01 M HNO_3 strip, and 0.1 M $H_2C_2O_4$ strip data for technetium all compare favorably to the distribution coefficients predicted by the GTM. Data for a TUCS strip and Na_2CO_3 solvent wash were not obtained.

Overall, the GTM efficiently predicted the behavior of U, Am, Pu, Np, and Tc with SBW solution. However, the GTM predicts extraction distribution coefficients for U, Am, and Pu that are consistently higher than were obtained experimentally. The overall extraction behavior predicted by the GTM is in agreement with the experimental results. The one major exception is with technetium. Experimental results indicate that technetium is readily extracted from the SBW simulant whereas the GTM predicts that technetium will be largely inextractable. Experimental distribution coefficients can be input into the GTM for those species for which the GTM is not accurate.

C. POTENTIAL FLOW SHEET FOR THE TREATMENT OF ICPP SBW

The distribution coefficients generated during the tracer experiments have been used in conjunction with the GTM to develop a potential process flow sheet for the large-scale treatment of ICPP SBW. The GTM allows input of user-specified distribution coefficients to develop and compare potential flow sheets. A flow sheet for the treatment of SBW is given in Figure 2.[8] The flow sheet consists of five stages of extraction, four stages of 0.2 M HNO_3 scrub, five stages of 0.25 M TUCS strip, two stages of 0.25 M Na_2CO_3 solvent wash, and one stage of 0.1 M HNO_3 rinse. The O/A ratios in the extraction, scrub, strip, carbonate wash, and acid rinse sections are 0.333, 3.0, 1.0, 5.0, and 10.0, respectively. The average composition of the ICPP SBW was used as the feed for the flow sheet; this composition is given in Figure 2.

The concentrations of the nonradioactive waste components in the flow sheet were generated by the GTM using the existing data base of distribution coefficients. Experimental distribution coefficients, obtained from the tracer studies with SBW, were input into the GTM for americium, neptunium, plutonium, uranium, and technetium. Where experimental data have not been obtained or where the experimental distribution coefficients are in very good agreement with the GTM, distribution coefficients as predicted by the GTM were used.

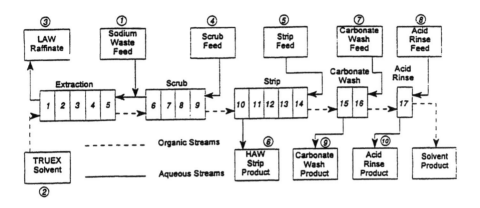

Component	Sodium Waste Feed (Molar) 1	TRUEX Solvent Feed 2	LAW Raffinate (Molar) 3	Scrub Feed (Molar) 4	TUCS Strip Feed (Molar) 5	HAW Strip Product (Molar) 6	Carbonate Wash Feed (Molar) 7	Acid Rinse Feed (Molar) 8	Carbonate Wash Product (Molar) 9	Acid Rinse Product (Molar) 10
HNO3	1.5		1.25	0.2	0.05	0.05		0.1		0.09
Na2CO3							0.25		0.25	
HEDPA					0.25	0.25				
Fe	2.45E-02		2.18E-02			3.66E-06			2.62E-16	4.52E-19
Al	0.545		0.485			3.09E-10			1.24E-20	2.07E-23
Na	1.327		1.182			7.91E-10			3.02E-20	5.03E-23
Ca	4.35E-02		3.87E-02			2.46E-11			9.91E-22	1.85E-24
Cd	1.84E-03		1.64E-03			1.42E-08			5.73E-19	9.55E-22
Ag	1.76E-06		1.57E-06			9.97E-15			4.01E-25	6.67E-28
F	6.32E-02		5.62E-02			7.92E-06			4.86E-05	6.25E-09
SO4	3.23E-02		2.88E-02			7.41E-12			2.98E-22	3.60E-25
PO4	9.44E-03		8.41E-03			1.52E-09			6.12E-20	7.40E-23
B	1.56E-02		1.04E-02			1.06E-02			1.88E-13	3.12E-16
Hg	2.03E-03		1.79E-03			4.47E-06			4.43E-08	6.12E-08
Cr(VI)	5.74E-03		9.94E-04			1.23E-02			4.96E-13	8.25E-16
Radionuclides										
Sr	6.23E-06		5.55E-06			3.53E-15			1.42E-25	2.36E-28
Cs	3.10E-04		2.76E-04			1.79E-13			7.06E-24	1.17E-26
Ba	4.59E-05		4.09E-05			2.60E-14			1.05E-24	1.74E-27
Eu	1.57E-08		5.26E-15			4.19E-08			8.75E-19	1.42E-22
Ce	2.08E-10		3.37E-17			5.55E-10			2.88E-20	3.94E-24
Tc	1.01E-06		9.85E-07			1.30E-06			5.47E-06	2.38E-07
ACTINIDES										
Am	8.82E-06		1.15E-12			2.35E-07			6.13E-18	1.79E-21
Np	3.16E-06		1.59E-06			3.66E-06			1.09E-16	5.45E-17
Pu	2.29E-05		2.41E-14			6.11E-06			8.87E-15	1.85E-20
U	3.77E-04		4.44E-14			1.01E-03			2.08E-14	1.91E-19
Activity nCi/g	348.5		0.22			1109				
Volume	400.0	150.0	448.3	50.0	150.0	150.0	30.0	15.0	30.0	15.8

Figure 2 TRUEX flow sheet for the treatment of SBW.

Based on the GTM, greater than 99.99% of the americium, plutonium, and uranium are removed from the SBW feed. However, only 43.4% of the Np(V) is removed from the SBW feed due to the low distribution coefficient of Np(V) ($D_{Np} = 1.9$). Reduction of the Np(V) to Np(IV) in the SBW feed would greatly increase the neptunium distribution coefficient in the extraction section. Even with Np(V) in the SBW feed, the total activity of the actinides is reduced from 348.5 to 0.22 nCi/g. The TUCS strip section removed essentially all of the actinides from the TRUEX solvent and only a very small fraction of the actinides is stripped in the carbonate wash and acid rinse sections.

The proposed flow sheet indicates 89% of the technetium is extracted from the SBW feed. Approximately 54% of the technetium extracted from the SBW feed is stripped from the TRUEX solvent in the TUCS strip section and the remaining technetium is stripped in the carbonate wash section.

The process flow sheet developed using the GTM with the experimental distribution coefficients obtained from the tracer studies indicates that large scale treatment of SBW with the TRUEX process is viable.

IV. APPLICATION OF THE GTM FOR TRUEX PROCESSING OF DISSOLVED HANFORD NCRW SLUDGE

A. FLOW SHEET DESCRIPTION

A flow sheet for TRUEX processing of dissolved Hanford neutralized cladding removal waste (NCRW) sludge is being developed and tested by workers at Pacific Northwest Laboratory (PNL).[11,12] As part of the current study, the NCRW flow sheet was modeled using the GTM, and the calculated distribution coefficients were compared to those obtained experimentally at PNL. The NCRW flow sheet involves washing and dissolution of the sludge, extraction of uranium utilizing PUREX technology, and TRUEX processing of the actinides. Ascorbic acid is added to the solvent extraction feed solution to adjust the Pu valence state to (III) in order to prevent the extraction of Pu in the TBP extraction cycle. This will allow the Pu to be separated from the U. The TRUEX portion of the flow sheet, as shown in Figure 3, consists of three stages of extraction, a two-stage 0.15 M $H_2C_2O_4$ scrub, a four-stage 5.0 M HNO_3 scrub, a two-stage 0.01 M HNO_3 scrub, a two-stage 0.01 M HNO_3 strip, a four-stage 0.01 M $H_2C_2O_4$ strip, and a single-stage 0.25 M Na_2CO_3 wash. O/A ratios for the flow sheet are 0.29 in the extraction section, 0.75 in the 0.15 M $H_2C_2O_4$ scrub section, 1.0 in the 5.0 M HNO_3 scrub section, 1.5 in the 0.01 M HNO_3 scrub section, 0.75 in the 0.01 M HNO_3 strip section, 3.0 in the 0.01 M $H_2C_2O_4$ strip section, and 9.6 in the 0.25 M Na_2CO_3 wash section. Design basis experiments have been performed to test this flow sheet.[11,12]

B. EXPERIMENTAL DESCRIPTION

The design basis experiment was performed using a series of batch contacts to simulate the countercurrent extraction flow sheet.[11,12] For the TRUEX portion of the flow sheet, raffinate from the uranium extraction section of the flow sheet feeds the TRUEX extraction section. A blend of this feed and the three scrub streams was used as a feed solution for the first extraction contact (stage 3) with fresh organic (0.2 M CMPO and 1.4 M TBP in a normal paraffin hydrocarbon). Raffinate from this contact was again contacted with fresh organic (stage 2). This was repeated for the third extraction contact (stage 1). Blends of the successive scrub streams were used as scrub feed solution for the scrub contacts. Blends of the successive strip streams were used as strip feed solution for the strip contacts. Organic from each previous stage was used for the scrub and strip contacts. O/A ratios equivalent to those of the flow sheet were used.

C. COMPARISON OF DESIGN BASIS EXPERIMENTAL RESULTS TO GTM RESULTS

Results of the design basis experiment are compared to results predicted by the GTM in Table 4 for uranium, americium, and plutonium distribution coefficients. Distribution

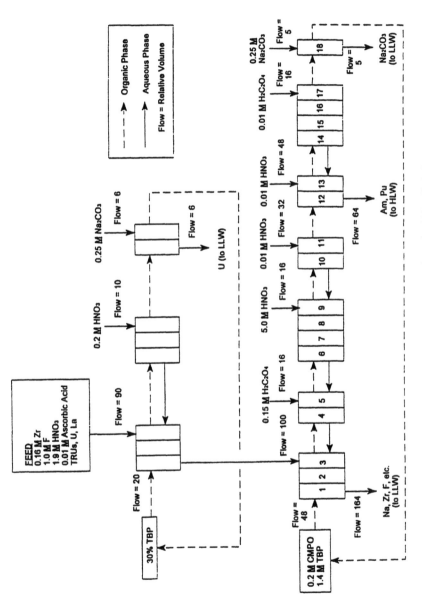

Figure 3 Solvent extraction flow sheet for NCRW sludge.

Table 4 Experimental vs. Estimated (GTM) Distribution Coefficients for TRUEX Extractions from Dissolved Hanford NCRW Sludge

Sample description		D_U		D_{Am-241}		D_{Pu}		
		Exptl.	GTM	Exptl.	GTM	Exptl.	GTM Pu(IV)	GTM Pu(III)
Forward	Stage 1	—	372	17	30.6	10	19.5	30.6
extraction	Stage 2	—	422	14	32.1	12	23.7	32.1
	Stage 3	34	485	11	32.7	17	29.4	32.7
0.15 M H$_2$C$_2$O$_4$ scrub section	Stage 5	54	212	14	26.6	14	19.5	26.6
5.0 M HNO$_3$ scrub section	Stage 9	400	734	21	29.9	200	1040	29.9
0.01 M HNO$_3$ scrub section	Stage 11	86	148	2.6	05.51	22	14.7	5.51
0.01 M HNO$_3$ strip section	Stage 13	2.5	0.259	0.007	0.0037	0.006	0.001	0.0037
0.01 M H$_2$C$_2$O$_4$ strip section	Stage 17	1.6	0.001	—	0.001	—	0.001	0.001

coefficients for the extraction section and the last stage of each of the scrub and strip sections are given. The GTM results were generated by modeling each individual batch contact performed in the design basis experiment. Version 3.1 of the GTM was used to perform the modeling.

As can be seen from Table 4, the uranium distribution coefficients predicted by the GTM in the extraction section are an order of magnitude higher than the experimental distribution coefficient. However, both indicate that uranium is readily extractable. The experimental scrub distribution coefficients compare fairly well to those predicted by the GTM. The 0.15 M H$_2$C$_2$O$_4$, 5.0 M HNO$_3$, and 0.01 M HNO$_3$ scrub distribution coefficients predicted by the GTM are higher than the experimental distribution coefficients by a factor of 4, 1.8, and 1.7, respectively. The GTM predicts that uranium is stripped in both the 0.01 M HNO$_3$ and 0.01 M H$_2$C$_2$O$_4$ strip sections, whereas the experimental results indicate that uranium is not stripped.

The americium distribution coefficients predicted by the GTM in the extraction section are about two times higher than the experimental distribution coefficients. However, both indicate that americium is readily extractable. The scrub experimental distribution coefficients are slightly lower than predicted by the GTM. Both the GTM and the experimental data indicate very low distribution coefficients for americium with a 0.01 M HNO$_3$ strip. Experimental data were not available for the 0.01 M H$_2$C$_2$O$_4$ strip section.

The plutonium distribution coefficients predicted by the GTM in the extraction section are two to three times higher than the experimental distribution coefficients, depending on the valence state of the Pu. It is expected that the plutonium is primarily present as Pu(III) due to the addition of ascorbic acid to the feed solution. The experimental 0.15 M H$_2$C$_2$O$_4$ scrub data is slightly lower than predicted by the GTM. The experimental 5.0 M HNO$_3$ and 0.01 M HNO$_3$ scrub distribution coefficients are slightly higher than predicted by the GTM. Both the GTM and experimental data indicate very low distribution coefficients for the 0.01 M HNO$_3$ strip. Data were not available for the 0.01 M H$_2$C$_2$O$_4$ strip section.

Overall, the GTM predicts extraction and scrub distribution coefficients for U, Am, and Pu that are consistently higher than were obtained experimentally. However, the overall extraction behavior predicted by the GTM is in agreement with the experimental results. The one major discrepancy between the experimental and GTM results was with uranium in the strip sections. The experimental data indicate that uranium is not stripped by 0.01 M HNO_3 or 0.01 M $H_2C_2O_4$, whereas the GTM predicts that uranium will be stripped, particularly with a 0.01 M $H_2C_2O_4$ strip ($D_U = 0.001$).

V. CONCLUSIONS

The generic TRUEX model can be a very useful tool for the development of TRUEX processes for the separation of transuranic elements from nuclear waste solutions. Overall, the GTM efficiently predicted the behavior of most of the components evaluated in the ICPP sodium-bearing waste and dissolved Hanford NCRW sludge solutions. The one trend noted for both waste solutions is that the extraction distribution coefficients predicted by the GTM for Am, Pu, and U are consistently higher than distribution coefficients obtained experimentally.

It is apparent that the behavior of some components in specific waste streams will vary from the behavior predicted by the GTM. Experimental data generated for the specific waste stream will identify where discrepancies exist. When developing TRUEX flow sheets, distribution coefficients for these components can be input directly into the GTM.

REFERENCES

1. **Mitchell, A. D.,** "SEPHIS-MOD4: A User's Manual to a Revised Model of the Purex Solvent Extraction System," ORNL-5471, Oak Ridge National Laboratory, 1979.
2. **Bendixsen, C. L.,** "A User's Manual for the Pulse Column Mathematical Model of Purex Extraction Systems," WIN-302, June 1989.
3. **Scotten, W. C.,** "SOLVEX — A Computer Program for Simulation of Solvent Extraction Processes," DP - 1391, September 1975.
4. **Gonda, K., Oka, K., and Fukuda, S.,** "Calculation Code Revised MIXSET for Purex Process," PNCT 841-79-26, Power Reactor and Nuclear Fuel Development Corporation, 1979.
5. **Mills, A. L.,** Computation and Modeling, in *Science and Technology of Tributyl Phosphate*, Schulz, W. S., Burger, L. L., and Navratil, J. D., Eds., CRC Press, Boca Raton, FL, 1990, chap. 5.
6. **Horwitz, E. P., Kalina, D. G., Diamond, H., and Vandegrift, G. F.,** "The TRUEX Process — A Process for Extraction of the Transuramic Elements from Nitric Acid Waste Utilizing Modified PUREX Solvent," *Solvent Extr. Ion Exch.* 3(1&2), 75, 1985.
7. **Vandegrift, G. F., Chamberlain, D. B., Conner, C., Copple, J. M., Dow, J. A., Everson, L., Hutter, J. C., Leonard, R. A., Nunez, L. Regalbuto, M. C., Sedlet, J., Srinivasan, B., Weber, B., and Wygmans, D. G.,** "Development and Demonstration of the TRUEX Solvent Extraction Process," WM'93, Tucson, AZ, February 1993.
8. **Herbst, R. S., Brewer, K. N., Law, J. D., Tranter, T. J., and Todd, T. A.,** "TRUEX Partitioning Studies Applied to ICPP Sodium-Bearing Waste," WINCO-1206, May 1994.
9. **Horwitz, E. P., Diamond, H., Gatrone, R. C., Nash, K. L., and Rickert, P. G.,** "TUCS: A New Class of Aqueous Complexing Agents for Use in Solvent Extraction Processes," Argonne National Laboratory, Chicago, IL, ANL/CP-71262, 1990.
10. **Lumetta, G. J. and Swanson, J. L.,** "Evaluation of 1-Hydroxyethane-1,1-diphosphonic Acid and Sodium Carbonate as Stripping Agents for the Removal of Am(III) and Pu(IV) from TRUEX Process Solvent," *Sci. Technol.* 28, 43, 1993.
11. **Lumetta, G. J. and Swanson, J. L.,** "Pretreatment of Neutralized Cladding Removal Waste Sludge: Status Report," PNL-8558, March 1993.
12. **Lumetta, G. J.,** "Pretreatment of Neutralized Cladding Removal Waste Sludge: Results of the Second Design Basis Experiment," PNL-9747, May 1994.

Chapter 3

Liquid Extraction, New Extraction Agents — Crown Ether Extractants

Mikhail K. Beklemishev and Chien M. Wai

I. INTRODUCTION

Macrocyclic polyethers (crown ethers) are a class of specific chelating agents capable of forming stable complexes with metal ions in aqueous solutions selectively. Since Pedersen first published his studies of alkali metal extraction with crown ethers in 1967, many reports have appeared in the literature describing the selectivity and general behavior of various synthetic crown ethers, particularly with respect to complexation with the alkali and the alkaline earth metal ions.[1] These reports have attracted the attention of separation chemists because selective extraction of the alkali and the alkaline earth metal ions is difficult to accomplish with other types of reagents. According to the principle of soft and hard acids and bases, oxygen-containing ligands such as crown ethers are hard bases that prefer to associate with hard acids including the alkali metals, the alkaline earth metals, and the f-block elements (lanthanides and actinides). Most of the metal ions of interest to nuclear waste management belong to these groups, thus predetermining the range of crown reagents applicable to their extraction; these reagents are oxygen-containing macrocycles, i.e., crown ethers, rather than nitrogen- or sulfur-containing macrocyclic reagents.

Table 1 shows the cavity sizes of some commonly available crown ethers and the ions of the Group IA, IIA, and the lanthanium series elements with matching ionic diameters.[2] As shown in Table 1, the 14-crown-4 host with a cavity diameter in the range of 1.2 to 1.5 Å is selective for Li^+ (ionic diameter 1.36 Å) and the 15-crown-5 host with a cavity diameter of 1.7 to 2.3 Å is selective for Na^+ (ionic diameter 1.90 Å), etc. This is the so-called ionic radius–cavity size compability concept for selective extraction of metal ions by crown ethers. This concept, though not working for most transition metals,[2] is nevertheless a fairly valuable principle for designing crown ethers for selective extraction of the alkali metal, the alkaline-earth metal, and the lanthanide ions.

The complex formed by a neutral crown ether and a metal cation is naturally charged. Lipophilic counteranions such as picrate, thiocyanate, and organosulfonic acids are often used to neutralize the charges carried by the crown–metal complexes and to facilitate their transport into organic solvents. Modification of crown structure by attaching negatively charged functional groups to the macrocyclic host can eliminate the need for specific counteranions. Crown ethers with attached proton-ionizable groups may function both as a cation exchanger and a coordinator. In recent years, a variety of proton-ionizable crown ethers have been synthesized and tested for selective extraction of the alkali metals, the alkaline metal ions, and the f-block elements.[3]

The scope of this chapter is to describe promising macrocycle solvent extraction systems for potential nuclear waste management applications. The materials of this chapter are classified by metals extractable by macrocyclic compounds. For cesium and

0-8493-4876-5/96/$0.00+$.50
© 1996 by CRC Press, Inc.

Table 1 Diameters of Crown Ether Cavities
Calculated from Crystallographic Data and
Diameters of Alakli and Alkaline Earth Metal Ions

Crown	Cavity diameter (Å)	Cation	Cation diameter (Å)
12C4	1.2	Li^+	1.20
15C5	1.72–1.84	Na^+	1.90
		Ca^{2+}	1.98
		Sr^{2+}	2.20
18C6	2.67–2.86	K^+	2.66
		Rb^+	2.96
		Ba^{2+}	2.70
21C7	3.4–4.3	Cs^+	3.34

Data taken from Reference 2, page 82.

strontium, where many papers dealing with crown ether complexation exist in the literature, we consider mostly the papers devoted to nuclear waste problems. For technetium and actinides, where there are fewer published papers dealing with crown ether extraction, we tried to cover all of the important ones.

II. EXTRACTION OF CESIUM

A comprehensive study of dicyclohexano-18-crown-6 (DC18C6) for cesium extraction was reported by McDowell and co-workers in 1980–81.[4,5] Liquid cation exchangers such as di(2-ethylhexyl)-phosphoric acid (HDEHP) were chosen as the counteranions in their study. HDEHP allows the extraction of metal ions at lower pHs, and, being highly soluble in organic solvents, HDEHP is not lost with the aqueous phase in solvent extraction. Cesium can be extracted by this system at pH as low as 1, the extraction mechanism being cesium–proton ion exchange. At pH > 4.5, where most of HDEHP is converted to its sodium salt, the dominant process is the exchange of Na^+ for Cs^+. To attain high cesium uptake (D_{Cs} = 50 at pH 4 to 5), quite high amounts of the reagents were necessary (0.1 to 0.25 M HDEHP and DC18C6). Benzene was found to be an efficient solvent for this extraction system.

DC18C6

btBB18C6 and analogues n = 1, btBB18C6;
n = 2, btBB21C7; n = 3, btBB24C8.

Substituents in the macrocyclic ring have a large effect on extraction of cesium. In acidic solutions benzo-substituted crowns are generally more efficient. For instance, D_{Cs} are 30 and 3.8 for bis(tert-butylbenzo)-18-crown-6 (btBB18C6) and DC18C6, respectively, in 0.2 M HNO$_3$, 0.1 M di-n-dodecylnaphthalene sulphonic acid (HDDNS), and 0.05 M crown ether.[6]

Myasoedova et al. studied a series of crowns starting with DC18C6 and ending up with dicyclohexano-24-crown-8 (DC24C8) for cesium and showed that the extraction efficiency increases at least four times in this row.[7] Nevertheless, the cavity of 24C8 is too large to fit cesium ion, and the selectivity of Cs^+ by crowns based on this host unit is not so high. For dibenzo-24-crown-8, a tolerance ratio Cs/Na equal to 20 was found while extracting cesium in the presence of 0.5 g/l sodium.[8] It was also possible to separate cesium from Zr, Hf, Th, Pb, Bi, and other ions present in fission products using this crown ether.

In one of the early publications, various 24-membered crowns with alkyl groups attached to benzene and cyclohexano rings were evaluated.[9] These reagents were designed to gain lipophilicity necessary to work in a PUREX-compatible nonpolar solvent, which also contained HDEHP and 25 vol% TBP in kerosene. However, the distribution coefficients of cesium were low (D_{Cs} = 1.5 at pH 5.5) even for the most efficient reagent *bis(tert*-butylbenzo)-24-crown-8 (btBB24C8).

Blasius and Nilles reported the extraction of cesium with crown ethers with different cavity sizes, varying from 18C6 to 30C10.[10] Dibenzo-21-crown-7 (DB21C7) was shown to be the most efficient reagent for extraction of cesium from medium-active waste (MAW) simulant solutions that contained 1 M HNO_3, 0.5 M $NaNO_3$, 1 g/l U, 0.8 g/l Pb, 0.2 g/l Ca and Mg, 0.15 g/l Fe, and also Cu, Zn, Cr, K, Mn, Ni, Zr, Ru, Ce, Sr, Sb, and Nb. For extraction into nitrobenzene with 0.012 M crown ether, the following D_{Cs} values were obtained:

Crown ether	DB18C6	DC18C6	DB21C7	DC21C7	DB24C8	DB27C9	DB30C10
D_{Cs}	0.02	0.02	1.03	0.14	0.09	0.11	0.19

The selectivity of DB21C7 may be explained by macrocycle size fit and by rigidity of the dibenzo-substituted molecule, which does not allow it to form complexes with the other (smaller) cations.[10] Among the solvents other than nitrobenzene, cyclohexanone and methyl isobutyl ketone (MIBK) are more or less efficient (D_{Cs} = 0.17 and 0.08, respectively), while 1,2-dichloroethane, dichloromethane, and chloroform are considered weaker extractants.

To increase the recovery, an appropriate anion partner is required, and the approach to its chosing is the most remarkable issue in the series of publications by Blasius and Nilles.[10] The authors proposed to use well-known reagents selective for cesium such as tetraphenylborate, heteropolymolybdate, heteropolytungstate, hexachloroantimonate, etc. These reagents form water-soluble complexes with cesium, which are not extractable by organic solvents. However, as counteranions they can enhance the D values of cesium, as extracted from MAW simulant with DB21C7, by one to three orders of magnitude (Table 2).

The adducts of crown ethers with the corresponding anions were obtained prior to extraction by these authors.[10] Of these adducts, the ones shown in Table 3 were found to be the most suitable for the extraction of cesium from MAW.

As shown in Table 3, excellent recovery of cesium from the simulant solution can be achieved by use of the DB21C7 adduct approach (up to 99% with DB21C7/NaSbCl$_6$). Solvents other than nitrobenzene are also applicable, for example, chloroform yields D_{Cs} = 20 as compared to 82 in nitrobenzene (for hexachloroantimonate), and those can be increased by raising the adduct concentration.

Alkali metals render a serious effect on extraction of cesium, and the presence of sodium in the simulant solution is considered to be the main interfering factor along with the presence of nitric acid. That can be demonstrated by comparing the D values of cesium as extracted from various media by 0.004 M DB21C7/$H_3[PMo_{12}O_{40}]$ in nitrobenzene:

Aqueous phase	D_{Cs}
Water	> 1000
1 M HNO$_3$	290
1 M HNO$_3$ + 0.5 M NaNO$_3$	26
The same, + 50 ppm K$^+$	22
MAW	22

For another solution simulating a different kind of nuclear waste and containing seven times more sodium (3.4 M NaNO$_3$), the D_{Cs} under the same conditions as in the above table is only 2.1.[10]

Attempts to create more efficient ligands for cesium binding had led to the synthesis of a lipophilic analogue of DB21C7 by introducing *tert*-butyl substituents into its benzene rings. The product, *bis*[4,4'(5')-*tert*-butylbenzo]-21-crown-7 (btBB21C7), was supposed to work in nonpolar solvents (conceivably, in a PUREX-compatible medium) and would not require that exotic and expensive counteranions as with DC21C7.[11,12] Theoretically, the DB21C7 could also be used in these cases but it is not soluble enough in nonpolar solvents. The new ligand btBB21C7 justified the hopes associated with its synthesis: it was a powerful and selective reagent for cesium in the presence of di-(*n*-dodecyl)-naphthalenesulfonic acid (HDDNS) as the source of counterions, superceding other crowns such as btBB18C6, DC21C7, DC18C6, and *tert*-butylcyclohexano-15C5. For 0.1 M HDDNS and 0.05 M btBB21C7 in toluene, the D values of cesium varied from 3×10^2

Table 2 Synergistic Factors for Extraction of Cs from MAW Simulant (0.012 M DB21C7 with the anions shown in the table)

Anion	Solvent	Synergistic factor[a]
BPh$_4^-$	Nitrobenzene	15
BPh$_3$CN$^-$	Nitrobenzene	3
PW$_{12}$O$_{40}^{3-}$	Nitrobenzene	30
SiW$_{12}$O$_{40}^{3-}$	Nitrobenzene	24
HgI$_3^-$	1,2-Dichloroethane	7400 (from pure water)
SbCl$_6^-$	1,2-Dichloroethane	1200
SbCl$_6^-$	Nitrobenzene	9

[a] SF = $D_{1+2}/(D_1 + D_2)$, where D_{1+2} is D value as obtained for the combined reagent (adduct), and D_1 and D_2 are D values for the separate constituents.

Table 3 Distribution Ratios of Cs Extracted from MAW Simulant with DB21C7 Adducts

Adduct in nitrobenzene	Concentration (M)	D_{Cs} from MAW simulant
DB21C7/H$_3$[PW$_{12}$O$_{40}$]	0.004	29
DB21C7/H$_4$[SiW$_{12}$O$_{40}$]	0.003	25
DB21C7/H$_3$[PMo$_{12}$O$_{40}$]	0.004	30
DB21C7/NaSbCl$_6$[a]	0.018	82

[a] Or NH$_4^+$SbCl$_6$, K$^+$SbCl$_6$.

Table 4 Separation Factors for Cs as Extracted with
0.025 M btBB21C7 and 0.025 M HDDNS in Toluene

Metal	Nonloading conditions	Loading and competitive conditions
Na	294	192
K	5.6	6.4
Rb	1.2	1.5

in 0.1 M nitric acid down to 40 in approximately 0.9 M nitric acid. The mathematical modeling showed that the main extracted species could be CsL(DDNS), (HDDNS)$_3$·L, CsL(DDNS)·HDDNS, and possibly Cs(DDNS)· HDDNS, where L stands for the crown ligand. The main form of HDDNS itself was (HDDNS)$_7$ or (HDDNS)$_8$. The achieved separation factors D_{Cs}/D_{Metal} for 0.025 M of both crown ether and HDDNS in toluene are given in Table 4. The loading and competitive conditions in Table 4 denote that 0.01 M of every metal were mixed together for extraction. These results meant that under competitive conditions the separation factors remained fairly high, while the D_{Cs} itself was naturally lower.[11]

This crown ether was recently tested for extraction of cesium from a simulated high-sodium acidic nuclear waste (INEL) that contained 1.8 M nitric acid, 1.25 M NaNO$_3$, 0.55 M Al^{3+}, and other components, including $7 \times 10^{-5} M$ cesium.[13] The composition of this type of waste solution is given in other chapters. In the presence of HDDNS or other organosulfonic acid anions, cesium extraction from the simulant solution was very low. A more efficient counteranion was needed to increase the extraction efficiency. The SbCl$_6^-$ anion was found as an effective counteranion as it was in the case of unsubstituted DB21C7. The range of solvent effects on extraction of cesium from the simulant using SbCl$_6^-$ differs from that for DB21C7. In this system, the nonpolar toluene was the most effective solvent; this could be attributed to the role of hydrophobic interactions of the *tert*-butyl groups of btBB21C7. From pure solutions, cesium was recovered by btBB21C7 almost quantitatively (D = 50 to 70). However, the maximum D value of cesium as extracted from nondiluted solution of the simulant was about 1.2, and increase in concentration of the crown or the anion did not raise this value significantly. This value is of the same order as obtained by Blasius and Nilles for DB21C7 for the simulant solution containing 3.4 M sodium.[10] For btBB21C7, repeated extractions (four times) in the presence of SbCl$_6^-$ helped to extract as much as 90% cesium. As shown by special interference studies, Al and Na played a significant role in diminishing cesium distribution ratios along with those of other components in the simulant.[13]

Dibenzo-21-crown-7 (in the form of its adduct with molybdophosphoric acid) was also used in extraction of cesium from HNO$_3$ solutions that were formed during HNO$_3$ leaching of soils and the acidification of natural waters.[14]

The 18-membered crown with *tert*-butyl groups (btBB18C6) was also studied for Cs extraction, but it could not compete with its 21-membered analogue.[6] In the 1980s, other crown ethers with substituents in their benzene rings were proposed, among them *bis*{4.4'(5')-[1-hydroxyheptyl]-benzo}-18-crown-6 and *bis*{4,4'(5')-[1-hydroxy-2-ethylhexyl]-benzo}-18-crown-6 ($n = 1$, Crown XVII).[15-17] The *bis*{4.4'(5')-[1-hydroxyheptyl]-benzo}-18-crown-6 in the presence of 5 vol% HDDNS in TBP–kerosene mixture (27:68 v/v) extracted cesium from 3 M HNO$_3$ with D value equal to 2, while strontium and lanthanum did not extract significantly. In the presence of hydrophobic sulphonic acids rather than HDEHP cesium extraction was found to be less sensitive to nitric acid concentration.[15] The most effective crown ether was found to be Crown XVII; it was capable of extracting cesium from a synthetic fission product solution simulating acidic high-activity waste. Its composition was as follows (concentrations given in mol/l): nitric acid, 3; Zr, 0.013; Nd, 0.01; Ru, 0.007; Cs, 0.006; Ce, 0.006; Sr, Ba, La, 0.003 to 0.004; Te, Rh, Y, Pr, Sm, Rb. 0.001 to 0.003. The highest D_{Cs} value was 5.6, the conditions being 0.1 M both of the crown and nonylnaphthalene sulfonic acid (NNS) in 27% TBP–68% kerosene.[16,17] Under these conditions, the distribution of all the other components of the simulant solution was also studied. Remarkably, only $Zr(D = 2)$, Rb^+ (D = 0.8), and Ru (D = 0.2) were also extracted; the other metal D values were within the range 0.00 to 0.08. Cesium could be stripped from the solvent by 0.5 to 1 M HNO$_3$ using a number of stripping steps (Figure 1).[16]

Crown XVII (n = 1) and its 21- and 24- membered analogues (n = 2 and 3, respectively).

Crown XVII was considered a valuable PUREX-compatible extractant, recommended by the authors for cesium recovery from high-activity wastes.[15-17] The analogs of Crown XVII with cyclohexano rather than benzo substituents or with another side chain were also tested and proved to be less efficient. A question remained, why was this structure more efficient for cesium than even btBB21C7, though it had an "improper" macrocycle size? Actually, there would be no contradiction if one took into account that the high-activity simulant solution does not contain any sodium (though it contains 3 M nitric acid). Meanwhile, the simulant solutions in the above mentioned papers in which the 21-membered crowns were used contained 0.5 to 1.25 M NaNO$_3$.[10,13]

A direct comparison of extraction capabilities of btBB21C7, Crown XVII, and its 21- and 24-membered analogues was performed in the presence of highly lipophilic anions (n-dodecylbenzene sulphonate, n-dodecyl sulfate, and DDNS$^-$).[18] Solvent extraction of cesium (though made from pure solution containing only 0.5 M HNO$_3$) was approximately the same for btBB21C7 and bis(octylOHBzo)21C7 (a 21-membered analogue of Crown XVII), and lower for Crown XVII and bis(octylOHBzo)24C8 (Table 5).

These data indicate that 21C7 analogues (rather than 18-membered Crown XVII) still have the proper ring size, and the OH group is a quite unnecessary feature, be it in Crown XVII, or in its larger ring analogues.

Another family of ligands tested for solvent extraction of cesium is crown ethers with ionizable groups. The most selective reagent in the series of crown ether carboxylic acids was a 22-membered reagent sym-dibenzo-22-crown-7-oxyacetic acid (DB22C7-COOH).[19] For this extractant, cesium was the best extracted species while Na$^+$ was the poorest one, Cs/Na selectivity factor being equal to 15.[19] These crowns also have a potential for application to the waste solutions, though they are not able to extract in acidic media.

n = 1, sym-dibenzo-16-crown-5-oxyacetic acid; n = 2, sym-dibenzo-19-crown-6-oxyacetic acid; n = 3, sym-dibenzo-22-crown-7-oxyacetic acid.

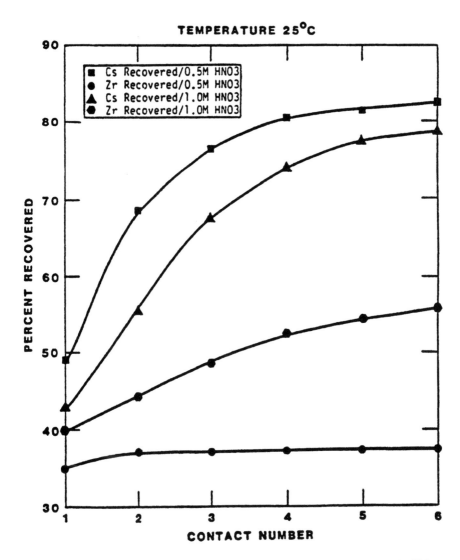

Figure 1 Percentage of cesium recovered on successive back-contacts from 0.05 M Crown XVII in 5 vol% NNS, 27 vol% TBP, and 68 vol% kerosene by use of 0.5 M or 1.0 M HNO$_3$ as a back-extractant.[16] (From Shuler, R. G. et al., Solv. Extr. Ion Exch., 3(5), 589, 1985. With permission from Marcel Dekker Inc., NY.)

Table 5 Distribution Coefficients of Cesium as Extracted from 0.5 M Nitric Acid by 5×10^{-3} M Crown Ethers in CHCl$_3$, $C_{anion} = 5 \times 10^{-3}$ M

Crown	n-Dodecylbenzene sulfonate	n-Dodecyl sulfate	DDNS$^-$
btBB21C7	0.30	0.38	0.24
Crown XVII	—	0.13	0.17
bis(OctylOHBzo)21C7	0.26	0.32	0.32
bis(OctylOHBzo)24C8	0.10	0.08	0.08

Some attempts were made toward selective separation of cesium by use of liquid membranes with a crown ether as the carrier. For the supported liquid-membranes (SLM) technique, crown ether n-decylbenzo-21-crown-7 was recently found to be the most appropriate.[20] Even with this compound, cesium cannot be quantitatively removed owing to the competition of the very high content of sodium in the waste solution.

III. EXTRACTION OF STRONTIUM

Strontium ion has an ion size (2.2 Å) intermediate between 15-crown-5 (1.72 to 1.84 Å) and 18-crown-6 type (2.67 to 2.86 Å) cavities (Table 1).[2] Similar to K+ and Na+, some 18-membered crowns give high stability constants and exhibit high selectivity for Sr^{2+}. A large number of papers is available in the literature dealing with extraction of strontium from pure solutions or in the presence of other alkali/alkaline-earth metals.[21-24]

Nonsubstituted 18-crown-6 was used sometimes in the procedures of strontium separation.[25] Dinaphthyl derivative of 18-crown-6 was also selective enough to separate strontium from $Ba(OH)_2$ matrix.[26] A possibility to extract strontium with derivatives of 24-crown-8 with alkyl-substituted benzo and cyclohexo rings was also reported.[9] In this case, extraction occurred only in the presence of large amounts of HDEHP, while the disadvantage of this anion supplier for nuclear waste treatment was a limited pH range of extraction (pH > 3).

The most selective and efficient among conventional (and commercially available) crowns for strontium is dicycolhexano-18-crown-6 (DC18C6). A facile technique of separation of Sr from Mg, Ca, and Ba from aqueous solutions by a one-step extraction with DC18C6 in $CHCl_3$ uptakes 97% Sr leaving behind up to 1.5 g calcium.[27] Determination of ^{89}Sr and ^{90}Sr in milk can be done with the aid of the same crown ether.[21] All of these procedures require a lipophilic counteranion like picrate, because in the presence of inorganic anions the extraction is low.

In acidic media, DC18C6 also results in good extraction of strontium. Nitrate, being present in large concentrations, plays the role of the counteranion, and the extraction reaches its maximum at 2 M HNO_3 when $CHCl_3$ or 1,2-dichloroethane are used as the solvents,[22,23] and at 1 M HNO_3 in the case of 1,1,2,2-tetrachloroethane (Figure 2).[24] DC18C6 was used for ^{90}Sr extraction from acidic media (soil extracts and others) with subsequent radiometric measurements.[14]

Application of solvent extraction of ^{90}Sr to nuclear waste treatment was reviewed by Shulz and Bray.[28] Among the crown ether extractants, DC18C6 was used to extract strontium from a solution simulating medium-activity waste (MAW).[24] The simulant solution contained: nitric acid, 1.0 M; NaNO_3, 0.5 M; U, 1000 ppm; Pb, 800 ppm; Ca and Mg, 200 ppm; Fe, 150 ppm; Ru, K, Zn, Cu, Mn, Cr, and Ni, 10 to 70 ppm; Zr, Cs, Ce, Sb, and Nb, 1 to 5 ppm; and Sr, 2 ppm. The main interfering elements were supposed to be Na, Pb, and K, and separation was possible only because of the larger extraction constant of the strontium complex. Other crown ethers such as benzo-15-crown-5, DB21C7, DB24C8, DB27C9, DB30C10, and bis-[4-phenylbutyl]-18-crown-6 were tested, all of them being less efficient than DC18C6 for the extraction of strontium from MAW. The following results show the solvent effect for DC18C6 (mixture of cis-isomers, 0.014 M):

Solvent	1,1,2,2-Tetrachloroethane	CH$_2$Cl$_2$	CHCl$_3$	(CH$_2$Cl)$_2$
D$_{Sr}$	8.7	4.5	4.4	2.0

Other solvents including nitrobenzene, alcohols, ketones, halogenated benzene derivatives, and hydrocarbons showed D values of Sr less than 1. Extraction of strontium was the highest in tetrachloroethane, and equilibrium was established in less than 1 min. Nitric acid concentration had a great effect on extraction efficiency, the optimum being 1.25 M HNO$_3$. At lower concentrations, increase in NO$_3^-$ ion concentration acts favorably because that ion plays the role of an ion partner for extraction of strontium. But, at concentrations higher than 1.25 M, extraction of nitric acid by the crown ether becomes very significant (species like DC18C6·nHNO$_3$·mH$_2$O are extracted, where n can be up to 40),[29] and, consequently, the effective concentration of the crown is diminished. This can qualitatively explain the cupola-like shape of the Sr–DC18C6 pH dependence curve.

Using higher concentration of the crown or applying multiple extraction steps, high decontamination factors for strontium (~100) can be obtained. The ligand can be easily regenerated by stripping with water. The procedure is recommended by the authors for separation of strontium from MAW.[24] There are some publications describing application of DC18C6 to strontium extraction from real samples.[30,31]

For separation of strontium from high-level waste (HLW) several crowns with *tert*-butyl, 1-hydroxy-*n*-heptyl, and other alkyl or hydroxy–alkyl groups attached to benzene

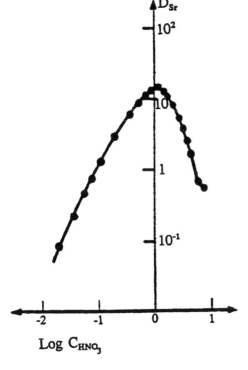

Figure 2   Distribution ratio of strontium (2 mg/l) as a function of nitric acid concentration when extracted with 1.37 × 10^{-3} M DC18C6 in 1,1,2,2-tetrachloroethane.[28] (From Blasius, E. et al., J. Radioanal. Nucl. Chem. 89(2), 395, 1985. With permission from Elsevier Science, Lausanne, Switzerland.)

or cyclohexane rings were tested.[16,32] A patented procedure of combined strontium and transuranics recovery (Np, Pu, and Am) from 1 to 3 M nitric acid employed binary extractant, a mixture of 0.2 M n-octyl(phenyl)-N,N-diisobutylcarbamoylmethylphosphine oxide and 0.2 M bis-4,4'(5')-[tert-butylcyclohexano]-18-crown-6; the diluent was 1.2 M TBP in isoparaffinic hydrocarbons. In this mixture, the strontium-active reagent is the crown ether. After extraction, scrubbing of iron, barium, and fission products is done by using 1 to 3 M nitric acid. Strontium, along with Np, Pu, and Am, can then be selectively stripped into 0.1 to 0.5 M complexone solution (tetrahydrofuran-2,3,4,5-tetracarboxylic acid).[32]

Other substituted 18-membered crowns tested for strontium extraction are Crown XVI and Crown XVIII.[15]

Crown XVI   Crown XVIII

Extraction of strontium with these crowns was investigated from the simulant solution that is described in Section II.[16] Crown XVI and Crown XVIII differ only by OH groups in the side hydrocarbon chains; however, their extraction abilities are quite different: with 0.05 M reagents in kerosene–TBP (68:27 v/v) in the presence of 0.1 M nonylnaphthalene sulfonic acid (NNS), the D_{Sr} values were 3.8 and 0.14 for Crown XVI and Crown XVIII, respectively. This may be caused by favorable hydrogen bonding of Crown XVI complex in the organic phase. However, no such effect of the hydroxyl group was observed for cesium, as described in the previous section. Extraction is selective for Sr, e.g., for 0.02 M Crown XVI, D_{Sr} = 2.0, and D_{others} < 0.6 except D_{Zr} = 2 and D_{Ba} = 2.4. The extractant (Crown XVI + NNS) was sufficiently stable in high radiation fields.[16] Strontium could be stripped from the extract with HNO_3, using multiple steps. The conditions shown above were recommended by the authors for strontium recovery from the HLW prior to its neutralization.[16]

Ionizable crown ethers were also studied with respect to strontium extraction from aqueous solutions, but not from waste simulants. Reagents with small cycle size (sym-dibenzo-14-crown-4 and -16-crown-5-oxyacetic acids) were moderately efficient for strontium, with D_{Sr} of the order of 10 for 0.01 M crowns in $CHCl_3$, and only small differences among the alkaline metal ions were observed.[33] Some 14- to 24-membered phosphorus-containing ionizable crowns were tested for alkaline-earth metals; they showed selectivity toward Ca or Ba rather than toward Sn.[34] Separation of ^{90}Sr from ^{90}Y is possible by use of the derivatives of sym-dibenzo-16-crown-5-oxyacetic acid.[35] Supported liquid-membrane technique was applied to extract strontium from evaporation concentrates of nuclear fuel with DC18C6 and bis-[tert-butylcyclohexano]-18-crown-6.[20,36]

IV. EXTRACTION OF URANIUM AND TRANSURANICS

A. EXTRACTION FROM NITRIC ACID SOLUTIONS

Like alkali and alkaline-earth metal ions, actinides can be extracted by macrocyclic compounds. The difference is that the size–cavity concept does not seem to work well in this case, and the main factors governing the actinide extraction are the oxidation state of the metal, and the type of substituents in the crown ether ring. For example, in one of the earliest works, it was shown that Pu(III) and Np(V) practically did not extract with any conventional crown ether, while Pu(IV) and Np(IV) were extracted from nitric acid solutions with high distribution ratios.[37] Uranium(VI) and other actinides in 6+ oxidation state took the intermediate position. It was noted that the sequence Pu(IV) > Pu(VI) >> Pu(III) parallels the general complexation tendency of these ions.[38] Thorium(IV) extraction by crown ethers from nitrate solutions is poor.[39] Many elements, including those having long-lived isotopes, such as ^{144}Ce, ^{137}Cs, and ^{106}Rh, show only negligible extraction from HNO_3 media when toluene is used as a solvent for selective extractions of some of the actinides by crown ethers.[38,40]

When extracted from nitric acid with crowns, the maximum of actinides uptake usually lies somewhere within the range 2 to 10 M HNO_3 (Figure 3). Thus, recovery of the macrocyclic reagents can be accomplished by dilution with water, as well as with dilute sulfuric or perchloric acid as a metal strippant. Anions of these acids have a stronger hydrophilic character than nitrate, and they are therefore unable to operate as counteranions. Complexing stripping agents such as oxalic acid can also be used.[40]

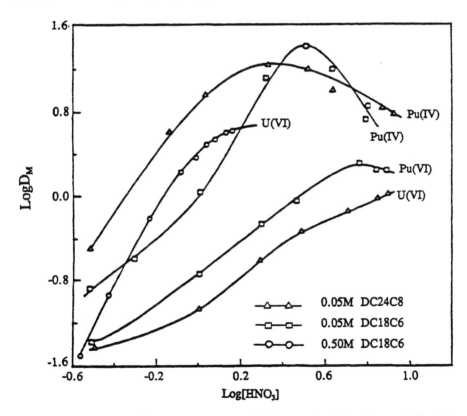

Figure 3 Plots of extraction of U(VI), Pu(IV), and Pu(VI) from varying nitric acid concentrations into benzonitrile by 0.05 M DC24C8 (triangles), 0.05 M DC18C6 (squares), and 0.5 M DC18C6 (circles). (From Shukla, J. P. and Lohithankshan, K. V., Chem. Scr., 29(4), 343, 1989. With permission from Royal Swedish Academy of Sciences.)

For extraction of transuranic elements, an important factor appears to be the composition of the complex formed in the organic phase. Extraction is always found to be more efficient when a disolvate with a crown can be formed. These complexes are probably sandwiches in which the metal ion can interact with a twice greater number of oxygen atoms. For example, U(VI), Pu(VI), and Np(VI) are extracted with various crowns in the form of 1:1 complexes, while Pu(IV) and Np(IV) form $M(NO_3)_4 \cdot 2L$ species, and extraction of these latter ions is usually considerably higher.[2,37,40]

In nitric acid solutions, the most efficient extractants for actinides (independent of their oxidation states) were dicyclohexano crowns such as DC18C6 and DC24C8, and the weakest ones were dibenzo-substituted crowns like DB18C6 and DB24C8.[2,37,40,41] This phenomenon can be explained by a higher basicity of dicyclohexano crowns.[42] Unsubstituted crowns take an alternating position. For example, actinide extraction decreases in the following sequences:[37]

> Pu(IV): DC24C8 > DC18C6 > 15C5 > 18C6 > DB24C8 > DB18C6
> Np(IV): DC24C8 > DC18C6 > 18C6 > 15C5 > DB24C8 ≈ DB18C6
> U(VI): DC18C6 > DC24C8 > DB24C8 > DB18C6 > 15C5 > 18C6

Similar trends for U(VI) and Pu(IV) were also observed by Shukla and Lohithakshan.[40] The following sequence was obtained by Yonezawa and Choppin for both UO_2^{2+} and Pu(IV):[42]

DC18C6 >> DB24C8 > DB18C6 ≈ btBB21C7 ≈ 1,10-dithia-18C6 > B15C5 ≈ C15C5

The introduction of lipophilic groups (btBB21C7) did not improve extraction from 0.5 to 8.0 M nitric acid ($D_{Pu} = 0.03$), i.e., lipophilic character is not crucial for this extraction mechanism.

Thus, DC18C6 was found to be one of the most powerful extractants for transuranic elements. Its advantage over DC24C8 is its higher partition value in the system $CHCl_3$–nitric acid.[41] Some attempts were made to use this crown ether as a potential partial substitute for TBP in PUREX process, namely for extraction of Pu(IV) from high- and medium-activity waste.[41,43] Using 0.67 M DC18C6 in chloroform[41] or benzonitrile,[43] extraction of Pu(IV), U(VI), and the sum of gamma-emitting fission products (Pf) was carried out from real waste solutions (Table 6).

Results from Table 6 show that a single extraction of the high-activity solution with a chloroform solution of DC18C6 permits the separation of plutonium, leaving about a half of uranium and most of the fission products in the aqueous phase. Washing the organic layer with 1 M HNO_3 improves the separation, and back-extraction using neutral water allows the recovery of plutonium. Eventually, 93% of plutonium can be recovered, while the use of 30% TBP in dodecane leads to an aqueous solution containing only 12% Pu, 34% U, and 0.01% fission products.[42] If benzonitrile is used as a diluent, washing of the organic phase in order to remove the fission products is done using 4 M nitric acid, and the back-extraction of uranium (with H_2O) precedes the back extraction of plutonium (with 0.5 M H_2SO_4).[43] The advantage of DC18C6 as a selective extractant of plutonium is that it is able to separate this element from uranium without the need for valence change as required in the PUREX process.[41,43]

Selectivity of DC18C6 extraction of transuranic elements was studied using Pu(IV) as an example.[40] Plutonium (1.2 ppm) was extracted from 2 M HNO_3 medium into benzonitrile

Table 6 Extraction of Plutonium from High-Activity
Solutions Using DC18C6 in Chloroform

| Elements | Initial conc. | Percentage found in aqueous phase after: | | |
		1st extn.	1 M HNO$_3$ wash	Water back-extn.
Pu	1.1 g/l	6	1.6	93
U	267 g/l	42	21	38
Pf[a]	1.1 TBq/l	92	1.1	0.14

[a] Pf = sum of fission products.

with 0.2 M DC18C6. The following quantities did not interfere in the extraction (g/l): Al, Co, Ni, Zn, SO$_4^-$ (30); Cu (20); F$^-$ (5); Fe^{3+}, Cr^{3+} (1 to 2); Zr (0.4); and UO$_2^{2+}$ (0.2). Thus, the extraction is highly selective with respect to many common transition metals.

Dicyclohexano-18-crown-6 has many conformational isomers, among which *cis-syn-cis* and *cis-anti-cis* usually prevail in the commercial product. By extracting plutonium(IV) from medium activity solutions, it was shown that *cis-syn-cis* isomer was somewhat more efficient than the other one: the D values of plutonium were two to three times higher while extraction of U(VI) and fission products remained the same.[41] In the same paper, radiation stability of DC18C6 solutions was studied by irradiation with ^{137}Cs gamma source or by storing the DC18C6 solution in the contact with a high-activity waste solution for 8 months. After that, no changes could be noticed in UV, IR, and NMR spectra or in HPLC, and the effectiveness and selectivity (Pu/U) of the crown was barely, if at all, affected. Under similar conditions, TBP would produce significant amount of mono- and dibutyl phosphate, which dramatically decrease its selectivity in the PUREX process.

Sometimes extraction efficiency can be improved by adding polar solvents to the organic phase. Thus, extraction of U(VI)[44] and Pu(IV)[45] with DC18C6 from 1 to 8 M HNO$_3$ increased several times (up to seven times for U(VI)) in the presence of up to 30 vol% acetonitrile, dioxane, or propylene carbonate as introduced into the aqueous phase (Figure 4). The extraction decreased with the addition of acetone, isobutanol, or methanol.

B. EXTRACTION FROM CHLORIDE SOLUTIONS

The chloride system is studied mostly for uranium. The composition of the species extracted from HCl media corresponds to the general formula nUO$_2$Cl$_2 \cdot m$Crown$\cdot p$HCl, which implies that actually we deal with the extraction of the anionic species of uranium. For example, for DB18C6 (L) the complexes UO$_2$Cl$_2 \cdot$2HCl$\cdot$2L and UO$_2$Cl$_2 \cdot$2KCl$\cdot$2L were found in the organic phase, which corresponded to [(DB18C6·H$_3$O)$_2$]$^{2+}$[UO$_2$Cl$_4$] and [(KDB18C6)$_2$]$^{2+}$[UO$_2$Cl$_4$].[46] DC18C6 extracts U(VI) as UO$_2$Cl$_2 \cdot$2L$\cdot$2HCl in case of cis-syn-cis and as (UO$_2$Cl$_2$)$_2 \cdot$3L$\cdot$4HCl in case of cis-anti-cis, the latter D value being 100 times less.[47-49] In nitrobenzene, the composition of the species extracted with DC18C6 was found to be UO$_2$Cl$_2 \cdot$2L$\cdot$2HCl, while the complexes with DB24C8, DB18C6, DB30C10, and B15C5 contained only one molecule of the crown.[50] The sequence of efficiency of various crown ethers with respect to uranium(VI) extraction was the following:

$$\text{DC18C6} > \text{B15C5} > \text{DB30C10} > \text{DB24C8} > \text{DB18C6}$$

which corresponds with the tendencies of extraction from nitric acid solutions.

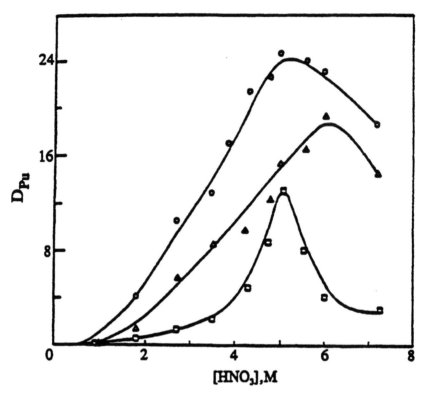

Figure 4 Effect of nitric acid concentrations on the extraction of Pu(IV) with 0.1 *M* DC18C6 in toluene from: pure aqueous solutions (squares), solutions containing 20 vol% acetonitrile (triangles), and solutions containing 20 vol% propylene carbonate (circles). (From Shukla, J. P. et al., Radiochim. Acta, 60, 104, 1993. With permission from R. Oldenbourg Verlag Gmbtl, Munchen, Germany.)

Extraction of uranium(IV) is unknown in nitric acid solutions, and it is extracted more poorly than U(VI) from chloride media. The composition of the complexes is usually $UCl_4 \cdot 2L \cdot 2HCl$.[46,50,51] DC18C6 in 1,2-dichloroethane is able to separate U(VI), leaving U(IV), in 6 *M* HCl with the separation factor of 10^3.[51] The sequence of efficiency of various crown ethers with respect to U(IV) extraction from HCl solutions is as follows:[51]

DC27C9 > DC24C8 > DC18C6 > DB24C8 > B18C6 > DB18C6

Thorium(IV) is almost not extracted from HCl media, and it can be separated from U(VI) in 6 *M* HCl with DC18C6 with a separation factor of approximately 10^3. If extracted in the presence of picrate ion, the separation gets much worse.[49]

C. EXTRACTION FROM BUFFER MEDIA

From the above discussion, it is apparent that noticeable actinides extraction with conventional crowns starts at concentrations of nitric acid >1 mol/l. However, there exists a class of crown ethers capable of extracting those elements in less acidic media. These are the ionizable crown ethers with attached carboxylic or hydroxamic groups such as *sym*-dibenzo-

16-crown-5-oxyacetic acid (DB16C5-COOH) and *sym*-dibenzo-16-crown-5-oxyacetyl-hydroxamic acid (DB16C5-CONHOH)[52] (Figure 5). The pH ranges of extraction of actinides from buffer solutions with 0.005 M DB16C5-COOH in chloroform are given below:

Ion	Th(IV)	U(VI)	Pu(VI)	Am(III)	Cm(III)	Np(VI)
pH	3–8	4.5–7	5–7	> 5.5	> 5	> 6
Maximal extn., %	100	100	95	100	98	40

Unlike non-ionizable crowns, Pu(IV) was poorly extracted, while Pu(VI) exhibited good extraction. It was shown that Th(IV) was extracted as $Th(OH)_2L_2$ (where L is the ionized DB16C5-COOH), while the most probable extraction species of uranium(VI) was found to be $UO_2(OH)L$, which corresponds with the bi-logarithmic analysis of D vs. [HL] dependences as well as with the pH ranges of stability of the corresponding hydroxo complexes. With the use of DB16C5-CONHOH, the extraction curves were slightly shifted to lower pH ranges, and the maximal distribution ratios of metals increased.[52]

D. EXTRACTION IN THE PRESENCE OF CHELATING AGENTS

Good extraction was observed for actinides in the presence of both crown ether and a chelating agent (the latter can also be considered as the source of efficient counteranions). The most widely investigated chelating agent was thenoyltrifluoroacetone (HTTA).[53-58] Using HTTA, high extraction efficiencies can be achieved from moderately acidic solutions: U(VI) with DC18C6 in benzene at pH 2;[53] Am(III) with DC18C6, DB18C6, and 18C6 in benzene at pH 3.5;[55] Am(III) and Cm(III) with a nitrogen-containing macrocycle 4,13-didecyl-1,7,10,16-tetraoxa-4,13-diazacyclooctadecane in $CHCl_3$ at pH 4.8;[54] Am(III) with different common crowns in benzene at pH 3 to 5.[58] The extracted species of trivalent metals were found to agree with the formula $M(TTA)_3 \cdot nCrown$, where n = 1 or 2. The order of synergism (DC18C6 > DB18C6 > 18C6) generally coincided with that observed in the absence of HTTA, in nitric acid solutions.[55]

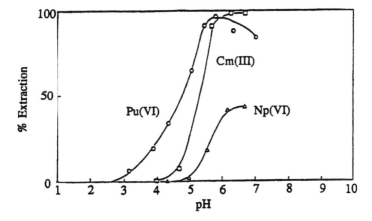

Figure 5 Solvent extraction of Np(VI), Pu(VI), and Cm(III) with 0.005 M *sym*-dibenzo-16-crown-5-oxyacetic acid in chloroform as a function of pH. (From Wai, C.M. et al., Anal. Sci., 7 (Suppl.), 41–44, 1991. With permission from the Japan Society of Analytical Chemistry, Tokyo, Japan.)

Among the other chelating agents used for synergistic extraction of actinides were acylpyrazolones. Thus, 1-phenyl-3-methyl-4-benzoyl-pyrazolone-5 with 12C4, 15C5, 18C6, DB18C6, and DC18C6 was studied; sometimes the effect was antagonistic.[42] 1-Phenyl-3-methyl-4-trifluoroacetyl-pyrazolone-5 was studied with B15C5 and DC18C6, and the composition of the extracted species was determined.[59] Di(2-ethylhexyl)-phosphoric acid caused a synergistic effect while extracting U(VI) from HNO_3, HCl, and H_2SO_4 with DC18C6 in nitrobenzene. The species $UO_2(DEHP)_2 \cdot HDEHP \cdot Crown$ was found, and extraction from nitric acid was the highest.[60]

Joint use of chlorinated cobalt dicarbollide with crown ethers allowed improvement of separation factors for Am(III)/Eu(III) up to the value of 3 to 5 by use of 18C6. For 15C5 and other (noncyclic) reagents, separation factors were not higher than 1.9. The extraction of Am(III) was on the order of 10 to 40%. More efficient extraction was achieved in the presence of picrate, dipicrylaminate, and carboxylic acids, but the selectivity dropped.[61,62]

E. EXTRACTION FROM OTHER MEDIA

From sulfuric or from perchloric acid, U(VI) is poorly extracted. From picric acid, extraction is good; the sequence of the crowns as they extract U(VI) may depend on the nature of the solvent used. Thus, in benzene, 4-methyl-B15C5 > DC18C6 > 15C5, but in 1,2-dichloroethane 18C6 > DC18C6 > 4-methyl-B15C5. The last two crowns extract U(VI) in the form of M:L = 1:2 complex in the presence of picrate.[39] Several extractants were tested for neptunium(VII) in alkaline media in the form of $NpO_4(OH)_2^{3-}$, and DC18C6 in neat TBP was found to be the most efficient among the neutral extractants.[63] Extraction of U(VI) from thiocyanate solutions was studied with the use of 15C5 and DC18C6.[53] From 0.1 M NH_4SCN + 0.01 M HSCN, U(VI) is extracted with 15C5 in nitrobenzene or 1,2-dichloroethane in the form $UO_2(SCN)_2 \cdot 3L$, which is consistent with the log D_U–log C_L plots and with the composition of the isolated uranium complex. No measurable extraction of Np(IV) and Am(III) was observed under these conditions. Rather, a possibility to extract Am(III) with 15C5 from 0.04 M picric acid was shown.[53]

There are only a few publications concerning the use of cryptands for actinide extraction. It was shown that cryptand [2.2B.2B] with two benzo rings was a more efficient synergist than 15C5 while extracting trivalent actinides with HTTA.[56] A tris-bipyridine cryptand was only slowly complexing with Am(III), though analytical separation of Am(III) from Eu(III) was possible.[64,65]

V. EXTRACTION OF TECHNETIUM

A. EXTRACTION OF TECHNETIUM (IV)

The usual oxidation state of technetium in aqueous solutions is 7+, which corresponds to the anion pertechnetate (TcO_4^-). However, a number of papers are available in the literature dealing with extraction of technetium(IV) in the presence of various reducing agents. In a series of papers, DB18C6 and 18C6 were studied for solvent extraction of technetium in the presence of $NaBH_4$[66,67] and ascorbic acid (HAsc).[68-71] The main extracted species are attributed to the formulas: $TcO_2 \cdot nH_2O \cdot mCrown$;[69] $TcO_2 \cdot nH_2O \cdot mCrown \cdot p(HAsc)$[71] (with probable solvation with the diluents used in these systems), and $TcO(OH)_2 \cdot Crown$.[67] The latter formula is hardly probable because it is in a contradiction with the pH dependence of the extraction: the acidity has no noticeable effect on the extraction in the range of pH 1 to 11 in this case (in the presence of tetraphenylborate).[67] A similar absence of a pH dependence of extraction of reduced forms of technetium was observed in the presence of ascorbic acid.[68] Both systems utilized nitrobenzene as the extraction solvent.[67,68] As opposed to that, in acetonitrile–benzene (3:1 v/v) mixture, D_{Tc} increased in the range of pH 3 to 5 by a factor of 10 per pH unit. The slope of log D_{Tc}–log [Crown] curves was usually 0.5 to 0.7.[66,67,71] Therefore, the actual mechanism of technetium extraction remains obscure.

Reduced forms of technetium are extracted with crown ethers preferentially into polar or mixed solvents. Thus, for 0.005 M 18C6 in nitrobenzene in the presence of 1 M ascorbic acid and pH range 2.5 to 5.5, log D_{Tc} = 1.8 to 1.9; for DB18C6 log D_{Tc} = 1.5 to 1.6.[68] In the presence of NaBH$_4$, log D_{Tc} = 1.4 to 1.6 for both of the crown ethers in nitrobenzene in a wide pH range.[67] Several mixtures of non-polar and water-soluble solvents were studied (e.g., acetone or acetonitrile with benzene).[69,71] The distribution ratios of technetium were not higher than those for regular polar solvents.

Extraction of technetium from reducing media was also studied in the presence of TTA. In this case crown ethers caused a significant synergistic effect at pH over 8.[66] Extraction of cationic forms of technetium allowed for separation of this element from molybdenum.[70,71]

Technetium(IV) can also be extracted from halogenide and pseudohalogenide media. For example, DC18C6 in 1,2-dichloroethane (0.01 M) was evaluated as an extractant for a number of elements including Tc(IV) from fluoride solutions. For 0.1 to 0.001 M HF + 0.1 M KF, log D_{Tc} = 2.0 to 2.5. Only Ta, Sb, and Re are also extracted under these conditions, while Co, Zn, Fe, Zr, Hf, Sn(IV), Nb, Pa, As(V), Mo, and W are not extracted.[72] Extraction of technetium along with Mo and W was studied from HCl–KSCN media in the presence of Sn(II) or ascorbic acid.[73]

B. EXTRACTION OF TECHNETIUM (VII)

Extraction of metal-containing anions extends the potential of macrocyclic compounds. One of the ions that is easily extracted with various crown ethers in the presence of a suitable cation is the TcO$_4^-$ anion. For 18-membered crowns the most suitable cations are K$^+$ and Na$^+$. For example, [K(Crown)]$^+$TcO$_4^-$ species are extracted with DC18C6 and 18C6.[29,74] Extraction is possible in highly alkaline solutions: DB18C6 extracts pertechnetate from 0.1 to 5.0 M NaOH (D_{Tc} ~ 0.4); under these conditions Ba, Ce, Nb, Nd, Np, Ru, and Zr do not extract.[75] These results suggest that technetium can be extracted by properly chosen crown ethers from high-sodium alkaline waste solutions. In another study, Jalhoom also reported the extraction of TcO$_4^-$ by use of crown ethers from sulfuric acid solution.[76]

Aza crown ethers were not widely studied with respect to technetium extraction. Extraction of pertechnetate from basic solutions by 1,10-diaza-18-crown-6 in 1,2-dichloroethane (DCE) was studied.[77] It was found that freshly prepared organic solutions gave almost negligible extraction, whereas aged solutions revealed very high levels of extraction as compared with other crown ethers (D_{Tc} = 7.7 × 10^2 in 0.18 M CsOH, and 5.3 × 10^2 in 4 M LiOH). Techniques of IR, UV, and mass spectrometry indicated that 1,10-diaza-18-crown-6 reacted slowly with DCE to form a cryptand [2.2.0] having the 2 nitrogen atoms bridged by the ethylene group of DCE; 2 mol HCl were liberated.[77]

REFERENCES

1. **Pedersen, C. J.** *The discovery of crown ethers.* Science, 241, 536, 1988.
2. **Zolotov, Yu A. and Kuzmin, N. M.** *Macrocyclic Compounds in Analytical Chemistry,* Nauka, Moscow, Russia, 1993, 320 pp. (in Russian).
3. **Wai, C. M.** in *Preconcentration Techniques for Trace elements,* Alfassi, Z.B. and Wai, C.M., Eds., CRC Press, Boca Raton, FL, 1991, 106–111.
4. **Kinard, W. F., McDowell, W. J., and Shoun, R. R.** *Studies of the size-selective extraction of alkali metal ions by the synergistic extraction system, crown ether–di(2-ethylhexyl) phosphoric acid - benzene.* Sep. Sci. Technol., 15, 1013, 1980.
5. **Kinard, W. F. and McDowell, W. J.** *Crown ethers as size-selective synergists in solvent extraction system: a new selectivity parameter.* J. Inorg. Nucl. Chem., 43(11), 2947–2953, 1981.
6. **McDowell, W. J., Moyer, B. A., Case, J. N., and Case, F. I.** *Selectivity in solvent extraction of metal ions by organic cation exchangers synergized by macrocycles: factors relating to macrocycle size and structure.* Solv. Extr. Ion Exch., 4(2), 217–236, 1986.
7. **Myasoedova, T. G., Ikonnikov, M. I., Ponomarev, A. V., Zagorets, P. A., and Filippov, E. A.** *Extraction of strontium(2+), cesium(+), and thallium(1+) with macrocyclic polyethers from nitric acid solutions.* Radiokhimiya, 29(2), 180–183, 1987.
8. **Vibhute, R. and Khopkar, M. J.** *Solvent extraction separation of cesium with dibenzo-24-crown-8 from picrate solution.* J. Radioanal. Nucl. Chem., 152(2), 487–496, 1991.
9. **Gerow, I. H., and Davis, M. W., Jr.** *The use of 24-crown-8's in the solvent extraction of cesium nitrate and strontium nitrate.* Sep. Sci. Technol., 14(5), 395–414, 1979.
10. **Blasius, E. and Nilles, K.-H.** *The removal of cesium from medium-active waste solutions. I. Evaluation of crown ethers and special crown-ether adducts in the solvent extraction of cesium.* Radiochim. Acta, 35(3), 173-182, 1984; **Blasius, E. and Nilles, K.-H.** *The removal of cesium from medium-active waste solutions. II. Solvent extraction using adducts of dibenzo-21-crown-7 with 12-heteropoly compounds and hexachloroantimonates(V) in an organic solvent: continuous extraction, preparation, capacities, bleeding and regeneration of the extractants.* Radiochim. Acta, 36, 207, 1984; **Blasius, E. and Nilles, K.-H.** Separating cesium ions from aqueous solutions by using an addition compound consisting of a macrocyclic polyether and an inorganic heteropolyacid. Eur. Pat. Appl. EP 73262 A1 830309, 37 pp.; **Blasius, E. and Nilles, K.-H.** Separating cesium ions from aqueous solutions. Eur. Pat. Appl. EP 73263 A1 830309, 49 pp.
11. **McDowell, W. J. and Case, G. N.** *Selective extraction of cesium from acidic nitrate solutions with didodecylnaphthalenesulfonic acid synergized with bis(tert-butylbenzo)-21-crown-7.* Anal. Chem., 64, 3013, 1992.
12. **McDowell, W. J.** *Crown ethers as solvent extraction reagents: where do we stand?* Sep. Sci. Technol., 23, 1251–1268, 1988.
13. **Wai, C. M., Beklemishev, M. K., and Elshani, S.** *Solvent Extraction and Sorption of Cesium from Tank Farm Waste Simulant with bis(tert-butylbenzo)-21-Crown-7 and Hexachloroantimonate(V).* unpublished.
14. **Yakshin, V. V., Vilkova, O. M., and Laskorin, B. N.** *Extractive separation of radionuclides of cesium and strontium with crown ethers.* Dokl. Akad. Nauk SSSR, 325(5), 967–969, 1992 (in Russian); Chem. Abstr., 118:72610.
15. **Gerow, I. H., Smith, G. E., Jr., and Davis, M. W., Jr.** *Extraction of cesium(1+) and strontium(2+) from nitric acid solution using macrocyclic polyethers.* Sep. Sci. Technol., 16(5), 519–548, 1981.
16. **Shuler, R. G., Bowers, C. B., Jr., Smith, G. E., Jr., Van Brunt, V., and Davis, M. W., Jr.** *The extraction of cesium and strontium from acidic high activity nuclear waste using a PUREX process compatible organic extractant.* Solv. Extr. Ion Exch., 3(5), 567–604, 1985.
17. **Davis, M. W., Jr. and Bowers, C. B., Jr.** *Extraction of cesium and strontium from nuclear waste.* Eur. Pat. Appl. EP 216473 A1 870401, 33 pp.; Appl. EP 86-306012 860805; Chem. Abstr., 107:185702.
18. **Beklemishev, M. K., Wai, C. M., and Elshani, S.** *Solvent Extraction of Cesium with Hydroxyl Groups Containing Benzo Crown Ethers.* unpublished.

19. **Walkowiak, W., Charewicz, W. A., Kag, S. I., Yang, I. W., Pugia, M. J., and Bartsch, R. A.** *Effect of structural variations within lipophilic dibenzocrown ether carboxylic acids on the selectivity and efficiency of competitive alkali metal cation solvent extraction into chloroform.* Anal. Chem., 62, 2018–2021, 1990.

20. **Casas i Garcia, J.** *Selective cesium and strontium decontamination of evaporation concentrates from nuclear fuel reprocessing plants with crown ethers by transport through supported liquid membranes.* Comm. Eur. Communities, [Rep.] EUR, EUR 14465, 319 pp., 1993; Chem. Abstr., 119:16658.

21. **Kimura, T., Iwashima, K., Ishimori, T., and Hamada, T.** *Separation of Sr-89 and -90 from calcium in milk with a macrocyclic ether.* Anal. Chem., 51(8), 1113–1116, 1979.

22. **Gloe, K., Muehl, P., Kholkin, A. I., Meerbote, M., and Beger, J.** *Extraction of metal salts by using macrocyclic crown ethers.* Isotopenpraxis, 18(5), 170–175, 1982.

23. **Filippov, E. A., Yakshin, V. V., Abashkin, V. M., Fomenkov, V. G., and Serebryakov, I. S.** *Extraction of alkaline earth metals from nitric acid solutions by the crown ether dicyclohexyl-18-crown-6.* Radiokhimiya, 24(2), 214-216, 1982; Chem. Abstr., 96:206290.

24. **Blasius, E., Klein, W., and Schoeon, U.** *J. Separation of strontium from nuclear waste solutions by solvent extraction with crown ethers.* J. Radioanal. Nucl. Chem., 89(2), 389–398, 1985.

25. **Mohite, B. S. and Khopkar, S. M.** *Solvent extraction separation of strontium as 18-crown-6 complex with picrate ion.* Anal. Chem., 59(8), 1200–1203, 1987.

26. **Helgeson, R. C., Timko, J. M., and Cram, D. J.** *Structural requirements for cyclic ethers to complex and lipophilize metal cations or alpha-amino acids.* J. Am. Chem. Soc., 95(9), 3023–3025, 1973.

27. **Kimura, T., Iwashima, K., Ishimori, T., and Hamaguchi, H.** *Separation of strontium ions from a large amount of calcium ion by the use of a macrocyclic ether.* Chem. Lett., No. 5, 563–564, 1977.

28. **Schulz, W. W. and Bray, L. A.** *Solvent extraction recovery of byproduct ^{137}Cs and ^{90}Sr from HNO_3 solutions — a technology review and assessment.* Sep. Sci. Technol., 22, 191–214, 1987.

29. **Rozen, A. M., Nikolotova, Z. I., Kartasheva, N. A., Luk'yanenko, N. G., and Bogatskii, A. V.** *Extraction of actinides and nitric acid by crown ethers.* Dokl. Akad. Nauk. SSSR, 263(5), 1165–1169. 1982.

30. **Saas, A.** *Quality control for low and medium active waste. A series of final reports (1985-89).* No. 42. Comm. Eur. Communities, [Rep.] EUR, EUR 13878, 134 pp., 1991; Chem. Abstr., 116:138418.

31. **Horwitz, E. P. and Dietz, M. L.** *Process for the Recovery of Strontium from Acid Solutions.* U.S. patent 5100585 A 920331, 11 pp.; Chem. Abstr., 117:35370.

32. **Horwitz, E. P. and Dietz, M. L.** *Combined Transuranic-Strontium Extraction Process.* U.S. patent 5169609.

33. **Uhlemann, E., Bukowsky, H., Dietrich, F., Gloe, K., Muehl, P., and Mosler, H.** *Extraction of alkali and alkaline earth metal ions with (sym-dibenzo-14-crown-4-oxy)- and (sym-dibenzo-16-crown-5-oxy)carboxylic acids.* Anal. Chim. Acta, 224, 47-53, 1989.

34. **Bartsch, R. A.** *Effect of structural variation within proton-ionizable crown ethers upon the selectivity and efficiency of solvent extraction of alkali metal and alkaline earth cations.* Solv. Extr. Ion Exch., 7(5), 829-854, 1989.

35. **Wood, D. J., Elshani, S., Du, H. S., Natale, N. R., and Wai, C. M.** *Separation of ^{90}Y from ^{90}Sr by solvent extraction with ionizable crown ethers.* Anal. Chem., 65, 1350-1354, 1993.

36. **Dozol, J. F., Eymard, S., Gambade, R., La Rosa, G., Casas i Garcia, J.** *Decontamination of evaporator concentrates of cesium, strontium, and transuranic elements.* J. Comm. Eur. Communities, [Rep.] EUR, EUR 13887, 217 pp. (French) 1992; Chem. Abstr., 117:57410.

37. **Yakshin, V. V., Filippov, E. A., Belov, V. A., Arkhipova, G. G., Abashkin, V. M. and Laskorin, B. N.** *"Crowns" in the extraction of uranium and actinides from nitric acid solutions.* Dokl. Akad. Nauk SSSR, 241(1), 159–162, 1978 (in Russian).

38. **Shukla, J. P., Singh, R. K., and Kumar Anil.** *Extraction of uranium(VI) and plutonium(IV) into toluene by crown ethers from nitric acid solution.* Radiochim. Acta, 54(2), 73–77, 1991.

39. **Wang, W.-J., Chen, B., Jin, Z., and Wang, A.** *Extraction of several lanthanide and actinide radionuclides with crown ethers.* J. Radioanal. Chem., 76(1), 49–62, 1983.

40. **Shukla, J. P. and Lohithakshan, K. V.** *Selective liquid-liquid extraction of uranium(VI) and plutonium(IV) from aqueous nitric acid media by crown ethers.* Chem. Scr., 29(4), 341–345, 1989.

41. **Lemaire, M., Guy, A., Chomel, R., and Foos, J. J.** *Dicyclohexano-18-crown-6 ether: a new selective extractant for nuclear fuel reprocessing.* Chem. Soc. Chem. Commun., (17), 1152–1154, 1991.

42. **Yonezawa, C. and Choppin, G. R.** *Synergistic extraction of americium(III), thorium(IV) and uranium(VI) by PMBP and crown ethers.* J. Radioanal. Nucl. Chem., 134(1), 233–239, 1989.

43. **Lemaire, M., Guy, A., Foos, J., Chomel, R., Doutreluigne, P., Perez, M., Thierry C. L., Guyon, V., and Le Roy, H.** *Separating plutonium and uranium from fission products using crown ethers, in the first stages of treating irradiated nuclear fuel.* Eur. Pat. Appl. EP 434513 A1 910626, 21 pp.; Chem. Abstr., 115:242254.

44. **Shukla, J. P., Kumar, Anil, Singh, R. K.** *Solvent extraction of uranium(VI) into toluene by dicyclohexano-18-crown-6 from mixed aqueous-organic solutions.* Talanta, 40(8), 1261–1266, 1993.

45. **Shukla, J. P., Kumar, Anil, and Singh, R. K.** *Liquid-liquid extraction of plutonium(IV) by dicyclohexano-18-crown-6 from aqueous-organic solutions.* Radiochim. Acta, 60(2-3), 103–107, 1993.

46. **Yakshin, V. V., Abashkin, V. M., and Korshunov, M. B.** *Extractive separation of mercury from cadmium and zinc using dicyclohexyl-18-crown-6.* Zh. Anal. Khim., 37(5), 938-940, 1982.

47. **Wang, W.-J., Sun, Q., and Chen, B.** *Extraction equilibrium of uranium(VI) with dicyclohexano-18-crown-6 and its application for separating uranium and thorium.* J. Radioanal. Nucl. Chem., 98(1), 11-16, 1986.

48. **Wang, W.-J., Chen, B., Zheng, P., Wang, B., and Wang, M.** *Solvent-extraction complex of uranium(VI) with cis,syn,cis-isomer of dicyclohexano-18-crown-6.* Inorg. Chim. Acta, 117(1), 81-82, 1986.

49. **Wang, W.-J., Sun, Q., and Chen, B.** *Further study of extraction equilibrium of uranium(VI) with dicyclohexano-18-crown-6 and its application to separating uranium and thorium.* J. Radioanal. Nucl. Chem., 110(1), 227–233, 1987.

50. **Lin, Z., Shaojin, C., and Yijuin, F.** *Solvent extraction of uranium(IV) with dibenzo crown ethers.* ISEC'83: Int. Solv. Extr. Conf., Part 1, Abstracts, Denver, CO, 1983, 453–454.

51. **Wang, W.-J., Jie, L., Hong, S., Peiji, Z., Ming, W., and Boyi, W.** *Extraction equilibrium and crystal structure of hydronium-uranium(IV) dicyclohexano-18-crown-6 isomer A complex.* Radiochim. Acta, 40(4), 199–202, 1986.

52. **Wai, C. M., Du, H. S., Meguro, Y., and Yoshida, Z.** *Selective extraction and separation of actinides with ionizable crown ethers.* Anal. Sci., 7 (Suppl., Proc. Int. Congr. Anal. Sci., 1991, Pt. 1), 41–44, 1991.

53. **Godbole, A. G., Thakur, N. V., Swarup, Rajendra, and Patil, S. K.** *Some studies on the extraction of actinides with crown ethers.* J. Radioanal. Nucl. Chem., 108(2), 89–98, 1986.

54. **Ensor, D. D., Nicks, M., and Pruett, D. J.** *Synergistic extraction of trivalent actinides and lanthanides using HTTA and an aza-crown ether.* Sep. Sci. Technol., 23(12–13), 1345–1353, 1988.

55. **Nair, G. M. and Prabhu, D. R.** *Synergism of crown ethers in the thenoyltrifluoroacetone extraction of americium(III) in benzene medium.* J. Radioanal. Nucl. Chem., 121(1), 83–90, 1988.

56. **Ensor, D. D. and Shah, A. H.** *Synergistic extraction of trivalent actinides and lanthanides using macrocyclic compounds.* J. Radioanal. Nucl. Chem., 127(4), 235–242, 1988.

57. **Shehata, F. A., Khalifa, S. M., and Aly, H. F.** *Synergic extraction of trivalent gadolinium, europium and americium radionuclides by 15-crown-5 or 18-crown-6 mixed with thenoyltrifluoroacetone from perchlorate medium.* J. Radioanal. Nucl. Chem., 159(2), 353–361, 1992.

58. **Adachi, T., Aoyagi, H., Gunji, K., Kato, Y., Kimura, T., Kohno, N., Meguro, Y., Muto, H., and Nakahara, Y.** *Solution chemistry and analytical chemistry of actinide elements.* Nippon Genshiryoku Kenkyusho, [Rep.] JAERI-M, JAERI-M 92–036, 31–57, 1992; Chem. Abstr., 118:204198.

59. **Mathur, J. N. and Khopker, P. K.** *Use of crown ethers as synergists in the solvent extraction of trivalent actinides and lanthanides by 1-phenyl-3-methyl-4-trifluoroacetyl-5-pyrazolone.* Solv. Extr. Ion Exch., 6(1), 111–124, 1988.

60. **Xu, S. C., Jin, J., Liu, M., and Pan, G.** *The synergistic extraction of uranyl ion by di(2-ethylhexyl)phosphate (HDEHP) and DC18C6.* J. Nucl. Radiochem., 6(1), 18–19, 1984.

61. **Kudryavtsev, E. G., Lazarev, L. N., Romanovskii, V. N., Proyaev, V. V., and Romanovskii, V. V.** *Extraction of TPE and RE from nitric acid solutions by mixtures of chlorated cobalt dicarbollide with polyoxocompounds.* Waste Manage., 1, 965–968, 1991.

62. (a) **Proyaev, V. V., Romanovskii, V. V., and Romanovskii, V. N.** *Extraction of trivalent f-elements by polyoxo compounds in the presence of chlorinated cobalt dicarbollide. Radiokhimiya,* 33(1), 46–48, 1991 (in Russian), Chem. Abstr., 115:36528. (b) **Proyaev, V. V. and Romanovskii, V. V.** *Extraction of trivalent f-elements by macrocyclic polyethers 18-crown-6 and 15-crown-5 in the presence of different hydrophobic anions.* Radiokhimiya, 34(1), 156–161, 1992 (in Russian); Chem. Abstr., 118:46711.

63. **Rozen, A. M., Nikoforov, A. S., Kartasheva, N. A., Nikolotova, Z. I., and Tananaev, I. G.** *Extraction of neptunium(VII) from alkaline media. Dokl. Akad. Nauk SSSR,* 312(4), 897–900, 1990 (in Russian); Chem. Abstr., 113:199082.

64. **Manchanda, V. K. and Mohapatra, P. K.** *Extraction studies of americium(III) and europium(III) with a tris-bipyridyl macrobicyclic ligand.* J. Incl. Phenom. Mol. Recognit. Chem., 15(2), 121–130, 1993.

65. **Manchanda, V. K. and Mohapatra, P. K.** *Complexation of a tris-bipyridine cryptand with americium(III).* Polyhedron, 12(9), 1115–1117, 1993.

66. **Le Tuong Minh and Lengyel, T.** *Synergistic behavior of crown ethers in the thenoyltrifluoroacetone extraction of technetium.* J. Radioanal. Nucl. Chem., 136(3), 225–230, 1989.

67. **Le Tuong Minh and Lengyel, T.** *Extraction of technetium by DB18C6 and 18C6 into nitrobenzene and acetylacetone from an aqueous phase containing sodium tetrahydroborate.* J. Radioanal. Nucl. Chem., 136(5), 363–369, 1989.

68. **Le Tuong Minh and Lengyel, T.** *Extraction of technetium from ascorbic acid medium with crown ethers.* J. Radioanal. Nucl. Chem., 128(5), 417–422, 1988.

69. **Le Tuong Minh and Lengyel, T.** *Extraction of technetium from aqueous medium with the aid of crown ethers dissolved in benzene-acetone mixture.* J. Radioanal. Nucl. Chem., 135(3), 223–229, 1989.

70. **Le Tuong Minh and Lengyel, T.** *Separation of molybdenum and technetium. Crown ether as extraction agent.* J. Radioanal. Nucl. Chem., 135(6), 403–407, 1989.

71. **Le Tuong Minh and Lengyel, T.** *The effect of solvent mixture composition on the extractive separation of technetium using a crown ether.* J. Radioanal. Nucl. Chem., 136(3), 219–224, 1989.

72. **Caletka, R., Hausbeck, R., and Krivan, V.** *The distribution of elements between polyether-type polyurethane foam, cyclic polyethers and hydrofluoric acid solution.* Talanta, 33(3), 219–224, 1986.

73. **Caletka, R., Hausbeck, R., and Krivan, V.** *Extraction of molybdenum, tungsten, and technetium with polyurethane foam and with cyclic polyether from thiocyanate/hydrochloric acid medium.* Talanta, 33(4), 315–320, 1986.

74. **Korpusov, G. V., Krylov, Yu. S., and Tsalon, S. I.** *Extraction of technetium by dicyclohexyl 18-crown-6.* Radiokhimiya, 26(3), 408–410, 1984 (in Russian); Chem. Abstr., 101:98621.

75. **Jalhoom, M. G.** *Extraction of technetium by crown ethers and cryptands.* J. Radioanal. Nucl. Chem., 104(3), 131–139, 1986.

76. **Jalhoom, M. G.** *Extraction of TcO₄⁻ ions by use of crown ethers from sulfuric acid solution.* Radiochim. Acta, 39(3), 195–197 (1986).

77. **Jalhoom, M. G.** *Studies on the extraction of TcO₄⁻ from solutions of mono- and bivalent metal hydroxides by use of Kryptofix-22 in 1,2-dichloroethane.* Proc. Int. Conf. Sep. Sci. Technol., 2nd, Vol. 2, 661–666. Editor(s): Baird, M. H. I. and Vijayan, S., Eds., Can. Soc. Chem. Eng.: Ottawa, Ont., 1989, 661–666; Chem. Abstr., 112:86415.

Chapter 4

Liquid Extraction,
Other Liquid Extraction Processes

Thomas E. Carleson and Arlin Olson

I. INTRODUCTION

This chapter concerns extraction processes other than TRUEX based (treated in Chapter 2) and other than crown ether based (treated in Chapter 3). There has been considerable research and development in other organophosphorus extractants, and even those that are not phosphorus based, as indicated below.

Metal chelation is produced by electron pair sharing between groups in the ligand and the metal ion. Ligands that can share more than one pair of electrons are called multidentate chelating agents. For many of these agents, there is at least one acidic group (–OH) as well as one or more basic donor atoms, such as nitrogen or phosphorus. The acidic group loses a proton, becoming an anion donor, and the basic nitrogen or phosphorus–oxygen (P=O) group donates a pair of electrons to the metal ion. The acid and basic groups often occur together on ligands. Examples of such ligands are ortho diphenols (two hydroxyl acid groups and an alkene group), ortho dithiols (two sulfuryl acid groups and an alkene group), ortho hydroxylaldoximes, glycine, beta-diketones, isonitrosoketones, mandelic acid, alpha acyloinoxime, oxine, oxinates, hyroxyquinolines, tropolones, alpha dioximes, ferroine, cuproine, and napthols.[1,2] Some of the carbamoyl-phosphoryl extractants (i.e., CMPO) have phenyl groups that improve americium (III) extraction.

Liquid extraction operations have been traditionally used for spent-fuel reprocessing. For some laboratories (Hanford) they are the preferred method over extraction chromatography, sorption, and precipitation processes for TRU removal.[3]

II. AGENTS FOR REMOVAL OF TRANSURANICS

A. ORGANOPHOSPHORUS EXTRACTANTS

Historically, the monofunctional agent tributyl phosphate (TBP) was developed for the extraction of actinides Am(III), Pu(IV,VI), and Np(IV,VI) from acidic media.[2] Organophosphorus chelating agents replace the water molecules attached to the metal and form a complex that is less hydrophilic and can be easily extracted by an organic solvent. Extraction of actinides with tributyl phosphate in a hydrocarbon diluent later came to be called the plutonium–uranium extraction (PUREX) process, which was used for recovery of uranium and plutonium from dissolved nuclear fuel elements. Tributyl phosphate is one of a series of organophosphorus compounds that have been evaluated for transuranic element (+4 and +6 actinides) extraction. Other mono- and bifunctional organophosphorus agents are bis- or also called di-(2-ethylhexyl) phosphoric acid (HDEHP), dibutylbutyl phosphonate (DBBP), butyl dibutylphosphonate (BDBP), di-isodecyl phosphoric acid (DIDPA), trioctylphosphine oxide (TOPO), dihexyl-*N,N*-diethylcarbamoyl methylenephosphonate (DHDECMP, or just CMP), and diphenyl-*N,N*-dibutylcarbamoylmethyl phosphine oxide (DPCMPO). The TRUEX monofunctional extractant octyl(phenyl)-*N*,*N*-diisobutylcarbamoyl methylphosphine oxide (CMPO) is another member of this carbamoyl-phosphoryl family of extractants.[4] It has been shown to be a better extractant than DHDECMP, although DHDECMP coextracts less Zr(IV) and Fe(III) and it is easier to strip actinides from DHDECMP than CMPO.[3]

The extractant DBBP was used with marginal success at the Hanford site in the 1970s to extract americium and plutonium from acidic wastes.[5] The monofunctional agent DBBP is a powerful extractant for americium(III) from low acid (pH >2) solutions, with high nitrate concentrations. This extractant was used with carbon tetrachloride diluent to extract americium and plutonium from plutonium finishing plant (PFP) wastes at Hanford. The process was designed to recover americium for further purification. Problems with the process concerned the difficulty in pH control of the unbuffered waste with 50% sodium hydroxide. The process was discontinued after an explosion of an americium-loaded cation–ion-exchange resin bed (used to further purify the americium) that had been stored for several months.

A process using HDEHP was developed by Swedish scientists (CTH actinide separation process) for removing TRU elements from PUREX process high-level liquid wastes.[6,7] The extractant was dissolved in Nysolvin 75A, an aliphatic solvent. The extractant HDEHP was able to extract protactinium, uranium, neptunium, and plutonium from 6 *M* nitric acid waste. The aqueous raffinate from the first extraction was stripped of excess nitric acid by tributyl phosphate. The extract phase was stripped with 0.5 *M* nitric acid. Although tested as a bench-scale process by both the Swedes and later the Germans, the HDEHP process was never evaluated on either pilot plant or production scale.

Processes using DIDPA have been extensively evaluated by the Japanese for removing americium and curium from PUREX process high-level liquid waste.[8-13] The process involves the following steps:

1. Acidity adjustment to 2 *M* nitric acid and extraction with 30% TBP in dodecane to remove traces plutonium and uranium left over
2. Denitration of raffinate to about 0.5 *M* nitric acid
3. Filtration of precipitated zirconium and molybdenum
4. Extraction of americium(III), curium(III), and rare earths with 0.5 *M* DIDPA–0.1 *M* TBP–paraffin hydrocarbon solvent from the near 0.5 *M* nitric acid solution
5. Stripping of the trivalent actinides and lanthanides with 4 *M* nitric acid

A hot pilot-scale test was conducted with mixer-settlers at the Tokai reprocessing plant. It appears that the extractant is fairly stable to decomposition. The process was reported to remove more than 98% of the plutonium and 99.99% of the americium and curium. In a Hanford report, the disadvantages of the process were stated as: need for two solvent extraction steps, need to adjust feed acidity, and need to filter the feed to the DIDPA extraction step.[3]

The Talspeak process proposed by scientists at Oak Ridge uses the extractant HDEHP.[14] It is used to separate lanthanides from trivalent actinides in slightly acidic solutions (pH above about 2.5). The feed is conditioned by adding lactic and diethylenetriaminepentacetic (DTPA) acids. In the first part of the process, trivalent lanthanides are extracted by HDEHP leaving the trivalent actinides in the aqueous phase. Potential disadvantages of the Talspeak process are the buildup of degradation products and their effect upon the operation.

Chinese investigators have been evaluating mixed trialkylphosphine oxides (TRPO) as an extractant for uranium ore leach material.[15-17] This extractant extracts +3, +4, and +6 actinides from 1 to 2 *M* nitric acid solutions. A bench-scale evaluation of TRPO on actual PUREX high-level liquid waste was conducted by German scientists.[18] The feed was adjusted to 1 M in nitric acid and contacted with 30% (volume) TRPO in kerosene solvent. More than 94% of the americium(III) and curium(III) were extracted. The extract was sequentially stripped of americium and curium (by 5.5 M nitric acid), plutonium (by 0.5 M oxalic acid), and uranium (0.5 M sodium carbonate solution). It was noted that zirconium, technetium, and molybdenum were more strongly extracted than for a CMPO (TRUEX) solution process.

The agent CMP has been proposed as an alternative to CMPO. It is a less powerful extractant than CMPO and subsequently extracts less Fe(III) and Bi(III) from nitric acid solutions than do solutions of CMPO or diamides.[3] Considerable preliminary tests evaluating CMP have been performed.[19-23] It is also easier to strip actinides from CMP solutions than from CMPO solutions. However, because the CMPO TRUEX process is further developed than the CMP (or diamide) process, it is preferred over these others.[3,24] The Chinese have also looked at this extractant, calling it DHDECMP.[25-27] With 30% (volume) DHDECMP in diethyl benzene, actinides were successfully extracted from acidic solution (2 to 6 M nitric acid). Experiments with simulated waste for six contacts resulted in more than 99.9% of the uranium, plutonium, and americium recovered.[26] About 99.4% of the neptunium was recovered. However, 99.7% of the lanthanides (gadolinium) were also extracted. Stripping (0.05 M nitric acid) efficiencies for plutonium, americium, and gadolinium were above 99.5%.

Russian scientists have been investigating DPCMPO as an extractant for actinides from nitric acid solutions.[28] In the presence of a fluorocarbon diluent, there appears to be no tendency toward third-phase formation even with metal saturated extractant. It is also claimed that DPCMPO is cheaper to synthesize than CMPO and that the degradation products are innocuous, so no alkaline washing step is needed.[3] However, it appears that DPCMPO has only limited solubility in normal paraffin hydrocarbon solvents.[29] In addition, the use of a fluorocarbon diluent of potential toxicity could be a problem.

The agent TOPO has been evaluated for thorium removal by BNFL.[30] A comprehensive review summarizing the carbamoyl-phosphoryl extractants has been prepared by scientists and engineers at Hanford.[3]

B. AMINE AND AMIDES

Extraction of actinides with triaklyamines or tetraalkyl ammonium salts (Tramex process) has been investigated for americium and curium removal.[31] The actinides (III and IV) are separated from lanthanides and other fission products by conditioning the feed to a high chloride concentration and conducting the extraction. However, these extra steps require additional chemicals and could produce additional potential waste products.

The French laboratories have investigated a series of diamide agents that extract +3, +4, and +6 actinides from acidic solutions (3 to 6 M nitric acid).[32,33] These agents contain two functional groups in the molecule — tertiary amine groups and ketone groups. The extraction of actinides is somewhat less with diamides than with CMPO (TRUEX) or CMP.[3] A miniature mixer-settler test of the diamide process was recently performed by French scientists.[34] More than 99.9% of the americium and plutonium were extracted from a concentrated waste of 5 M nitric acid. In addition, over 99% of the iron was extracted (no oxalic acid was added). A dilute nitric acid hydroxylamine nitrate solution was used to strip 99.9% of the americium and plutonium.

In the diamide process, the +3 actinides (i.e., americium) are less efficiently extracted than +4 and +6 actinides. It also appears that Fe(III) and Zr(IV) are somewhat extracted at high acid concentrations about the same as uranium and plutonium, with D being between 50 and 100.[35] Attempts to alleviate this coextraction of inerts with the actinides by addition of oxalic acid does not appear successful.[3] The extractant is in hydrogenated propylene tetramer that sometimes leads to problems with third-phase formation.[36] The structure seems to affect third-phase formation. Longer-chain alkyl groups appear to be more stable to third-phase formation in acidic solutions. Another problem with diamides is hydrolysis when in contact with nitric acid solutions. One proposed advantage of the triamide process over one using phosphorus-based extractants is the incineration of spent extractants to innocuous compounds (carbon dioxide, water, and nitrogen). Other claimed advantages are that degradation products are not deleterious to process performance, routine cleanup would not therefore be required, and the agents cost less than for CMPO.[32,37] Further work is progressing with these agents.[35] A recent report by the Hanford laboratory summarizes the advantages and disadvantages of the diamide process.[3] Amides have also been investigated in Italy for metals removal and amines have been evaluated by BNFL for thorium removal.[38,39]

C. BETA-DIKETONES

The most important beta-diketones are acetylacetone, benzoylacetone, dibenzoylmethane, and thenotrifluoracetone.[2] These compounds have a hydrogen atom replaceable by a metal and a ketonic oxygen, which can complete a chelate ring. Good separation of trivalent actinides from lanthanides has been found using thio-derivatives of beta-diketones.[40] Beta-diketones, however, are fairly corrosive and may cause process problems.

III. CESIUM AND STRONTIUM

A review of liquid extraction agents for removal of cesium and strontium from alkaline waste solutions was published recently by Schulz and Bray.[41] A similar review of processes for removal of cesium and strontium from acidic solutions was published by Kolarik.[42]

A. DICARBOLLIDES (COBALT) H+{[PI-(3)-(1), 2-B9C$_2$H$_{11}$CL$_2$]$_2$CO

Eastern European countries (Czechoslovakia) and Russia have been investigating cobalt dicarbollide dissolved in a polar diluent such as nitrobenzene for extraction of cesium-137 from acidic waste solutions (2 to 3 M nitric acid).[43-47] Cesium forms an ion pair with the dicarbollide and high acid concentrations (>3 M nitric acid) are required to strip it. The extractant has been synthesized commercially in Russia and applied to extract megacuries of cesium-137 and strontium-90 from actual PUREX waste. (With the addition of polyethylene glycol, strontium-90 can also be extracted from 0.5 M nitric acid solutions. The strontium-90 can be stripped with 3 M or higher concentrations of nitric acid.) With appropriate feed conditioning (nitric acid concentration of 0.5 M or below), it is reported that the extractant can also extract +3 valence state actinides (americium and curium) and +3 lanthanides.[47] A recent Los Alamos report summarizes much of the current published information on cobalt dicarbollides.[48] Several problems are evident with the dicarbollide process.[49] These include the toxic properties of the diluent, compatibility with other processes (such as TRUEX), degradation products of the solvent and extractant and possible enhancement of corrosion due to chlorides, and effects of other species (Ca[II] and Zr[IV]) on the extraction.

B. PHENOLS

At the Hanford site, 4-*sec*-butyl-2-(alpha-methyl-benzyl)phenol (BAMP) was used to extract cesium from alkaline solutions.[41]

REFERENCES

1. **Cheng, K.L., K.U. Ueno, and T. Imamura, Eds.,** *CRC Handbook of Organic Analytical Reagents,* CRC Press, Boca Raton, Florida, 1982.
2. **Stary, J.,** *The Solvent Extraction of Metal Chelates,* Pergamon Press, New York, 1964.
3. Westinghouse Hanford Company (WHC), "Alternative Pretreatment Technologies for Removal of Transuranium Elements from Hanford Site Wastes," WHC-EP-0577, Westinghouse Hanford Company, Richland, Washington, January, 1993.
4. **Horwitz, E.P., H. Diamond, and D.G. Kalina,** "Carbamoylmethylphosphoryl Derivatives as Actinide Extractants. Their Significance in the Processing and Recovery of Plutonium and Other Actinides," in *Plutonium Chemistry,* W.T. Carnall and G.R. Choppin, Eds., ACS Symposium Series 216, American Chemical Society, Washington, D.C., 27, 1983.
5. **Schulz, W.W.,** *The Chemistry of Americium,* U.S. Energy Research and Development Administration Technical Information Center, Oak Ridge, Tennessee, 1976.
6. **Liljenzin, J.O., I. Hagstrom, G. Persson, and I. Svantesson,** "Separation of Actinides from PUREX Waste," Proc. Int. Solv. Extraction Conf., ISEC '80, Vol. 3, Paper 80–180, 1980.
7. **Liljenzin, J.O., G. Persson, I. Svantesson, and S. Wingefors,** "Experience from Cold Tests of the CTH Actinide Separation Process," in *Transplutonium Elements. Production and Recovery,* ACS Symposium Series 161, American Chemical Society, Washington, D.C., 1981.
8. **Kubota, M., H. Nakamura, S. Tachimori, T. Abe, and H. Amano,** "Removal of Transplutonium Elements from High-Level Waste," IAEA-SM-246/24, International Atomic Energy Agency, Vienna, Austria, 1981.
9. **Kubota, M.I., I. Yamaguchi, K. Okada, Y.Morita, and H. Nakamura,** "Development of Partitioning Method: Partitioning Test with Nuclear Fuel Reprocessing Waste Prepared at PNC 1," JAERI-M-83-011, Tokai, Japan, 1983.
10. **Kubota, M., I. Yamaguchi, K. Okada, Y. Morita, K. Nakano, and H. Nakamura,** "Partitioning of High Level Waste as Pretreatment in Waste Management," in *Proceedings Scientific Basis of Nuclear Waste Managment,* Volume VII, G.L. McVay, Ed., Elsevier, Boston, Massachusetts; *Mat. Res. Soc. Symp. Proc.,* 26, 551, 1984.
11. **Tachimori, S., A. Sato, and H. Nakamura,** "Extraction of Lanthanides (III) with Diisodecyl Phosphoric Acid from Nitric Acid Solution," *J. Nucl. Sci. Technol.,* 15, 421, 1978.
12. **Yamaguchi, I., M. Kubota, K. Ohada, Y. Morita, and H. Nakamura,** "Development of Partitioning Method: Partitioning Test with Nuclear Fuel Reprocessing Waste Prepared at PNC 2," JAERI-M-84-070, Tokai, Japan, 1984.
13. **Morita, Y. and M. Kubota,** "Extraction of Neptunium with Di-isodecyl Phosphoric Acid from Nitric Acid Solution Containing Hydrogen Peroxide," *Solv. Extn. Ion Exch.,* 6, 233, 1988.
14. **Weaver, B., F.A. Kappelmann, and F.A. Talspeak,** "A New Method of Separating Americium and Curium from Lanthanides by Extraction from an Aqueous Solution of Aminopolyacetic Acid Complex with a Monoacidic Phosphate or Phosphonate," Oak Ridge National Laboratory report ORNL-3559, August, 1964.
15. **Zhu, Y.R., R. Jiao, S. Wang, S. Fan, B. Liu, H. Zeng, S. Zhou, and S. Chen,** Proc. Int. Solv. Extraction Cong. ISEC '83, Denver, Colorado, 1983.
16. **Jiao, R., S. Wang, S. Fan, B. Liu, and Y. Zhu,** "Trialkyl (C6-C8) Phosphine Oxide for the Extraction of Actinides and Lanthanides from High Active Waste," *J. Nucl. Radiochem.,* 7, 65, 1985.
17. **Zhu, Y. and C. Song,** "Recovery of Np, Pu, and Am from Highly Active Waste; Triakyl Phosphine Oxide Extraction," in *Transuranium Elements; A Half Century,* L.R. Morss and J. Fuger, Eds., ACS Books, Washington, D.C., 1992, chap 2.
18. **Siddall, T.H., III,** "Bidentate Organophosphorus Compounds as Extractants: I. Extraction of Cerium, Promethium and Americium Nitrates," *J. Inorg Nucl Chem,* 25, 883, 1963.
19. **Apostolidis, R., L. DeMeester, R. Molinet, J. Liang, and Y. Zhu,** "The Extraction of Actinides and Other Constituents from Highly Active Waste (HAW) by Trialkyl Phosphine Oxide (TRPO)," in *New Separations Chemistry Techniques for Radioactive Waste and Other Specific Applications,* Elsevier, New York, 1991.

20. **Schulz, W.W.,** "Bidentate Organophosphorus Extraction of Americium and Plutonium from Hanford Plutonium Reclamation Facility Waste," ARH-SA-203, Atlantic Richfield Hanford Company, Richland, Washington, 1974.

21. **McIsaac, L.D., J.D. Baker, and J.W. Tkachyk,** "Actinide Removal from ICPP Wastes," ICP-1080, Allied Chemical Company, Idaho Falls, Idaho, 1975.

22. **Shoun, R.R., W.J. McDowell, and B. Weaver,** "Bidentate Organophosphorus Compounds as Extractants from Acidic Waste Solutions: A Comparative and Systematic Study," Proc. Int. Solv. Extraction Conf., ISEC '77, Toronto, Canada, 1977.

23. **Marsh, S.F. and S.L. Yarbro,** "Comparative Evaluation of DHDECMP and CMPO as Extractants for Recovering Actinides from Nitric Acid," LA-11191, Los Alamos National Laboratory, Los Alamos, New Mexico, 1988.

24. **Nenni, J.A.,** "Identification of Alternative Technologies for the Treatment of Sodium-Bearing Waste at the ICPP," Westinghouse Idaho Nuclear Company (letter report), JAN-06-92, September 1992.

25. **Zhao, H. Y. Ye, X. Yang, and Z. Lin,** "Extraction of Np(IV), Pu(IV), and Am(III) by Bidentate Organophosphorus Extractant," *Inorg. Chim. Acta,* 94, 189, 1984.

26. **Zhao, H. L. Fu, X. Wei, S. Liu, G. Ye, L. Yang, and J. Jiang,** "Separation of Actinides and Lanthanides from Nuclear Power Reactor Fuel Reprocessing Waste by Bidentate Organophosphorus Extractant," *At. Energy Sci. Technol.,* 24, 66, 1990.

27. **Zhao, H.,** "Progress on the Separation of Actinides and Lanthanides from High Level Waste," *At. Energy Sci. Technol.,* 27, 277, 1993.

28. **Dzekun, E.G., G.M. Gelize, I.V. Smirox, A. Ya. Shadrin, V.V. Milyutin, B.G. Myasoedov, and M.K. Chumtova,** "Use of Bidentate Organophosphorus Compounds and Ampholites for Recovery of Transplutonium Elements (TPE) from Highly Active Waste (HAW)," Waste Management '92, University of Arizona, Tucson, 1992.

29. **Horowitz, E.P., K.A. Martin, H. Diamond, and L. Kaplan,** "Extraction of Am from Nitric Acid by Carbamoyl-Phosphoryl Extractions: The Influence of Substituents on the Selectivity of Am over Fe and Selected Fission Products," *Solv. Extr. Ion Exch.,* 4, 449, 1986.

30. **Eccles, H. and A. Rushton,** "Separative Technologies for the Removal of Thorium from an Acidic Waste Liquor," *Sep. Sci. Technol.,* 28, 59, 1993.

31. **Lloyd, M.H.,** "An Anion Exchange Process for Americium–Curium Recovery from Plutonium Process Waste," *Nucl. Sci. Eng.,* 17, 452, 1963.

32. **Musikas, C. and H. Hubert,** Proc. Int. Solv. Extraction Conf. ISEC '83, Denver, Colorado, 1983.

33. **Musikas, C.,** "Potentiality of Monorganophosphorus Extractants in Chemical Separations of Actinides," *Sep. Sci. Technol.,* 23, 1211, 1988.

34. **Cuillerdier, C., C. Musikas, P. Hoel, L. Nigond, and X. Vitart,** "Malonamides as New Extractants for Nuclear Waste Solutions," *Sep. Sci. Technol.,* 26, 1229, 1991.

35. **Musikas, C. and W.W. Schulz,** "Solvent Extraction in Nuclear Science and Technology," in *Principles and Practices of Solvent Extraction,* Marcel Dekker, New York, 1992, chap 11.

36. **Cuillerdier, S.C., C. Musikas, and L. Nigond,** "Extraction of Actinides from a Chloride Medium Using Pentaalkypropanediamide," *Sep. Sci. Technol.,* 28, 1993.

37. **Bernard, C., P. Miquel, and M. Viala,** "Advanced PUREX Process for New Reprocessing Plants in France and Japan," *Proc. 3rd Int. Conf. Nuclear Fuel Reprocessing Waste Management,* Atomic Energy Society of Japan, Vol. 1, April, 1991.

38. **Gasparini, G.M. and G. Grossi,** "Long Chain Distributed Aliphatic Amides as Extracting Agents in Industrial Applications of Solvent Extraction," *Solv. Extr. Ion Exch.,* 4, 1233, 1986.

39. **Carswell, D.J. and J.J. Lawrence,** "Solvent Extraction with Amines 1. The System Th-HNO$_3$-Triocylamine," *J. Inorg. Nucl. Chem.,* 11, 69, 1959.

40. **Ensor, D.D., G.D. Jarvinen, and B.F. Smith,** "The Use of Soft Donor Ligands, 4-Benzoyl-2,4-dihydro-5-methyl-2-phenyl-3H-pyrazol-3-thione and 4,7-Diphenyl-1,10-phenanthroline for Improved Separation of Trivalent Americium and Europium," *Solv. Extr. Ion Exch.,* 6, 439, 1988.

41. **Schulz, W.W. and L.A. Bray,** "Solvent Extraction Recovery of Byproduct Cesium-137 and Strontium-90 from Nitric Acid Solution — A Technology Review and Assessment," *Sep. Sci. Technol.,* 22 (2&3), 191, 1987.

42. **Kolarik, Z.** "Separation of Actinides and Long-lived Fission Products from High Level Radioactive Wastes (a Review)," KfK 4945, Kernforshungzentrum, Karlsruhe, GmbH, November 1991.

43. **Rais, J., P. Selucky, and M. Kyra,** "Extraction of Alkali Metals into Nitro Benzene in the Presence of Univalent Polyhedral Borate Anions," *J. Inorg. Nucl. Chem.,* 38, 1376, 1976.

44. **Rais, J., S. Tachimori, and P. Selucky,** "Synergetic Extraction in Systems with Dicarbollide and Bidentate Phosphonate," *Sep. Sci. Technol.,* 29, 261, 1994.

45. **Kyrs, M. and J. Rais,** "Separation of Fission Products by Solvent Extraction with Dicarbolide in Radioanalytical Chemistry and Technology," Presentation at the Separation of Ionic Solutes, SIS, '85 Conference, Smolenice, Czechoslovakia, September, 1985.

46. **Galkin, B.Y., I.N. Lazarev, R.I. Lyubsev, V.N. Romanovskii, D.N. Shiskin, J. Rais, M. Kyrs, P. Selucky, and L. Kadlecova,** in Proc. 5th Symp., Marianske Lazne, Czechoslovakia Atomic Energy Commission, Prague, April, 1981.

47. **Esimantovskii, V.M., L.N. Lazarev, R.I. Lyubtsev, V.N. Romanovskii, and D.N. Shishikin,** "Technological Tests of HAW Partitioning with the use of Chlorinated Cobalt Dicarbolyde (CHCODIC); Management of Secondary Wastes", in Proc. Symp. Waste Management, Tucson, March 1992, p. 801.

48. **Reilly, S.D., C.F.V. Mason, and P.H. Smith,** "Cobalt (III) Dicarbollide, a Potential Cs-137 and Sr-90 Waste Extraction Reagent", LA-11695, Los Alamos National Laboratory, Los Alamos, New Mexico, 1992.

49. **Olson, A.L., W.W. Schulz, L.A. Burchfield, C.D. Carlson, J.L. Swanson, and M.C. Thompson,** "Evaluation and Selection of Aqueous-Based Technology for Partitioning Radionuclides from ICPP Calcine," WINCO Report No. 1171, Idaho Falls, Idaho, February, 1993.

Chapter 5

Adsorption and Ion-Exchange Processes

John R. FitzPatrick

I. INTRODUCTION

The cleanup or removal of radionuclides from various liquid waste streams using either adsorption or ion-exchange techniques is an area of high interest. Nuclear facilities must treat their liquid waste streams in some manner. Aqueous waste streams are formed during some nuclear operations that contain small quantities of radionuclides in addition to large quantities of various inorganic salts and possibly some organic constituents. The removal of radionuclides down to an acceptable level from waste streams, prior to further treatment, is probably more easily accomplished at the source rather than at a liquid waste treatment facility. This chapter is a review of the processes for the separation of radionuclides from liquid wastes generated during the recovery of plutonium and other actinides from scrap materials and for removal of the actinides and fission products from waste streams arising from the processing of nuclear fuel by ion-exchange or sorption techniques. The necessary liquid waste cleanup steps associated with nuclear power plants have been reviewed elsewhere. In addition to reviewing basic ion-exchange processes, this review includes chelating resins, solvent-impregnated resins, inorganic ion exchangers, and various other sorbents for radionuclides such as bone char, clays, and other naturally occurring products.

II. ION EXCHANGE

Basic ion-exchange theory and technique is described in numerous texts, articles, and in many manufacturer's publications.[1-5]

Actinides and other radionuclides are usually separated and purified by either solvent extraction or ion exchange. Anion exchange is commonly used for purification of Pu and separation of Pu from Am.[6-8] However, it is possible to use cation exchange to purify Pu from other species in either HCl or HNO_3 media.[9]

Uranium recovery can also be performed using anion exchange. The resins utilized for these processes usually contain either quaternary ammonium groups (strong base) or tertiary amine groups (weak base).[10]

A. CLEANUP OF WASTE STREAMS ASSOCIATED WITH PLUTONIUM PROCESSING

The waste streams resulting from the recovery of U, Pu, or Am from scrap processing and Pu/Am separation is discussed first. These waste streams are typical of those produced at either the Los Alamos Plutonium Facility (TA-55), Rocky Flats Plant (prior to the shutdown of that facility), or the plutonium finishing areas at either Hanford or Savannah River. The waste streams from anion-exchange processing of Pu, typically the column effluents and washes, contain 10 to 100 mg/l Pu and 0.1 to 1.0 mg/l Am.[11] Further removal of the actinide species from 6 to 8 M HNO_3 or HCl is difficult using anion-exchange

0-8493-4876-5/96/$0.00+$.50
© 1996 by CRC Press, Inc.

techniques; however, the high affinity of anion-exchange resins for Pu can give some additional removal of Pu from the HNO_3-containing waste streams prior to a thermoevaporation or some other concentrating step. Some additional removal can be obtained using small-particle size resins such as 300–400 mesh Dowex-1 × 4; however, the pressure drop across such columns is very high.[12] (Dowex resins are trademarked by Dow Chemical Co.) Additional removal of Pu from a low-level waste stream has also been demonstrated using an anion-exchange column containing 20–50 mesh Amberlite IRA-938.[11] (Amberlite resins are trademarked by Rohm and Haas Co.) The larger mesh size is preferable to reduce the possibility of column plugging by particulate matter in the waste stream. Marsh has reported a new macroporous polyvinylpyridine resin for Pu anion exchange that offers both high-capacity and rapid sorption kinetics.[13] This resin appears quite promising for providing some additional removal of the Pu from these lean waste streams. Additional removal of Am species requires the use of either chelating resins or extraction chromatography techniques.

At Los Alamos, the HNO_3-containing waste streams are evaporated in a thermosyphon evaporator, which vaporizes nitric acid, to reduce the volume of radionuclides and salts. The acid can then be reused or released to the liquid waste treatment facility. Hydrochloric acid solutions are neutralized with caustic (either potassium hydroxide [KOH] or sodium hydroxide [NaOH]), thus forming a metal hydroxide sludge containing both the radionuclides and many other easily formed metal hydroxides that are present as impurities in the waste stream. This sludge can be redissolved in order to recover the radionuclides or it can be mixed into a cement-based grout or some other suitable transuranic (TRU) waste matrix.

However, special chelating resins or extraction chromatography using CMP type extractants or the newer CMPO extractants provide much better removal of the actinides present in the waste stream. These are discussed in much more detail in their respective sections. There are also a variety of sorbents, such as bone char (calcium hydroxyapatite), activated alumina, and inorganic ion-exchange media, that can be used to remove metal species after the solutions have been adjusted to pH 7 or higher.[14] These adsorbents will be discussed further in the section on inorganic ion exchangers.

Currently under investigation at Hanford and Los Alamos, is the cleanup of the tank wastes generated by the Hanford Plutonium Finishing Plant (PFP). These wastes are quite similar to the waste streams generated at the Los Alamos Plutonium Facility. The goal of this research effort is to demonstrate processes that would reduce the actinide content of the sludge from Tank #102-SY. The sludge will first be slurried so that it can be pumped from the tank. It is possible but not necessary to treat the slurry by super critical water oxidation (SCWO) or some other oxidation process to decompose the organic material and ferrocyanides, as well as most of the nitrites and nitrates. It has been proposed to separate the sludge from the solution either as it comes directly from the tank or after an oxidation step, then dissolve it in HNO_3. The resulting solution can be treated by anion exchange[15] or by solvent extraction (the transuranium extraction [TRUEX] process)[16] to further reduce the actinide content. If anion-exchange is used, then the effluent from the anion-exchange column is evaporated using a thermosyphon evaporator, where the acid fraction is separated from the remaining actinides and the large quantity of other inorganic salts. The distilled acid can then be neutralized and treated by standard flocculation techniques. The solids formed are removed by filtration and placed into a suitable low-level waste (LLW) matrix, such as grout. The evaporator bottom fraction, containing the Am, U, and traces of Pu, as well as other salts, such as Al and Fe, are then neutralized to pH 2, and any Fe(III) present is reduced to Fe(II). The solution is passed through a cation-exchange column to further reduce the actinide content. The effluent from the cation-exchange column is neutralized, mixed with flocculants to form a solid that is placed in a suitable LLW matrix. The concentrated eluate from the cation column is neutralized, solidified, and placed into a suitable matrix and stored as TRU waste. This waste form is suitable for shipment to a repository such as the Waste Isolation Pilot Plant (WIPP).

The current wastewater treatment method for such waste streams at Los Alamos uses lime to make the solution basic. A flocculating agent containing iron is then added to the waste stream to coprecipitate the remaining actinides and some of the other heavy metals as a solid hydroxide sludge. This process requires one or more repetitions. Mixing of acidic and basic waste streams (arising from the neutralization of HCl solutions) to yield a basic solution for the iron-assisted precipitation is useful. The water leaving the waste water treatment facility has a radionuclide activity of ~50 pCi/l.[17]

A variety of mixed-bed ion-exchange materials and some inorganic ion-exchange materials will work very well for removing the radionuclides from neutral or basic solutions. This is discussed further in the section on inorganic ion exchangers/sorbents.

B. NUCLEAR FUEL PROCESSING/REPROCESSING

The chemistry of nuclear fuel reprocessing is much more complicated than the removal of only the actinides. This is due to the presence of both actinides and fission products. There have been several good reviews of nuclear fuel processing using either solvent extraction or ion exchange.[6,7,12,18-22]

The waste streams generated during the processing of nuclear fuel are complicated not only by the actinides present, but also by the high radioactivity levels associated with the various fission products, such as ^{90}Sr and ^{134,137}Cs. There are also some very long-lived fission products, such as ^{99}Tc, that could be removed and stored with the actinides. The primary separation of the actinides from the fission products is usually done by solvent extraction using tributyl phosphate (TBP) in a normal paraffinic hydrocarbon (NPH; PUREX technology). The actinides and a small fraction of the fission products are retained in the organic or TBP phase while the aqueous phase (raffinate) contains the majority of the fission products. This aqueous waste stream contains very small quantities of actinides and some organic material.

At Hanford it was this particular waste stream that was made strongly basic (pH $\geq$14) and then pumped out to the large underground tanks for long-term storage. The origin of the various wastes at the Hanford site is more fully described in a paper by Swanson and Lumetta.[23] Currently there are large programs underway at Hanford, Savannah River, Oak Ridge, and Idaho involving the solidification and stabilization of these tank wastes. Research at these various sites is being conducted to determine the best methods for remediating these waste streams. There are several options available, including *in situ* vitrification;[24] *in situ* sludge washing;[25] removal of the tank contents for sludge washing, separation of the ^{90}Sr, ^{137}Cs and separation of the actinides from the sludge.[26] The actinide content consists mostly of U (~7 g/l), Pu at ~10 mg/l, and Am at ~0.1 mg/l.[27] Removal of the ^{90}Sr, ^{134,137}Cs, ^{99}Tc, and the actinides from the tank waste sludges allows the remaining undissolved sludges to be disposed of into a suitable LLW form.[28] The actinides, which now comprise a very small volume, will be vitrified into glass, which is a suitable TRU waste form.

Separation of the ^{134}Cs ($t_{1/2}$ = 2.1 years), ^{137}Cs ($t_{1/2}$ = 30.2 years), and ^{90}Sr ($t_{1/2}$ = 28 years) into a separate waste stream from the one containing the actinides would allow for solidification and interim storage of these radioisotopes. This waste form must be relatively stable for at least 300 years, at which time ten half-lives will have elapsed and the residual activity from these two elements will be quite small. This waste form must also be able to withstand the high heat load resulting from the decay of these radionuclides.[28-30] The current preferred waste form is glass.

At Savannah River, it has been demonstrated that low levels of Sr and Cs can be removed from dilute waste streams using a sulfonic acid cation-exchange resin.[31] The removal of Cs from the waste stream has been demonstrated using a resorcinol–formaldehyde resin developed at Westinghouse Savannah River Company.[32-34] It should also be noted that the separation of the Sr/Cs from the actinides is not a necessity. Additional methods for the removal of the ^{134,137}Cs and ^{90}Sr are given in the sections on extraction chromatography and inorganic ion exchangers/absorbers.

III. CHELATING RESINS

These types of ion-exchange resins are typically commercially produced. However, there have been a large number of resins produced in small quantities by various researchers. The synthesis and use of this type of resin has been reviewed, both by Millar and co-workers[35] and by Eccles and Greenwood.[36] The use of chelating resins for analytical chemistry purposes was reviewed by Myasoedova, Savvin, and Vernadsky in 1986.[37] There are a number of desirable properties that form the basis for the synthesis of this type of resin. The major properties are high metal capacity, high selectivity, fast kinetics, and high mechanical strength. There are basically four groups of commercial resins available today, each containing one of the following groups: iminodiacetic acid, aminophosphonic acid, amidoxime, or thiol.[36] Typical commercial resins with these functional groups are Amberlite IRC-718 or Chelex-100®, which contain iminodiacetic acid groups (Chelex is supplied by Bio-Rad Chemical Division), Duolite® C-467 with aminophosphonic acid groups, Duolite C-346, which contains amidoxime groups, and Purolite® S-920 or Duolite GT-73 with thiol groups.[35] (Duolite is a trademark of DISA Limited, U.K. and Purolite is carried by Rohm and Haas Co.; resins are from Purolite Co.) The cost of synthesizing these types of resins is substantially higher than that of ordinary ion-exchange resins. Perhaps it is this higher cost that limits the number of different commercial chelating resins currently available.

Because there is not a large quantity of chelating resins available today, there are a very limited number of references on their use. Duolite C-467 has been used for the removal of uranium and thorium from phosphoric and/or nitric acid waste streams.[38,39] Chelex-100® also has been investigated for the removal of actinides from near-neutral waste solutions.[40] The behavior of Dowex® A-1 chelating resin for adsorption of both actinides and lanthanides has been reported by Mathur and Khopar.[41] Myasoedova and co-workers also have reported on chelating sorbents containing complexing ion exchangers in a fibrous matrix (Polyorgs).[42]

Eichrom Industries has recently begun to manufacture a new chelating resin called Diphonix.® This resin is based upon the work by Horwitz and co-workers at Argonne National Laboratory (ANL) in conjunction with Alexandratos and co-workers at the Department of Chemistry, University of Tennessee. This resin is a styrene-divinyl benzene, sulfonated ion-exchange resin containing gem-diphosphonic acid groups.[43] Several recently published articles indicate that this resin will be able to extract the actinides as well as various transition metal cations such as Fe(III) and Cr(III) from near-neutral to very acidic solutions.[44-48] This resin appears to be quite promising as a final polishing step for both nitric and hydrochloric acid waste streams.

IV. EXTRACTION CHROMATOGRAPHY

The area of extraction chromatography, although not as popular as ion-exchange chromatography, has been used at various times by a variety of researchers. There are examples of its use sprinkled throughout the literature. Extraction chromatography appears to have become much more popular in the last decade, perhaps due to the lack of commercial chelating resins available for removal of specific metal ions. These extraction chromatographic materials are sometimes called solvent-impregnated resins. The work in this field was reviewed in 1973.[49] The concept or need for ion-specific resins has not led to the development of ion-exchange resins that are specific for only one or two particular species. In fact, this may be almost impossible due to the similar chemistry exhibited by the various metal ions within a periodic family. The rationale behind the solvent-impregnated resins is that a specific extractant can be loaded onto a substrate; either a polymeric resin, or silica gel, or some other inert substrate. This solvent-impregnated resin now can be used to extract the specific metal ions without extracting many of the other ions that may be present in solution. There are many examples of their use in the literature over the past two decades.[50-61]

With the specificity of chelating extractants, it is now possible to treat a waste stream by pumping the solution through a bed of solvent-impregnated resin, largely removing only the desired components, typically the radioactive species. This has been demonstrated by Horwitz and co-workers at ANL[62] with resins that are now currently being produced by Eichrom Industries.[63] These resins were originally for analytical use but there have been reports on their use for relatively large-scale cleanup of both HNO_3 and HCl waste streams. The use of extraction chromatography for processing is not commonplace. However, in certain instances, columns containing these resins have been used for production-scale work.[64,65]

The use of this type of resin for cleanup and recovery of a larger fraction of the actinide or other radioactive species present in a particular waste stream is not only desirable, but is also quite practical. In particular, removal of radioactive components from the Hanford tank wastes has been studied using both actinide-specific resins and Sr-specific resins.[28,66,67] Another interesting possibility is the bonding of macrocyclic ligands to silica, thus eliminating the problems of the extracting agent washing off the support material.[68] This is a very new concept and has not been tested except for separations of various metal ions from relatively clean solutions, but not on simulated tank waste solutions. Current work at Los Alamos and other laboratories to demonstrate a more thorough removal of the Pu and Am from both the HNO_3 and HCl waste streams has also been reported.[69,70]

Other countries facing the problem of nuclear wastes, including Russia, India, France, Italy, Germany, and Japan, are currently investigating the use of CMPO-loaded resins for removal of the actinides from both high-level waste (HLW) and from intermediate-level or medium-activity waste (ILW). This minimizes the volume of HLW while producing a larger volume of easily handled LLW. Smaller volumes of HLW help to reduce the size of a HLW repository and also help to decrease the amount of shielding at the handling areas of the repository.[71-75]

The removal of some of the other elements in either high- or medium-level waste using extraction chromatography has also been demonstrated. These elements are typically the gamma emitters $^{134,137}Cs$, ^{125}Sb, ^{106}Ru, ^{144}Ce, and $^{154,155}Eu$. Although they are beta emitters, ^{99}Tc and ^{90}Sr could also be included in this listing. If the waste has aged for 10 years or more, most of the ^{106}Ru and ^{144}Ce will have decayed because at least ten half-lives will have elapsed. Separation of these elements has been studied and methods have been developed for their effective removal. Most of these removal methods are discussed in the section on inorganic ion exchangers/sorbents.

Separation of Sr from tank solutions/sludges has also been demonstrated using both solvent extraction[23,28,76,77] and extraction chromatography.[29,78,79] The extraction chromatography columns contain an inert resin loaded with a crown ether extractant, (di-t-butylcyclohexano-18-crown-6, DtBuCH18C6). This is the same extractant used in the solvent extraction process mentioned above. The results from the early tests indicate that this extraction chromatography material performs very well.[29]

Other specific methods for the removal of both [134,137]Cs and ^{90}Sr from waste solutions are discussed in the section on inorganic ion exchangers/sorbents.

V. INORGANIC ION EXCHANGERS/SORBENTS

The potential uses of inorganic materials as ion-exchange media have long been recognized. There are a number of reviews dating back to the 1960s. One of the more comprehensive reviews was done by Clearfield et al. in 1973.[80] Another good review, although not directly applicable to nuclear waste processing, was done by Shanks et al. on metallurgical applications.[81] Applications of inorganic sorbents to actinide separation processes was presented at a 1979 actinide separations symposium.[82] Recently, research has been published comparing 60 different absorbers (some of them inorganic) for removal of a variety of elements under both acidic and basic conditions.[83] A literature survey has also been recently published on partitioning of HLW from alkaline solutions.[84] There are a number of references made to inorganic ion-exchange media in this survey also.

During 1975 to 1978, removal of actinides using bone char and other inorganic absorbents was the focus of a study at Mound,[85-87] while similar studies on various titanates were ongoing at Sandia National Laboratory.[88,89] A study of the use of sodium titanates was also done at Hanford in this time period.[90] It should be noted that the sodium titanate material could be incorporated onto a resin much like the solvent-impregnated resins discussed earlier. Although the capacity for Cs was not as high as the pure material, the removal of Cs was adequate, and perhaps a higher loading of the titanate onto the resin would be feasible and improve the Cs capacity of this material. The impregnated resin was somewhat easier to handle. With the renewed interest in removal and stabilization of radionuclides from the Hanford and other high-level tank wastes, research on various inorganic ion exchangers has resurfaced. These materials include those studied earlier at Mound and Sandia as well as the newer silicotitanates and also phosphorous/molybdenum or tungsten compounds. These are discussed in more detail later in this section.

Bone char, which is formed by burning cattle bones, was the major focus of the work in the late 1970s at Mound. This coarse material (8–28 mesh) was produced by Stauffer Chemical. The research done at Mound indicated that bone char was quite effective for the removal of Pu (and Am to some extent) from neutral or basic solutions. In 1989, bone char was again evaluated along with some other resins and absorbents for the removal of Pu from low-level wastewater.[14] The conclusion from this study indicated that bone char was far superior to any of the other materials for removal of Pu from near-neutral solutions. This material is no longer produced by Stauffer, and a synthetic material, called bone ash, does not remove the radionuclides as thoroughly as bone char does. This bone ash appears to be a synthetic calcium hydroxyapatite that has been calcined. Perhaps it is still possible to obtain bone char as an import from Great Britain.

Modified peat was also investigated at Rocky Flats for the cleanup of wastewaters.[91] The peat, which is quite cheap, was treated with sulfuric acid to form a particulate material that was permeable to water yet resistant to leaching and could be used as a column material. It was not as effective as bone char for the removal of Pu from these solutions. However, it performed better than coconut carbon and, in some cases, was comparable to zeolite for removal of both Pu and Am from wastewaters.

Separation of Sr, Cs, and other fission products from ML- or ILW and HLW arising from the reprocessing of nuclear reactor fuel has been the main use for the sodium titanate materials. Ammonium 12-molybdophosphate (AMP) and the corresponding tungsten analog, ATP, have been suggested for the same separation. Faubel and Ali list a variety of reagents for removal of Cs from ILW that have been previously studied.[92] Capacities for Cs removal are given for ammonium hexacyano cobaltous ferrate, potassium hexacyano nickel ferrate, zirconium phosphate, titanium phosphate, antimony pentoxide, titanium dioxide, and AMP. All of these absorbers are restricted to use with pH 2 to 12 solutions, except AMP, which can be used with high nitric acid concentrations. AMP has one of the highest capacities for Cs, up to 1.5 mmol/g. The column containing microcrystalline AMP is not a fixed-bed column, but more of a resin in pulp or fluidized column. Once the column is loaded, the AMP is removed and dissolved in 1 to 10 M NaOH, thus yielding a much smaller volume of HLW.

Some of these inorganic materials, such as silicotitanates and a zeolite, have been investigated for removal of Cs from actual tank wastes.[30] Although the silicotitanates worked very well, they were rejected because they have not been well enough characterized and are still considered to be in the development stage. There are recent reports on both advanced titanate and silicotitanate exchangers, but the amount of information currently available about these materials is quite limited.[93,94] The zeolite was rejected because of the low capacity for Cs and the fact that there was not an easy way to remove the Cs from the zeolite, thus making regeneration of the column impossible. The cesium capacity in zeolite has been shown to be ~1 g/kg zeolite.[95] This is quite low compared to AMP or zirconium phosphate (100 g/kg).

Various materials, including some new, different inorganic absorber materials were reported in the proceedings from a technical seminar that was jointly organized by the Commission of the European Communities (CEC) and the Italian Commission for Nuclear and Alternative Energy Sources (ENEA) in 1991.[96] This book also contains a good review on hexacyanoferrate removal of Cs. It is possible to incorporate Cu hexacyanoferrate into macroporous resins, forming a solvent-impregnated resin, which was used to remove Cs from low-level waste streams.[97]

There are various other absorbing materials (tuff, clays, earth, and various synthetic zeolites) that have been studied for the removal of a variety of metal ions from solutions. Most of these materials have not been tested on any synthetic or actual waste stream solutions nor have they been thoroughly researched for their potential uses in the cleanup of waste streams. Most of these materials show some potential for waste stream cleanup. However, they are so new that it will be several years before they have been developed to the point where they can be tested and utilized for waste stream cleanup.

VI. CONCLUSIONS

We see that ion exchange using regular ion-exchange resins, chelating resins, solvent-impregnated resins, or inorganic ion-exchange materials has been demonstrated to be a viable method for the separation of radionuclides from radioactive waste streams. The minimization of the quantity of high-level or TRU waste requiring either vitrification or encapsulation in some other appropriate matrix material is desirable. These ion-exchange processes offer a usable, yet economically viable alternative to solvent extraction for the removal of specific metal ions from a large variety of these waste streams. The generation of large quantities of low-level waste, although not entirely desirable, allows for incorporation of a minimum amount of radionuclides into a waste form matrix that is economically feasible and is much easier to handle and store.

Although not all of the possible materials for ion exchange associated with waste stream separation were covered in depth, the references listed in this chapter will hopefully lead to other references that will, in turn, allow for a reasonable decision to be made about which process or method to choose for the treatment of a particular waste stream. At the current time, treatment of waste streams is an area of great interest in the United States and other countries. The knowledge in this area will continue to grow in the next 5 to 10 years as the nuclear industry expands the knowledge base concerning separation and decontamination of radioactive species from the large variety of waste streams. This growth will be associated with research at nuclear power plants, nuclear fuel fabrication/fuel-reprocessing facilities and a variety of other nuclear projects, including nuclear medicine and radioactive isotope production.

REFERENCES

1. Rohm and Haas Co., "Ion Exchange Resins — Laboratory Guide".
2. Bio-Rad Chemical Division, "Guide to Ion Exchange," Cat. No. 140-9997.
3. Dow Chemical, "Fundamentals of Ion Exchange," in IDEAS ± Exchange, Vol. 1, No. 1, Form No. 177-1196-83.
4. G. J. Moody and J. D. R. Thomas, "Inorganic Ion Exchange in Organic and Aqueous-Organic Solvents, A Review," *Analyst,* 93, 557–587 (1968).
5. J. J. Fardy, "Preconcentration of Trace Elements by Ion Exchangers," in *Preconcentration Techniques for Trace Elements,* Edited by Z. B. Alfassi, C. M. Wai, CRC Press, Boca Raton, FL, Chapter 7, 1992, pp. 211–241.
6. J. M. Cleveland, *The Chemistry of Plutonium,* American Nuclear Society, LaGrange Park, IL, 1979.
7. O. J. Wick, Editor, *Plutonium Handbook, a Guide to Technology,* Gordon and Breach, New York, 1967.
8. G. Choppin, "Separation Processes for Actinide Elements," *Sep. Sci. Technol.,* 19(11&12), 911–925 (1984–1985).
9. C. Keller, *The Chemistry of the Transuranium Elements,* Verlag Chemie, Weinheim , Germany, 1971, pp. 461–464.
10. M. Streat and D. Naden, "Ion Exchange in Uranium Extraction," in *Ion Exchange and Sorption Processes in Hydrometallurgy, Critical Reports on Applied Chemistry,* Vol. 19, Edited by M. Streat and D. Naden, John Wiley & Sons, New York, Chapter 1, 1987, pp. 1–55.
11. L. L. Martella, J. D. Navratil, and M. T. Saba, "Recovery of Pu and Am from Nitric Acid Waste Streams," in *Actinide Recovery from Waste and Low-Grade Sources,* Proceedings of the International Symposium cosponsored by the Divisions of Industrial and Engineering Chemistry and Nuclear Chemistry and Technology of the American Chemical Society, New York, August 24–25, 1981, Edited by J. D. Navratil and W. W. Schulz, Harwood Academic Publishers, New York, 1982, pp. 27–40.
12. Kjeller Report, "Reprocessing of Fuel from Present and Future Power Reactors," pp. 229–244, Report: KR-126, September 1967.
13. S. F. Marsh, "Reillex(TM) HPQ: A New, Macroporous Polyvinylpyridine Resin for Separating Pu Using Nitrate Anion Exchange," *Solv. Extr. Ion Exch.,* 7(5), 889–908 (1989).

14. G. S. Barney, A. R. Blackman, K. J. Lueck, and J. W. Green, "Evaluation of Adsorbents for Removal of Pu from a Low-level Process Wastewater," Report: WHC-SA-0533, February 1989.

15. S. L. Yarbro, W. A. Punjak, S. A. Schreiber, S. L. Dunn, G. B. Jarvinen, S. F. Marsh, N. G. Pope, S. Agnew, E. R. Birnbaum, K. W. Thomas, and E. M. Ortiz, "Tank 102-SY Remediation Project: Flow Sheet and Conceptual Design Report," Report: LA-12701-MS, January 1994.

16. G. J. Lumetta and J. L. Swanson, "Pretreatment of PFP Sludge, Report for the Period October 1990–March 1992," Report: PNL-8601, April 1993.

17. J. R. Covey, W. S. Midkiff, and F. Cadena, "Actinide Removal from Wastewater Applying Waste Minimization Techniques," presented at 47th Annual Purdue University Industrial Waste Conference, May 11–13, 1992, Report: LA-UR-92-1538.

18. I. L. Jenkins and A. G. Wain, "Ion Exchange in the Atomic Energy Industry with Particular Reference to Actinide and Fission Product Separations," *Rep. Prog. Appl. Chem.*, 57, 308–319 (1973).

19. W. L. McCready and J. A. Wethington, Jr., "Solvent Extraction and Ion Exchange in the Nuclear Fuel Industry," *Nucl. Technol.*, 53, 280–294 (1981).

20. I. L. Jenkins, "Ion Exchange in the Atomic Energy Industry with Particular Reference to Actinide and Fission Product Separation," *Solv. Extr. Ion Exch.*, 2(1), 1–27 (1984).

21. J. D. Navratil, "Ion Exchange Technology in Spent Fuel Reprocessing," *J. Nucl. Sci. Technol.*, 26(8), 735–743 (1989). (Note: same as report: RFP-3639.)

22. J. P. Bibler, "Ion Exchange in the Nuclear Industry," in *Recent Developments in Ion Exchange 2*, Edited by P. A. Williams and M. J. Hudson, Elsevier, New York, Proceedings from International Conference on Ion Exchange Processes(ION-EX '90), at The North East Wales Institute of Higher Education, July 9–11,1990, pp. 121–133. (Note: same as report: WSRC-RP-89-1206.)

23. J. L. Swanson and G. J. Lumetta, "Batch Testing of the TRUEX Process on Some Unique Hanford Wastes and of the TRUEX and SREX Processes on Commercial Fuel Reprocessing High-Level Waste," Presented at the 7th Symposium on Separation Science and Technology for Energy Applications, October 20–24, 1991, Knoxville, TN, Report: PNL-SA-19681. (Same as CONF-911049-3.)

24. S. S. Koegler, R. D. Gibby, and L. E. Thompson, "In-situ Vitrification of Radioactive Underground Tanks," Report: PNL-SA-19225, October 1991.

25. B. A. Hamm, R. E. Eibling, and J. R. Fowler, "Demonstration of In-Tank Sludge Processing. Part 1: Al Dissolution, Sludge Washing and Settling Results," Report: DPST-83-668, July 12, 1983.

26. J. L. Straalsund, J. L. Swanson, E. G. Baker, J. J. Holmes, E. O. Jones, and W. L. Kuhn, "Clean Option: An Alternative Strategy for Hanford Tank Waste Remediation," Report: PNL-8388, Vol. 1, December 1992; Vol. 2, September 1993.

27. G. J. Lumetta and J. L. Swanson, "Pretreatment of Neutralized Cladding Removal Waste Sludge: Status Report," Report: PNL-8558, March 1993.

28. G. J. Lumetta, M. J. Wagner, N. G. Colton, and E. O. Jones, "Underground Storage Tank Integrated Demonstration — Evaluation of Pretreatment Options for Hanford Tank Wastes," Report: PNL-8537, June 1993.

29. G. J. Lumetta, D. W. Wester, J. R. Morrey, and M. J. Wagner, "Preliminary Evaluation of Chromatographic Techniques for the Separation of Radionuclides from High-Level Radioactive Waste," *Solv. Extr. Ion Exch.*, 11(4), 663–682 (1993).

30. S. A. Barker, C. K. Thornhill, and L. K. Holton, "Pretreatment Technology Plan," Report: WHC-EP-0629, March 1993.

31. J. P. Bibler and R. M. Wallace, "Ion Exchange Processes for Cleanup of Dilute Waste Streams by the F/H Effluent Treatment Facility at the Savannah River Plant," from a paper proposed for presentation at Ion-Ex '87, Wrexham, Clwyd, Wales, UK, April 13–16, 1987. (Also as Report: DP-MS-86-187 and CONF-870485-1.)

32. D. E. Kurath, W. G. Richmond, E. O. Jones, L. A. Bray, and B. C. Bunker, "Cs Removal from Hanford Tank Waste," Presented at the AIChE Spring National Meeting, March 28–April 1, 1993, Houston, TX, Report: PNL-SA-21848.

33. **J. P. Bibler (SRL), R. M. Wallace (SRL), and L. A. Bray,** "Testing of a New Cs-Specific Ion-Exchange Resin for Decontamination of Alkaline-High Activity Waste," Presented at the 1990 Waste Management Meeting, Tucson, AZ, February 25, March 1, 1990. Waste Management '90, Edited by R. G. Post, Proceedings of the Symposium on Waste Management, Vol. 1, pp. 747–751. (Also as Report: PNL-SA-17370, 1989.)

34. **L. A. Bray, K. J. Carson, and R. J. Elovich,** "Cs Recovery Using SRL Resorcinol-Formaldehyde Ion Exchange Resin," Report: PNL-7273, 1990.

35. **J. R. Millar, D. Petruzzelli, and G. Tiravanti,** "Some Problems in the Use of Chelating Resins for Environmental Protection from Toxic Metals," in *Recent Developments in Ion Exchange 2,* Edited by P. A. Williams and M. J. Hudson, Elsevier, New York, Proceedings from International Conference on Ion Exchange Processes(ION-EX '90), at The North East Wales Institute of Higher Education, July 9–11, 1990, pp. 337–346.

36. **H. Eccles, and H. Greenwood,** "Chelate Ion-Exchangers: The Past and Future Applications, A User's View," *Solv. Extr. Ion Exch.,* 10(4), 713–727 (1992).

37. **G. V. Myasoedova, S. B. Savvin, and V. I. Vernadsky,** "Chelating Sorbents in Analytical Chemistry," *CRC Crit. Rev. Anal. Chem.,* 17(1), 1–63 (1986).

38. **S. Gonzalez-Luque and M. Streat,** " The Recovery of By-Product Uranium from Wet Process Phosphoric Acid Solution Using Selective Ion Exchange Resins," in *Ion Exchange Technology,* Edited by D. Naden and M. Streat, Ellis Horwood, Chichester, 1984, pp. 679–689.

39. **H. Eccles,** "The Removal of Th Isotopes from HNO_3 Liquors," in *Ion Exchange Technology,* Edited by D. Naden and M. Streat, Ellis Horwood, Chichester, 1984, pp. 698–708.

40. **F. H. El-Sweify, R. Shabana, N. Abdel-Rahman, and H. F. Aly,** "Distribution of Some Actinides and Fission Products between the Chelating Ion Exchanger, Chelex-100 and Certain Carboxylic Acid Solutions," *Radiochim. Acta,* 38, 211–214 (1985).

41. **J. N. Mathur and P. K. Khopkar,** "Ion Exchange Behavior of Chelating Resin Dowex A-1 with Actinides and Lanthanides," *Solv. Extr. Ion Exch.,* 3(5), 753–762 (1985).

42. **G. V. Myasoedova, I. I. Antokol'skaya, O. P. Shvoeva, M. S. Mezhirov, and S. B. Savvin,** "New Chelating Sorbents Based on Fibrous Materials Filled With Complexing Ion Exchangers," *Solv. Extr. Ion Exch.,* 6(2), 301–321 (1988).

43. **S. D. Alexandratos, A. Q. Trochimczuk, D. W. Crick, E. P. Horwitz, R. Chiarizia, and R. C. Gatrone,** "Macromolecules," submitted.

44. **E. P. Horwitz, R. Chiarizia, H. Diamond, R. C. Gatrone, S. D. Alexandratos, A. Q. Trochimczuk, and D. W. Crick,** "Uptake of Metal Ions by a New Chelating Ion-Exchange Resin. Part 1: Acid Dependencies of Actinide Ions," *Solv. Extr. Ion Exch.,* 11(5), 943–966 (1993).

45. **R. Chiarizia, E. P. Horwitz, R. G. Gatrone, S. D. Alexandratos, A. Q. Trochimczuk, and D. W. Crick,** "Uptake of Metal Ions by a New Chelating Ion-Exchange Resin. Part 2: Acid Dependencies of Transition and Post-Transition Metal Ions," *Solv. Extr. Ion Exch.,* 11(5), 967–985 (1993).

46. **K. L. Nash, P. G. Rickert, J. V. Muntean, and S. D. Alexandratos,** "Uptake of Metal Ions by a New Chelating Ion-Exchange Resin. Part 3: Protonation Constants via Potentiometric Titration and Solid State ^{31}P NMR Spectroscopy," *Solv. Extr. Ion Exch.,* 12(1), 193–209 (1994).

47. **R. Chiarizia, E. P. Horwitz, and S. D. Alexandratos,** "Uptake of Metal Ions by a New Chelating Ion-Exchange Resin. Part 4: Kinetics," *Solv. Extr. Ion Exch.,* 12(1), 211–237 (1994).

48. **E. P. Horwitz, R. Chiarizia, and S. D. Alexandratos,** "Uptake of Metal Ions by a New Chelating Ion-Exchange Resin. Part 5: The Effect of Solution Matrix on Actinides," *Solv. Extr. Ion Exch.,* 12(4), 831–846 (1994).

49. **A. Warshawsky,** "Extraction with Solvent Impregnated Resins," in *Ion Exchange and Solvent Extraction, A Series of Advances,* Vol. 3, Edited by J. A. Marinsky and Y. Marcus, Marcel Dekker, New York, 1973, pp. 229–310.

50. **J. S. Fritz and D. C. Kennedy,** "Extraction Chromatography of U with Dioctyl Sulphoxide," *Talanta,* 17, 837–843 (1970).

51. **K. Watanabe, Y. Nakagawa, and K. Sato,** "Mutual Separation of Pu, Th, and U by Extraction Chromatography," *Tokyo Gakugei Daigaku Kiyo, Dai-4-Bu,* 29, 158–162 (1977).

52. E. P. Horwitz, M. L. Dietz, D. M. Nelson, J. J. LaRosa, and W. D. Fairman, "Concentration and Separation of Actinides from Urine Using a Supported Bifunctional Organophosphorous Extractant," *Anal. Chim. Acta*, 238, 263–271 (1990).

53. C. Madic, C. Kertesz, R. Sontag, and G. Kowhly, "Application of Extraction Chromatography to the Recovery of Np, Pu and Am from an Industrial Waste," *Sep. Sci. Technol.*, 15(4), 745–762 (1980).

54. L. L. Martella and J. D. Navratil, "Recovery of U from Mixed Pu-U Residues by an Extraction Chromatography Process," Report: RFP-3289, 1982.

55. S. F. Marsh and O. R. Simi, "Applications of DHDECMP Extraction Chromatography to Nuclear Analytical Chemistry," Anal. Chem. Nucl. Technol., Proc. Conf. Anal. Chem. Energy Technol., 25th, Edited by W. S. Lyon, 1982, pp. 69–77.

56. N. V. Jarvis, "Small-Scale Recovery and Separation of Actinides Using TBP Extraction Chromatography," *Solv. Extr. Ion Exch.*, 7(6), 1077–1088 (1989).

57. T. Kimura, "Extraction Chromatography in the TBP-Nitric Acid System. I. Extraction Behavior of Th(IV) and U(VI) with TBP/XAD-4 Resin," *J. Radioanal. Nucl. Chem.*, 141(2), 295–306 (1990).

58. E. P. Horwitz, M. L. Dietz, and R. Chiarizia, "Application of Novel Extraction Chromatographic Materials to the Characterization of Radioactive Waste Solutions," *J. Radioanal. Nucl. Chem.*, 161(2), 574–583 (1992).

59. T. Kimura, "Extraction Chromatography in the TBP-Nitric Acid System. II. Characteristics of the TBP/XAD-4 Resin on the Separation of Actinide Elements," *J. Radioanal. Nucl. Chem.*, 141(2), 307–316 (1990).

60. T. Kimura and J. Akatsu, "Extraction chromatography in the DHDECMP–HNO₃ System. I. Extraction Behavior of Ce(III) and Am(III) with the DHDECMP/XAD-4 Resin," J. *Radioanal. Nucl. Chem.*, 149(1), 13–23 (1991).

61. T. Kimura and J. Akatsu, "Extraction Chromatography in the DHDECMP–HNO₃ System. II. Characteristics of the DHDECMP/XAD-4 Resin on the Separation of Trivalent Actinide Elements," *J. Radioanal. Nucl. Chem.*, 149(1), 25–34 (1991).

62. E. P. Horwitz, M. L. Dietz, R. Chiarizia, H. Diamond, A. M. Essling, and D. Graczyk, "Separation and Preconcentration of U from Acidic Media by Extraction Chromatography," *Anal. Chim. Acta*, 266(1), 25–37 (1992).

63. Eichrom Industries, Darien, IL 60561, produces a number of solvent impregnated resins, including: TRU-Spec for actinide removal, UTEVA-Spec for U and +4 actinide removal, Sr-Spec for Sr removal, RE-Spec for removal of the rare-earth elements, Ln-Spec for removal of the lanthanide elements..

64. W. I. Yamada, L. L. Martella, and J. D. Navratil, "Americium Recovery and Purification Using a Combined Anion Exchange-Extraction Chromatography Process," *J. Less Common Metals*, 86, 211–218 (1982).

65. J. Bourges, C. Madic, and G. Koehly, "Transplutonium Elements Production Program — Extraction Chromatographic Process for Pu Irradiated Targets," in Actinide Separations, Edited by J. D. Navratil and W. W. Schulz, a symposium sponsored by ACS Division of Industrial and Engineering Chemistry at the ACS/CSJ Chemical Congress (177th ACS National Meeting), Honolulu, HI, April 3–5, 1979, pp. 35–50.

66. G. S. Barney and R. G. Cowan, "Separation of Actinide Ions from Radioactive Waste Solutions Using Extraction Chromatography," Report: WHC-SA-1520-FP, 1992.

67. G. J. Lumetta, M. J. Wagner, D. W. Wester, and J. R. Morrey, "Evaluation of Solid-Based Separation Materials for the Pretreatment of Radioactive Wastes," Report: PNL-8596, May 1993.

68. D. M. Camaioni, N. G. Colton, and R. L. Bruening, "Investigation of the Potential of Silica-Bonded Macrocyclic Ligands for Separation of Metal Ions from Nuclear Waste," Report: PNL-7894, January 1992.

69. L. D. Schulte, J. R. FitzPatrick, D. A. Romero, R. R. Salazar, B. S. Schake, and T. M. Foreman, "Extraction of Pu and Am from HCl or HNO₃ Waste Streams Using Actinide-Specific Resins," Poster, P-79, Actinides-93, September 19–24, 1993, Santa Fe, NM.

70. A. Ramanujam, P. S. Dhami, V. Gopalakrishnan, J. N. Mathur, M. S. Murali, R. H. Iyer, L. P. Badheka, and A. Banerji, "Separation of Actinides from HNO₃ Solutions Using CMPO Adsorbed Chromosorb," Poster P-192, Actinides-93, September 19-24, 1993, Santa Fe, NM.

71. W. Faubel and S. A. Ali, "Partitioning of Nitric Acid Intermediate-Level Waste Solutions by Sorption," *Nucl. Technol.*, 86, 60–65 (1989).

72. W. Faubel, "Separation of Am, Eu and Ce from Liquid Wastes with CMPO by Extraction Chromatography," in *New Separation Chemistry Techniques for Radioactive Waste and Other Specific Applications*, Edited by L. Cecille, M. Casarci, and L. Pietrelli, Elsevier Applied Science, New York, 1991, pp. 73–79.

73. W. Faubel, R. Mehret, and P. M. Menzler, "Decontamination of MAW-Concentrate by a Chromatographic Method," Report: EUR-12911, 1990.

74. L. Pietrelli, A. Saluzzo, and F. Troiani, "Actinides Removal by Means of Octyl(phenyl)-*N,N*-diisobutyl Carbamoyl Methyl Phosphine Oxide (CMPO) Sorbed on Silica," in *New Separation Chemistry Techniques for Radioactive Waste and Other Specific Applications*, Edited by L. Cecille, M. Casarci, and L. Pietrelli, Elsevier Applied Science, New York, 1991, pp. 87–94.

75. J. P. Brunette and J. F. Dozol, "Application of CMPO Containing Gels to Metal Extraction," in *New Separation Chemistry Techniques for Radioactive Waste and Other Specific Applications*, Edited by L. Cecille, M. Casarci and L. Pietrelli, Elsevier Applied Science, New York, 1991, pp. 116–121.

76. E. P. Horwitz, M. L. Dietz, and D. E. Fisher, "SREX: A New Process for the Extraction and Recovery of Sr from Acidic Nuclear Waste Streams," *Solv. Extr. Ion Exch.*, 9(1), 1–25 (1991).

77. E. P. Horwitz, M. L. Dietz, H. Diamond, R. D. Rogers, and R. A. Leonard, "Combined TRU-Sr Extraction/Recovery Process," Topic E-7, at the International Solvent Extraction Conference, ISEC'93, York, England, September 9–15, 1993. (Same as CONF-930945—4.)

78. E. P. Horwitz, M. L. Dietz, and R. Chiarizia, "The Application of Novel Extraction Chromatographic Materials to the Characterization of Radioactive Waste Solutions," *J. Radioanal. Nucl. Chem.*, 161(2), 575-583 (1992).

79. M. L. Dietz, E. P. Horwitz, R. Chiarizia, and H. Diamond, "Novel Extraction Chromatographic Materials for the Separation and Preconcentration of Radionuclides," Topic J-4, at the International Solvent Extraction Conference, ISEC'93, York, England, September 9–15, 1993. (Same as: CONF-930945—3.)

80. A. Clearfield, G. H. Nancollas, and R. H. Blessing, in *Ion Exchange and Solvent Extraction, A Series of Advances*, Vol. 5, Edited by J. A. Marinsky and Y. Marcus, Marcel Dekker, New York, 1973, pp. 1–120.

81. D. E. Shanks, E. G. Noble, A. M. Pierzchala, and D. J. Bauer, "Application of Inorganic Ion Exchangers to Metallurgy," Report: BMRI-8816, 1983.

82. W. W. Schulz, J. W. Koenst, and D. R. Tallant, "Application of Inorganic Sorbents in Actinide Separations Processes," in Actinide Separations, Edited by J. D. Navratil and W. W. Schulz, a symposium sponsored by ACS Division of Industrial and Engineering Chemistry at the ACS/CSJ Chemical Congress (177th ACS National Meeting), Honolulu, HI, April 3–5, 1979, pp. 17–32.

83. S. F. Marsh, Z. V. Svitra, and S. M. Bowen, "Distributions of 14 Elements on 60 Selected Absorbers from Two Simulant Solutions (Acid-Dissolved Sludge and Alkaline Supernate) for Hanford HLW Tank 102-SY," Report: LA-12654, October 1993.

84. S. F. Marsh, "Partitioning High-Level Waste from Alkaline Solution: A Literature Survey," Report: LA-12528, May 1993.

85. G. L. Silver and J. W. Koenst, "A Study of the Reaction of U and Pu with Bone Char," Report: MLM-2384, January 1977. (Includes references to earlier reports in this series, MLM-2244, September 1975 and MLM-2371,m September 1976.)

86. W. R. Herald, R. C. Roberts, and M. K. Williams, "Development of Ultrafiltration and Inorganic Absorbents for Reducing Volumes of Low-Level and Intermediate-Level Liquid Waste: July–September 1978," Report: MLM-2566, October 1978. (Also includes references to other MLM reports in this series, MLM-2420, June 1977; MLM-2464, 1977; MLM-2499, February 1978; MLM-2503, February 1978; MLM-2513, April 1978; and MLM-2538, July 1978.)

87. **D. E. Blane and W. R. Herald,** "Removal of Pu from High-Level Caustic Waste Solutions Using Bone Char–Pilot Study," Report: MLM-2534, October 1978.

88. **R. G. Dosch,** "The Use of Titanates in Decontamination of Defense Waste," Report: SAND-78-0710, June 1978.

89. **R. L. Schwoebel and C. J. Northrup,** "Proceedings of the Sandia Laboratories Workshop on the Use of Titanate Ion Exchangers for Defense Waste Management," Report: SAND-78-2019, September 1978.

90. **W. W. Schulz,** "Decontamination of Hanford Pu Reclamation Facility Salt Waste Solution," Report: RHO-SA-23, June 1978.

91. **C. M. Smith, J. D. Navratil, and P. McCarthy,** "Removal of Actinides from Radioactive Wastewaters by Chemically Modified Peat," *Solv. Extr. Ion Exch.,* 2(7and8), 1123–1149 (1984).

92. **W. Faubel and S. A. Ali,** "Separation of Cs from Acid ILW-Purex Solutions by Sorption on Inorganic Ion Exchangers," *Radiochim. Acta,* 40, 49–56 (1986).

93. **R. G. Dosch, N. E. Brown, H. P. Stephens, and R. G. Anthony,** "Cs Separation Using Crystalline Silicotitanate Ion Exchangers," Report: SAND-92-2737C, 1993.

94. **R. G. Dosch, N. E. Brown, H. P. Stephens, and R. G. Anthony,** "Treatment of Liquid Nuclear Wastes with Advanced Forms of Titanate Ion Exchangers," Report: SAND-93-0428C, 1993.

95. **A. Marrocchelli and L. Pietrelli,** "Cs Adsorption with Zeolites from Nuclear High Salt Content Alkaline Wastes," *Solv. Extr. Ion Exch.,* 7(1), 159–172 (1989).

96. *New Separation Chemistry Techniques for Radioactive Waste and other Specific Applications,* Edited by L. Cecille, M. Casarci, and L. Pietrelli, Elsevier Applied Science, New York, 1991.

97. **I. J. Singh and M. Ramaswamy,** "Removal of Cs from Low Level Waste Solutions by Copper Hexacyanoferrate Loaded Resins," Report: BARC/1991/E/022, December 1991.

Chapter 6

Electrochemical Processes

Mark F. Buehler and Jeffrey E. Surma

I. INTRODUCTION

Electrochemical techniques have been used in the field of nuclear waste separations since the mid-1950s, but the applications were limited until research and development over the past few decades produced new electrode materials (i.e., Ebonex® and DSA®); ion-exchange membranes (i.e., Nafion®); and advancements in electrochemical engineering (i.e., dimensional analysis).[1,2] Also, manufacturers have recently made economic commerical electrochemical equipment/reactors readily available, which has provided a platform for scientists and practitioners to explore novel separation schemes.[3] These new developments have created opportunities to expand the use of electrochemical processes for nuclear waste separations.[4]

Electrochemical separations have many advantages for nuclear waste applications. The primary advantage of electroseparations is that little waste is generated upon separation. This is especially important for nuclear waste treatment where handling and disposal of secondary waste can control the economics of remediation. Typical operating conditions of electrochemical processes are also attractive because they operate at low temperatures (<100°C) and accommodate continuous flows, large pH variations (pH = 0 to 14), and suspended solids.

Electrochemical separation processes are supported by a large literature base of theoretical and experimental work. Theoretical modeling of electrochemical cells allows for process optimization, and existing experimental data reduce the need for excessive process screening. In addition, the thermodynamic stability of each chemical species is unique and controlled by the applied potential, thus providing high selectivity for separation applications.

This chapter provides an overview of electrochemical processes for separations, demonstrated applications, and emerging technologies for near-term applications. The techniques described here include electropolishing, electrodeposition, electrochemical ion exchange, transuranic dissolution, organic destruction, and nitrate destruction. Technologies showing promise for use in the near future include electrodialysis and electrically enhanced solid/liquid separations.

II. ELECTRON-TRANSFER MECHANISMS

Electrochemical separation technologies are based on the transfer of electrons driven by an applied potential.[5-8] The valence state of a material is altered, which results in a change of the solubility, phase, or chemical structure. At the positive electrode interface (anode), electrons are released and oxidation reactions occur; conversely, at the negative electrode interface (cathode), electrons are accepted and reduction reactions occur. For most of the technologies discussed here, electron-transfer reactions can be categorized into two distinct mechanisms: direct and indirect.[2] Operating parameters are chosen based on the mechanism dominating the separation process. When selecting an electrochemical separation process, it is important to note that electroactive species can also undergo chemical reactions. The combination of electron-transfer and chemical reactions creates a complex reaction scheme and, therefore, a complete understanding of reaction mechanisms is not always clear.

Direct electron-transfer reactions occur at the electrode/electrolyte interface. Processes discussed here that operate by direct oxidation include electropolishing and organic destruction; processes that operate by direct reduction include nitrate destruction and electrodeposition. There are many important intermediate steps that can occur during direct electron transfer, such as adsorption, diffusion, and multielectron transfers; however, these steps will be consolidated and interpreted as one step to describe the separation process. Because the electrons are transferred at an interface, the active electrode area is the key engineering parameter for economic considerations. Also, vigorous electrolyte mixing is typically preferred to minimize mass transfer limitations.

Indirect electron transfer occurs by way of a catalytic or intermediate electroactive species. This mechanism is widely used in industry for electrochemical organic synthesis and has been applied to nuclear waste treatment by many researchers.[7] Here, electrolysis is facilitated by an electroactive species either oxidizing or reducing another species in the bulk electrolyte phase. Processes using indirect oxidation include methods for transuranic dissolution and organic destruction. The primary advantage of this approach is that electron-transfer reactions occur in the bulk electrolyte phase and not at the electrode interface. This allows high conversions to be obtained relative to diffusion-controlled reactions occurring directly on the electrode surface. Processes using indirect reduction are not widely used in nuclear separations and are not described here.

III. EQUIPMENT

Electrochemical reactor designs are similar to those of chemical reactors and include batch, semibatch, continuously stirred tank, and plug-flow reactors. Extensive reviews of the design parameters and scale-up considerations have been described by several authors.[3,7,9-13] All electrochemical reactors or cells consist of two electron-conducting electrodes (anode and cathode) connected by a conducting electrolyte medium (usually a liquid). In certain applications, a charge-selective or inert-porous membrane is inserted between the electrodes to restrict or selectively control the transport of cations or anions.

Although several types of electrochemical cells exist,[7] a common cell found in nuclear separations is the "plate-and-frame" configuration. Here, the anode and cathode are opposing parallel plates with the waste stream passing between them acting as the electrolyte. The cell geometry is shown in Figure 1. The user has the option of inserting a separator for ion selectivity or turbulence promoter to increase mass transfer between the electrode materials. The cell is operated so that the electrolyte is continuously flushed by the surfaces of the electrodes. The cell has two outer frame ends that provide support so seals can be made by compression using threaded bolts. Other electrochemical cell and reactor configurations include fluidized beds, profiled compartments, packed-bed electrodes, rotating electrodes, moving-bed electrodes, and spiral wound electrodes.[7]

Desired properties of electrode materials include high physical, chemical, and electrochemical stability; high electrical conductivity; nonfouling, high side-reaction overpotentials; and low overpotentials for the reactions of interest. Typical electrode materials for the anode include platinized titanium, DSA®, Ebonex®, and lead oxide; for the cathode, materials include stainless steel, graphite, Hastelloys®, nickel, and lead. The

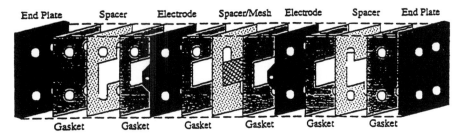

Figure 1 Plate-and-frame configuration for an electrochemical flow cell.

choice of anode and/or cathode material for a specific electrochemical application is based upon optimizing the reaction selectivity and minimizing costly side reactions.[14-16]

IV. DEMONSTRATED APPLICATIONS

Electrochemical separation techniques have been demonstrated on a variety of nuclear waste streams. This section illustrates representative applications on a bench or pilot scale. The oxidation and reduction processes described below provide separation by dissolution, decomposition, phase change, or ion exchange.

A. OXIDATION PROCESSES
1. Transuranic Dissolution
Indirect oxidation provides an attractive alternative to the standard HNO_3-HF technology that is commonly used to dissolve fired or calcined plutonium oxides. The advantages of indirect oxidation include minimization of secondary waste, use of an oxidant that is readily regenerated, and elimination of fluorine ion.

For indirect oxidation, the oxidation step is facilitated by one of three catalysts or mediators: silver, cobalt, or cerium. The oxidation potentials (in 4 M HNO_3) of these mediators that have been successfully applied in the CEPOD process are

$$Ce^{+3} \rightarrow Ce^{+4} + e^- \qquad E^0 = 1.61 \text{ V} \qquad (1)$$

$$Co^{+2} \rightarrow Co^{+3} + e^- \qquad E^0 = 1.82 \text{ V} \qquad (2)$$

$$Ag^+ \rightarrow Ag^{+2} + e^- \qquad E^0 = 1.98 \text{ V} \qquad (3)$$

Silver ion is the most kinetically favored for the dissolution of PuO_2. The oxidizing ions are continuously regenerated at an anode surface; therefore, no net consumption of the oxidizing ion occurs.

One example of this approach is the Catalyzed Electrochemical Plutonium Oxide Dissolution (CEPOD) process developed at Pacific Northwest Laboratory (PNL) for fuel reprocessing.[17-21] Here, insoluble plutonium oxide is oxidized from the (IV) state to the (VI) state, allowing for dissolution of PuO_2^{+2} ion in a nitric acid solution.

Figure 2 shows a schematic representation of an electrochemical cell separated by a porous membrane.[17] The electrocatalyst or mediating ion is pumped by the anode for continuous regeneration. The electrolyte used for plutonium dissolution is nitric acid, ranging in concentration from 4 to 6 M.

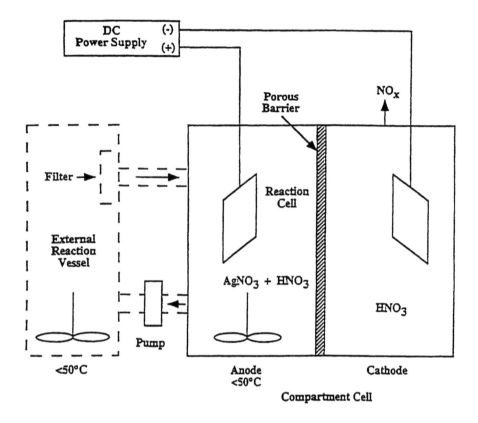

Figure 2 Schematic diagram for the experimental-scale Catalytic Electrochemical Plutonium Oxide Dissolution (CEPOD) process.

In addition to the dissolution of PuO_2, indirect oxidation has been demonstrated on many common solid materials encountered in plutonium processing and transuranic waste treatment. These examples range from incinerator ash to contaminated clothing.[20] This application is useful for concentrating the transuranic waste and ultimately reducing the volume of high-level waste by rendering the residue a low-level waste.

2. Organic Destruction

Both indirect and direct electron-transfer mechanisms have been used on a commercial scale for years to synthesize organic materials. Some examples of indirect oxidation include the synthesis of methoxylations, oxidation of glucose to gluconic acid, and oxidation of polynuclear aromatic hydrocarbons; while examples of direct oxidation for organic synthesis include conversion of sugars to acids, partial oxidation of benzene and methyl aromatics, and epoxidation of olefins.[7] These well-established methods can also be used to destroy or eliminate organics in wasteto streams containing radionuclides.

In particular, the U.S. Department of Energy (DOE) sites at Savannah River and Hanford have numerous storage tanks containing millions of gallons of mixed waste, some of which has accumulated over the past 50 years. The organic materials in the tank waste are primarily complexants that were added to precipitate specific radioactive metal ions such as cesium and the transuranics. Organics often must be removed from a waste stream before treatment because of various safety concerns. Specifically, radiolysis decomposition reactions of soluble organic materials can produce gaseous hydrogen. Although these reactions are kinetically slow, hydrogen gas can accumulate over time and reach explosive limits. Another concern is during vitrification of mixed waste for permanent immobilization, the carbon resulting from decomposition of organics can affect the structural integrity of the glass matrix. Apart from safety problems, other downstream treatment processes, such as ion exchange, are enhanced if organic complexants are removed prior to treatment. Indirect oxidation of organics in simulant tank waste has been demonstrated at several DOE laboratories, including PNL, Lawrence Livermore National Laboratory (LLNL), and Savannah River Laboratory (SRL).

An organic material can be electrochemically oxidized to nitrogen, carbon dioxide, and acid, depending on the organic functional groups present. The oxidation process occurs via an electrochemical–chemical mechanism. The catalytic cation, for example, Ag^{+2}, will oxidize the organic material and will react with water to produce oxidizing radicals such as $OH\cdot$ or $O\cdot$.[22] Oxidation reactions for benzene and ethylene glycol are shown below:

$$10AgNO_3^+ + (CH_2OH)_2 + 2H_2O \rightarrow 10Ag^+ + 2CO_2 + 10HNO_3 \qquad (4)$$

$$30AgNO_3^+ + C_6H_6 + 12H_2O \rightarrow 6CO_2 + 30Ag^+ + 30HNO_3 \qquad (5)$$

Studies on waste pretreatment at PNL demonstrated the destruction of toluene, phenol, chloroform, trichloroethylene, perchloroethylene, methyl isobutyl ketone, quinoline, and others.[23] A flow-through cell was used with an anionic membrane. The oxidizing intermediate was $1\ M$ $Ni(NO_3)_2$ or $Co(NO_3)_2$ in a HNO_3 supporting electrolyte at 22 to 50°C. Destruction rates of 2.7 to 3 g/h were obtained at an operating current of 6 A. Also, catalytic destruction of Trimsol® cutting oil was examined as an alternative to incineration at the Rocky Flats Plant (Golden, CO).[24] Destruction efficiencies of greater than 99% were obtained for the silver intermediate at concentrations of 0.25 and 0.5 M with a flow-through electrode configuration. The operating temperature ranged between 60 and 80°C, and the current efficiencies varied from 23 to 55%.

Farmer and co-workers at LLNL used silver, cobalt, and iron intermediates to oxidize a variety of organic materials including ethylene glycol, isopropanol, benzene, Trimsol® cutting oil, 2-monochloro-1-propanol, and 1,3-dichloro-2-propanol.[25-28] A rotating cylinder electrode with a cationic separator in a batch cell configuration was used to generate the oxidizing intermediate. Greater than 90% conversion of organics to carbon dioxide was obtained for all the organics studied. The current efficiencies varied from 20 to 84% and increased with decreasing current density. Concentrations of the catalytic intermediates were 0.5 M with a HNO_3 or H_2SO_4 supporting electrolyte. The operating temperature was varied from 27 to 50°C.

Almon et al. at SRL examined the decomposition of the following organic materials: benzene, tributylphosphate (TBP), methanol, propanol, tetraphenylborate, and cyclohexane.[22] The total organic destruction of TBP was 94% using a 0.5 M $AgNO_3$/3 M HNO_3 solution operating at 0.7 A and 27°C.

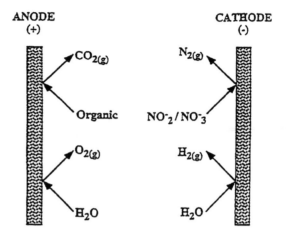

Figure 3 Schematic depicting typical direct electron-transfer mechanism for anodic and cathodic reactions. Anodic reactions include oxidation of organics and water, while the cathodic reactions include the reduction of nitrate and water.

In a similar study, Steel et al. from the Atomic Energy Authority Technology (United Kingdom) used the silver and cerium intermediates to decompose several organic materials.[29] Their pilot-plant facility treated TBP, kerosene, benzene, and many other organics. A Nafion® 324 fluoropolymer cationic membrane was used in a separated cell with $1 \, M$ AgNO$_3$/$4 \, M$ HNO$_3$. Destruction efficiencies varied between 80 and 90% for kerosene and TBP at current density of 1 A/cm^2.

Figure 3 shows a schematic for the electrochemical process of direct oxidation at the anode interface. As with indirect oxidation, the products generated are typically carbon dioxide and acids, depending on the organic material. The advantage of direct oxidation is the intermediate or catalytic oxidizing ion is not used, which eliminates a separation and regeneration step. Also, the technique is operationally simple and commercial flow-through cells are used. However, a disadvantage for large volumes of waste is the large electrode surface area or long residence times required for high conversion of organics. High surface area electrodes such as porous or three-dimensional electrodes have been developed to address this issue, but, because of pore plugging, these materials do not perform well with suspended solids.

SRL examined the direct oxidation of organic materials such as TBP, benzene, and cyclohexane.[22] A batch cell was used with an applied voltage of 3 V and a supporting electrolyte of $3 \, M$ HNO$_3$. The temperature was varied from 10 to 80°C, and the TBP destruction rate reached a maximum at 42°C. The corresponding destruction rates ranged from 0.4 to 0.77 ml/Ah.

Recently, researchers at PNL investigated the oxidation of tank waste simulants from the Hanford Site.[30,31] A flow-through cell operating with a recycle stream was used. The simulant processed contained EDTA as the organic surrogate along with 20 other chemical species and approximately 10% suspended solids. Various anode materials were evaluated, with platinized titanium providing the highest destruction rate of 0.022 g/Ah for carbon at an overall current efficiency of 22%. This application is important to establish direct oxidation as a viable method for eliminating organics from a waste slurry with complex chemistry and substantial solids content.

3. Metal Surface Decontamination by Electropolishing

Electropolishing is commonly practiced in industry and is used in the nuclear field to decontaminate metal surfaces.[32] In this process, a metal surface is corroded or oxidized until the contaminated material is dissolved into solution, leaving a clean surface. In certain applications, the dissolution process is accompanied by a mechanical abrasion to enhance erosion and increase the decontamination rate. Although the method is straightforward, understanding the complete reaction mechanisms is generally difficult because of the heterogenous nature and the complex geometry of contaminated surfaces.

As the contaminated metal is polarized anodically, oxidation and dissolution of metal yields a smooth, polished surface. The thickness of material removed usually ranges between 1 and 50 μm, depending on the original condition of the contaminated surface. Any radioactive material trapped on the surface or within the surface imperfections is removed and released into the electrolyte solution. The smooth surface of the decontaminated object minimizes electrolyte loss as the item is removed from the electrolyte solution. Demonstrated applications include the removal of beta and gamma emitters such as fission products, plutonium, and uranium.[32-37]

Two procedures can be used to decontaminate surfaces by electropolishing: immersion or *in situ* techniques.[32] In immersion electropolishing, a contaminated object is immersed into an electrolyte bath, and a potential is applied, driving it anodic. The primary advantage of this method is that equipment can be treated without prior disassembly and with rapid dissolution or decontamination rates. A disadvantage of immersion electropolishing is that only conducting surfaces can be treated, and, therefore, nonconducting material must be removed before treatment.

Allen et al. describe the use of electropolishing for decontamination of equipment used in nuclear reactors, such as, large reactor valves, hand tools, steel ducting, animal cages, vacuum system parts, core drill bits, and glove box waste.[32] Phosphoric acid was used as the electrolyte because of its stability and ability to solubilize a wide range of alloys. A typical reduction in dose rate of contaminated objects was from 1 million disintegrations per min · 100 cm^2 to background in less than 5 min.

In situ electropolishing is used to treat remote areas on an object or when objects are too large for immersion treatment. Examples of suitable uses of *in situ* electropolishing include interior surfaces of pipes or tubes, shipping casks, refueling cavity surfaces, or pool walls. Commercial devices and equipment include internal cathode, brush, contact, and pumped stream.[36]

B. REDUCTION PROCESSES

1. Nitrate Reduction

At several DOE sites, spent or waste nitric acid has been stored in numerous underground tanks. These tanks were originally designed to store the waste temporarily, but storage in these units became long term. Therefore, the waste acid had to be neutralized to prevent rapid corrosion of the carbon steel storage tanks. As a result, the tank waste is a radioactive, highly alkaline solution with high concentrations of nitrate and nitrite. Large volumes of this type of waste are located at SRL, Hanford, and Rocky Flats.

Electrochemical reduction and separation have been investigated at SRL as a means to reduce the total volume and radioactive level of the mixed waste, recover chemicals for recycling, and reduce the corrosive nature of the waste. For the other sites, an additional motivation is to minimize the generation of NO_x off-gas during vitrification of the waste. The electrochemical reduction reactions are shown in the following:

$$NO_3^- + H_2O + 2e^- \rightarrow NO_2^- + 2OH^- \tag{6}$$

$$2NO_2^- + 3H_2O + 4e^- \rightarrow N_2O + 6OH^- \tag{7}$$

$$2NO_2^- + 4H_2O + 6e^- \rightarrow N_2 + 8OH^- \tag{8}$$

$$NO_2^- + 5H_2O + 6e^- \rightarrow NH_3 + 7OH^- \tag{9}$$

Both separated and unseparated flow-through cells were used to evaluate SRL simulant waste.[38,39] For example, a lead cathode and a platinum anode were used in a separated flow-through cell operating at 70°C.[40] The current efficiency was high, ranging from 55 to 90%. The measured off-gas composition from the catholyte compartment primarily consisted of hydrogen (from water reduction) and ammonia with less than 1% nitrous oxide after 700,000 C had passed. Also, the efficiency of nitrate transport through the anionic membrane was low because of the co-migration of hydroxyl ions. However, precipitates formed within the anionic membrane, which could eventually lead to membrane failure.

2. Recovery and Purification of Plutonium and Uranium by Electrodeposition

Electrodeposition typically involves the complete reduction of an ionic species accompanied by a phase change. Each chemical species has a unique thermodynamic potential for complete reduction described by the Gibbs free energy. Therefore, it is feasible to selectively separate or deposit different ion species by controlling the applied potential to the cathode electrode. This approach has been used for both recovery and purification of plutonium and uranium and for selective removal of ^{99}Tc and ^{106}Ru.[39,41]

The use of electrodeposition for the recovery/purification of plutonium and uranium was introduced in 1930 and has been used extensively since then to obtain metal purities greater than 99.99%.[42-56] The approach typically uses a cadmium anode with an overlying halide molten salt blanket and an iron cathode.[41] The operating temperature of this system is limited to 500°C because of the high volatility of cadmium. Uranium and plutonium are separated selectively with the uranium deposits growing significant dendrites.

In the treatment of the supernate from the underground storage tanks at the Savannah River Site, nitrate destruction (discussed above) was accompanied by the removal of ^{99}Tc and ^{106}Ru.[39] A continuous-flow cell was used to selectively remove these radioactive metals with a lead cathode. The ruthenium deposit was suspected to be present in various oxidation states and the removal varied between 31 and 71%. The variation in deposits was attributed to the surface characteristics of the cathode. Electrodes with predeposited ruthenium removed larger amounts of ruthenium in the waste than those with no or little ruthenium on the surface.

C. ELECTROCHEMICAL ION EXCHANGE

Electrochemical ion exchange (EIX) was originally developed by researchers in the United Kingdom and focused on radioactive waste applications in the nuclear industry.[57-61] The technique combines an ion-exchange material and water electrolysis to selectively remove radioactive ions from waste streams. The absorption and elution characteristics of the ion exchanger are controlled by an applied potential. Specifically, the pH within the ion-exchange material is varied by water oxidation or reduction, allowing both acidic and basic conditions to be obtained. The primary advantage of this method is the ability to reversibly cycle the absorption and elution steps. This cycling reduces energy requirements and minimizes the elution volume and, thus, secondary waste generation.

The EIX process is begun by embedding an electrode into an ion-exchange material and applying a potential to drive either the oxidation or reduction of water. Typically, a weak acid cation exchanger or a weak basic anion exchanger is used. For a cationic exchanger, usually the site active group is a carboxylic acid, which is deprotanated by hydroxyl ions generated through water reduction. The reactions are listed below:

$$\text{Cathode:} \quad 2H_2O + 2e^- \rightarrow H_2 + 2OH^- \tag{10}$$

$$\text{Ion exchanger:} \quad R\text{--}COOH + OH^- \rightarrow R\text{--}COO^- + H_2O \tag{11}$$

After the sites have been activated, the exchanger is exposed to the waste stream and unwanted cations are absorbed by the exchanger. The negative polarity of the electrode aids in the migration of cations into the ion-exchange material.

The ion exchanger is eluted by changing the electrode polarity and driving it anodic. Hydrogen ions are generated inside the ion-exchange material by water oxidation. These ions exchange with the stripped cations and occupy the cationic sites in the exchanger. The reactions at the anode and exchanger are listed below:

$$\text{Anode:} \quad 2H_2O \rightarrow O_2 + 4H^+ + 4e^- \tag{12}$$

$$\text{Ion exchanger:} \quad R\text{--}COOM + H^+ \rightarrow R\text{--}COOH + M^+ \tag{13}$$

This technology has been demonstrated on a variety of nuclear waste streams at low concentration.[62] For example, pressurized-water nuclear reactor waste stream was treated by EIX with encouraging results. An anion absorber (IRN-78L) was used to recover borate, yielding decontamination factors (DF) of greater than 375. Also, fuel storage pond wastewater was treated by EIX to remove ^{137}Cs and ^{60}Co using a zirconium phosphate exchanger. The application provided cesium DFs of 100 to 5000 and cobalt/cesium DFs of 100 to 200. In addition, plutonium and americium were extracted from a fuel fabrication waste stream in a low-salt environment with DFs of 11 and 70, respectively.

Although EIX has not been widely incorporated into the nuclear industry, its application to a variety of waste streams makes it an attractive alternative. EIX offers high throughputs and high capacities that reduce capital and operational costs. The process is ideal for dilute contaminants because of its high selectivity and low energy requirements. Furthermore, the ability to extend the exchanger lifetime by recycling the absorption and elution steps reduces secondary waste costs by minimizing the volume of exchange material needed for immobilization and disposal.

V. EMERGING ELECTROCHEMICAL TECHNOLOGIES

Many other electrochemical methods have evolved as feasible for separation processes in the nuclear industries. A few of the techniques under development with near-term application are discussed below.

A. ELECTRODIALYSIS

Electrodialysis has been widely used in both industry and research to separate ionic species. Ion-selective membranes are inserted between the anode and cathode to selectively allow either positive or negative ions to pass.[63] Cationic membranes allow transport of positive ions, and anionic membranes allow transport of negative ions.

A simplified schematic diagram of a three-compartment electrodialysis cell is shown in Figure 4. The figure depicts cationic and anionic membranes sandwiched between two electrodes, creating three internal compartments. The solution flowing in the compartment on the anode side is the anolyte; the solution on the cathode side is the catholyte. Anions are transported toward the anode and cations toward the cathode. Production-scale units typically operate with 100 to 500 cationic/anionic membrane pairs per stack. The exposed electrode area ranges from 1 to 70 m^2.

Figure 4 also shows the generation of hydrogen ions (H^+) at the anode and hydroxyl ions (OH^-) at the cathode. This occurs if the applied potential is sufficiently large enough to reduce water at the cathode and oxidize water at the anode (Reactions 10 and 12). These reactions are very important because the free H^+ and OH^- ions can combine with ions transported from the center compartment to generate useful acid or base streams. Therefore, the overall electrodialytic process can desalinate a highly concentrated salt feed into an alkaline stream (catholyte), an acidic stream (anolyte), and a low-concentration salt effluent.

Applications of electrodialysis to nuclear waste treatment have been somewhat limited because of the radiological stability of the polymer-based, charge-selective membranes. Implementation of ceramic-based membranes is currently being evaluated as a replacement. Despite this drawback, separation of ^{106}Ru, ^{21}Na, ^{12}K, ^{90}Sr, and ^{137}Cs from waste streams has been demonstrated using this technology.[64-70]

B. ELECTROCHEMICALLY ENHANCED SOLID/LIQUID SEPARATIONS

The separation of solids and liquids can be improved or enhanced by taking advantage of electrochemical reactions. Two promising techniques currently under development for nuclear waste applications are electrokinetics and electroflotation.

1. Electrokinetics

Electrokinetics encompasses three distinct transport mechanisms occurring when a charged species is subjected to an electric field. These phenomena are electrophoresis (the movement of charged particles), electroosmosis (the movement of water), and electromigration (the movement of ions).[71-75] Figure 5 shows these three mechanisms for a porous soil. For nuclear separations, electrokinetics has been applied to soils or concrete where liquid is transported through a porous matrix and to slurries where charged particles are isolated at either the cathode or anode. The primary advantage of the technique is that charged contaminants can be transported or concentrated *in situ*, without soil excavation or concrete destruction. Also, secondary waste is minimized, along with the cost associated with decommissioning and decontaminating excavation equipment.

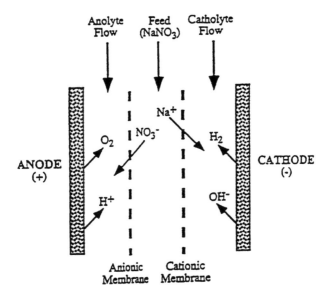

Figure 4 Simplified schematic representation of the electrodialytic separation of sodium nitrate.

Electrokinetic remediation of soil and concrete relies on the fact that many solids undergo surface ionization in the presence of water. The charge-bearing particles and dissolved ions are then attracted to either the positive or negative electrode as an electric field is applied. Positively charged species and water move toward the cathode (negative electrode), and negatively charged species move toward the anode (positive electrode). Typically, the contaminated soil and concrete are well packed, which prohibits the movement of solid particles and suppresses electrophoresis. Therefore, electroosmosis and electromigration are the dominant transport mechanisms. In addition to these electro-kinetic phenomena, electrolysis reactions occur at the cathode and anode. These reactions include the reduction and oxidation of water Reactions 10 and 12) and other ionic species near the electrode surface. Also, it should be noted that the operating conditions for successful contaminant transport are dependent on several soil and concrete parameters such as zeta potential, pore water composition, and pore size distribution. Therefore, each contaminated site must be evaluated independently to determine if the approach is applicable.

Electrokinetic soil or concrete remediation research has been conducted at several national laboratories, including Sandia National Laboratory (SNL), SRL, PNL, and Oak Ridge National Laboratory (ORNL). SNL demonstrated that organics and ions can be transported through unsaturated soils.[76,77] SRL focused on concentrating mercury from seepage basins using cylindrical electrodes and a sorbent material called Electrosorb®.[78] The transport and removal of radioactive materials in the soils has been examined at PNL by monitoring the movement of ^{137}Cs and ^{60}Co ions in saturated Hanford soil.[79] The ORNL study involved decontaminating several radioactive concretes by removing ura-nium and ^{99}Tc.[80] Also, Acar et al. determined the feasibility of uranium, thorium, and radium ion transport in kaolinite.[81-83]

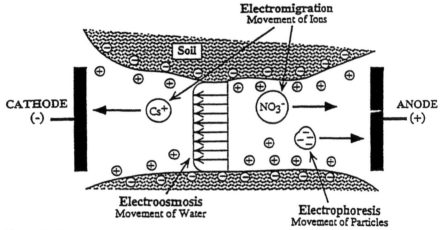

Figure 5 Electrokinetic phenomena of electroosmosis, electrophoresis, and electromigration.

Electrophoresis was used to purify sewage from a Russian radiochemical production facility.[84] This laboratory study examined removal of ^{90}Sr and ^{137}Cs adsorbed to ferric hydroxide colloidal agglomerates that were subsequently separated electrophoretically. Although this approach has not been widely used, electrophoresis is applicable to many nuclear situations such as enhancing settling rates or selectively separating colloidal particulates based on a difference in zeta potential.

2. Electroflotation

Flotation is a method to separate solids from a suspension by attaching bubbles to the solid particulates and allowing the less dense gas to elevate the solids to the top of the liquid solvent. The solids are then collected by skimming the solvent surface. Additives can be used to develop a stable foam at the air/suspension interface to allow easy collection of the solid particles. Flotation has been used extensively in the mining industries over the last 60 years with a high degree of success.[85,86]

In electroflotation, oxygen and hydrogen bubbles are produced by the electrolysis of water. Typically, these bubbles are very small, less than 30 μm in diameter. The advantage of this approach is that bubble production is easily initiated and controlled by the applied voltage. As the solid/gas bubble rises, coagulation can occur by collision and/or electrophoretically driven velocities. Electroflotation has been applied to nuclear waste separation in Russia on a laboratory scale.[84,87,88] The Russian researchers reported successful separation of particulates containing ^{90}Sr and ^{137}Cs using electroflotation. In these studies, detergents and ferrocyanides were added to aid in the solid/liquid separation.

VI. CONCLUSIONS

Electrochemical separation methods show promise as an alternative technology for nuclear waste treatment. Recently, studies have shown that these methods effectively remove or alter the waste species while minimizing the creation of secondary waste. A further benefit is that electrochemical techniques are inherently "clean" because only electrons are transported for remediation. Not all electrochemical separation methods have been developed to maturity, although pilot-scale operations have been conducted for some processes. The majority of existing data has been obtained from laboratory- or bench-scale experiments. However, the data measured at these levels have provided sufficient information to demonstrate the viability of electrochemical separations for radioactive materials.

ACKNOWLEDGMENTS

Pacific Northwest Laboratory is operated for the U.S. Department of Energy by Battelle Memorial Institute under Contract DE-AC06-76RLO 1830 and has the right to use and reproduce all or parts of this work for U.S. government purposes.

REFERENCES

1. **Benedict, M. and Pigford, T. H.,** *Nuclear Chemical Engineering*, McGraw-Hill, New York, 1957.
2. **Pletcher, D. and Weinberg, N. L.,** "The Green Potential of Electrochemistry Part 1: The Fundamentals," *Chem. Eng.*, August, 98, 1992.
3. **Pletcher, D. and Weinberg, N. L.,** "The Green Potential of Electrochemistry Part 2: The Applications," *Chem. Eng.*, November, 132, 1992.
4. **Weinberg, N. L.,** "Environmental Legislation in the U.S.A.: Opportunities for Electrochemical Technology," in *Electrochemistry for a Cleaner Environment*, Eds. J. D. Genders and N. L. Weinberg, The Electrosynthesis Company, Inc., New York, 1, 1992.
5. **Newman, J.,** *Electrochemical Systems*, Prentice Hall, New Jersey, 1991.
6. **Bard, A. J. and Faulkner, L. R.,** *Electrochemical Methods: Fundamentals and Applications*, John Wiley & Sons, New York, 1980.
7. **Pletcher, D. and Walsh, F. C.,** *Industrial Electrochemistry*, Chapman and Hall Ltd., New York, 1990.
8. **Pletcher, D.,** *A First Course in Electrode Processes*, The Electrochemical Consultancy, Hants, England, 1991.
9. **Kuhn, A. T., Ed.,** *Industrial Electrochemical Process*, Elsevier, New York, 1971.
10. **Pickett, D. J.,** *Electrochemical Reactor Design*, Elsevier, New York, 1979.
11. **Savinell, R. F.,** "Some Aspects of Electrochemical Reactor Design," *AIChE Symp. Ser. 229, 79*, 13, 1983.
12. **Scott, K.,** *Electrochemical Reaction Engineering*, Academic Press, New York, 1991.
13. **Walsh, F. C.,** *A First Course in Electrochemical Engineering*, Alresford Press Ltd., Hants, England, 1993.
14. **Tilak, B. V. and Weinberg, N. L.,** "Some Criteria for Selection of Electrodes for Electrochemical Processes," *AIChE Symp. Ser. 204, 77,* 60, 1981.
15. **Clarke, R. and Pardoe, R.,** "Applications of Ebonex® Conductive Ceramic Electrodes in Effluent Treatment," in *Electrochemistry for a Cleaner Environment*, Eds. J. D. Genders and N. L. Weinberg, The Electrosynthesis Company, Inc., New York, 331, 1992.
16. **Johnson, D. C., Chang, H., Feng, J., and Wang, W.,** "New Anodes for Electrochemical Incineration: Concepts of Atomic Engineering," in *Electrochemistry for a Cleaner Environment*, Eds. J. D. Genders and N. L. Weinberg, The Electrosynthesis Company, Inc., New York, 349, 1992.
17. **Ryan, J. L. and Bray, L. A.,** *ACS Symp. Ser.*, 117, 449, 1980.
18. **Bray, L. A. and Ryan, J. L.,** "Catalyzed Electrolytic Dissolution of Plutonium Dioxide," in *Actinide Recovery from Waste and Low-Grade Sources*, Eds. J. D. Navratil and W. W. Schulz, Harwood Academic Publishers, London, pp. 129-154, 1982.
19. **Ryan, J. L., Bray, L. A., and Boldt, A. L.,** U.S. Patent 4,686,019, 1987.

20. **Bray, L. A., Ryan, J. L., and Wheelwright, E. J.,** "Electrochemical Process for Dissolving Plutonium Dioxide and Leaching Plutonium from Scrap and Waste," *AIChE Symp. Ser. 254,* 83, 120, 1987.
21. **Wheelwright, E. J., Ryan, J. L., Bray, L. A., Bryan, G. H., Surma, J. E., and Matheson, J. D.,** "The Use of Catalyzed ELectrolytic Plutonium Oxide Dissolution (CEPOD) for Waste Treatment," in Proceedings from the First Hanford Separation Science Workshop, PNL-SA-21775, Richland, Washington, July 23–25, 105, 1991.
22. **Almon, A. C. and Buchanan, B. R.,** "Electrochemical Oxidation of Organic Wastes," in *Electrochemistry for a Cleaner Environment,* Eds. J. D. Genders and N. L. Weinberg, The Electrosynthesis Company, Inc., New York, 301, 1992.
23. **Molton, P. M., Fassbender, A. G., Nelsen, S. A., and Cleveland, J. K.,** Proceedings from 13th Annual Environ. Qual. R&D Symposium, Battelle, Pacific Northwest Laboratory, Richland, Washington, 1988.
24. **Surma, J. E., Geeting, J. G. H., Blanchard, D. A., Bryan, G. H., and Wheelwright, E. J.,** "Phase II-Catalyzed Electrochemical Oxidation (CEO) of Rocky Flats Contaminated Oils," PNL-8938, Pacific Northwest Laboratory, Richland, Washington, 1993.
25. **Farmer, J. C.,** "Electrochemical Treatment of Mixed and Hazardous Wastes," in *Environmental Oriented Electrochemistry,* Ed. C. A. C., Sequeira, Elsevier, Amsterdam, 1991.
26. **Farmer, J. C., Wang, F. T., Hawley-Fedder, R. A., Lewis, P. R., Summers, L. J., and Foiles, L.,** "Electrochemical Treatment of Mixed and Hazardous Wastes: Oxidation of Ethylene Glycol and Benzene by Silver (II)," *J. Electrochemical Society,* 139, 654, 1992.
27. **Farmer, J. C., Wang, F. T., Hawley-Fedder, R. A., Lewis, P. R., and Summers, L. J.,** "Destruction of Chlorinated Organics by Cobalt(III)-Mediated Electrochemical Oxidation," *J. Electrochem. Soc.,* 139, 3025, 1992.
28. **Farmer, J. C., Wang, F. T., Hawley-Fedder, R. A., Lewis, P. R., and Summers, L. J.,** "Electrochemical Treatment of Mixed and Hazardous Wastes: Oxidation of Ethylene Glycol by Cobalt(III) and Iron (III)," *Trans. Inst. Chem. Engrs.,* 70(B), 158, 1992.
29. **Steele, D. F., Richardson, D., Craig, D. R., Quinn, J. D., and Page, P.,** "Destruction of Industrial Organic Wastes Using Electrochemical Oxidation," in *Electrochemistry for a Cleaner Environment,* Eds. J. D. Genders and N. L. Weinberg, The Electrosynthesis Company, Inc., New York, 287, 1992.
30. **Pillay, G., Surma, J. E., Buehler, M. F., and Lawrence, W. E.,** "Electrochemical Treatment of Organic Tank Waste," Extended Abstracts, The Electrochemical Society, May 22–27, San Francisco, California, 94-1, 911, 1994.
31. **Lawrence, W. E., Buehler, M. F., Surma, J. E., Pillay, G., Gervins, K. L., and Schmidt, A. J.,** "Electrochemical Organic Destruction in Support of Hanford Tank Waste Pretreatment," PNL-10131, Pacific Northwest Laboratory, Richland, Washington, 1994.
32. **Allen, R. P., Arrowsmith, H. W., Charlot, L. A., and Hopper, J. L.,** "Electropolishing as a Decontamination Process: Progress and Applications," PNL-SA-6858, Pacific Northwest Laboratory, Richland, Washington, 1978.
33. **Allen, R. P. and Arrowsmith, H. W.,** *Mater. Perform.,* 18(11), 21, 1979.
34. **Allen, R. P. and Arrowsmith, H. W.,** *Trans. Am. Nucl. Soc.,* 38, 621, 1981.
35. **Spalaris, C. N.,** "Field Test and Evaluation of Electropolishing and Preoxidation Processes for Type 316 Stainless Steel Nuclear Grade Piping," EPRI NP-3832, Electric Power Research Institute, Palo Alto, California, 1985.
36. **Allen, R. P.,** "Electropolishing Applications in the Nuclear Industry," *AIChE Symp. Ser. 254,* 83, 156, 1987.
37. **Holton Jr., L. K., Elliott, M. L., Surma, J. E., Goles, R. W., Allen, R. P., Haun, F. E., Brouns, R. A., Klein, R. F., Bryan, G. H., and Peters, R. D.,** "Processing Summary Report: Fabrication of Cesium and Strontium Heat and Radiation Sources," PNL-6790, Pacific Northwest Laboratory, Richland, Washington, 1989.
38. **Hobbs, D. T. and Ebra, M. A.,** "Electrochemical Processing of Alkaline Nitrate and Nitrite Solutions," *AIChE Symp. Ser. 254,* 83, 149, 1987.

39. **Hobbs, D. T.,** "Electrochemical Treatment of Nuclear Waste in the Savannah River Site," in *Electrochemistry for a Cleaner Environment*, Eds. J. D. Genders and N. L. Weinberg, The Electrosynthesis Company, Inc., New York, 259, 1992.

40. **Westinghouse Savannah River Company,** "Final Report — Electrochemical Processing of Nitrate Waste Solutions," WSRC-TR-93-090, The Electrosynthesis Company, Inc., 1992.

41. **Burris, L., Steunenberg, R. K., and Miller, W. E.,** "The Application of Electrorefining for Recovery and Purification of Fuel Discharged from the Integral Fast Reactor," *AIChE Symp. Ser.* 254, 83, 135, 1987.

42. **Driggs, F. H. and Lilliendahl, W. C.,** "Preparation of Metal Powders by Electrolysis of Fused Salts. I. Ductile Uranium," *Ind. Eng. Chem.*, 22, 516, 1930.

43. **Marzano, C. and Noland, R.A.,** "The Electrolytic Refining of Uranium," ANL-5102, Argonne National Laboratory, 1953.

44. **Blumenthal, B.,** "Melting of High Purity Uranium," *J. Metals*, 7, 499, 1955.

45. **Boisde, G., Chauvin, G., Coriou, H., and Hure, J.,** "Mechanisms of Electrorefining of Uranium in Fused Salt Baths," *Electrochim. Acta*, 5, 54, 1961.

46. **Chauvin, G., Coriou, H., and Hure, J.,** "Electrolytic Refining of Certain Nuclear Metals in Fused Salt Baths," *Metaux*, 37, 112, 1962.

47. **Kolodney, M.,** "Preparation of the First Electrolytic Plutonium and of Uranium from Fused Chlorides," *J. Electrochem. Soc.*, 129, 2438, 1982.

48. **Brodsky, M. B. and Carlson, B. G. F.,** "Electrodeposition of Plutonium and Uranium from Molten Salt Solutions of the Di-Chlorides," *J. Inorg. Nucl. Chem.*, 24, 1675, 1962.

49. **Mullins, L. J., Leary, J. A., and Maraman, A. N.,** "Plutonium Electrorefining," *Ind. Eng. Chem. Process Design Develop.*, 2, 20, 1963.

50. **Mullins, L. J., Leary, J. A., Morgan, A. N., and Maraman, W. J.,** "Plutonium Electrorefining," LA-2666, Los Alamos Scientific Laboratory, 1962.

51. **Mullins, L. J., Morgan, A. N., Apgar, S. A., and Christensen, D. C.,** "Six-Kilogram-Scale Electrorefining of Plutonium Metal," LA-9469-MS, Los Alamos National Laboratory, 1982.

52. **Coops, M. S., Knighton, J. B., and Mullins, L. J.,** "Pyrochemical Processing of Plutonium. Technology Review Report," UCRL-88116, Lawrence Livermore National Laboratory, 1982.

53. **Baldwin, C. E. and Navratil, J. D.,** "Review of Plutonium Process Chemistry at Rocky Flats," RFP-3379, Rockwell International Corp., 1982.

54. **Niedrach, L. W. and Glamm, A.,** "Electrorefining for Removing Fission Products from Uranium Fuels," *Ind. Eng. Chem.*, 48, 977, 1956.

55. **Niedrach, L. W. and Glamm, A.,** "Electrofining of Uranium — A New Approach," *J. Electrochem. Soc.*, 103, 521, 1956.

56. **Niedrach, L. W., Glamm, A., and Fountain, G. R.,** "The Behavior of Representative Fission Products and Plutonium in the KAPL Electrorefining Process and a Review of the Status of the Process," KAPL-1692, Knolls Atomic Power Laboratory, 1957.

57. **Turner, A. D., Bridger, N. J., Neville, M. D., and Jones, C. P.,** "An Experimental Investigation of Electrochemical Ion-Exchange for the Treatment of Tihange PWR Low Level Liquid Waste," AERE-G5134, Atomic Energy Authority, U.K., 1989.

58. **Neville, M. D.,** "Electrochemical Options for Effluent Treatment — the Harwell Experience," *Institute of Chemical Engineers Environmental Protection Bulletin* 005, March 1990.

59. **Turner, A. D., Bridger, N. J., Jones, C. P., Neville, M. D., and Junkison, A. R.,** "The EIX Process for Radioactive Waste Treatment," in *Proceedings of the CEC/ENEA Seminar on New Separation Chemistry Techniques for Radioactive Waste and Other Specific Applications*, Rome, May 16–18, 1990.

60. **Bridger, N. J., Jones, C. P., and Neville, M. D.,** "Electrochemical Ion-Exchange," *J. Chem. Technol. Biotechnol.*, 50, 469, 1991.

61. **Hirose, Y., Turner, A. D., Junkison, A. R., Bridger, N. J., and Neville, M. D.,** "Electrochemical Processing of Low Level Liquid Radioactive Waste," The Third International Conference on Nuclear Fuel Reprocessing and Waste Management (RECOD'91), Sendai, Japan, April 14–18, 1991.

62. **Jones, C. P., Neville, M. D., and Turner, A. D.,** "Electrochemical Ion Exchange," in *Electrochemistry for a Cleaner Environment,* Eds. J. D. Genders and N. L. Weinberg, The Electrosynthesis Company, Inc., New York, 207, 1992.

63. **Strathmann, H.,** *Membrane Handbook,* Eds. W. S. W. Ho and K. K. Sirdar, Van Nostrand Reinold, New York, 1991.

64. **Rauzen, F. V., Kuelshov, N. F., Trushkov, N. P., and Dudnik, S. N.,** *Atom. Energ.,* 45, 49, 1978.

65. **Sugimoto, S.,** *J. Nucl. Sci. Technol.,* 15, 753, 1978.

66. **Moskvin, L., Gurskii, V. S., and Grigorev, G. L.,** *Radiokhimiyia,* 22, 438, 1980.

67. **Gurskii, V. S. and Moskvin, L. N.,** *Zh. Prikl. Khim.,* 60, 2203, 1987.

68. **Abdul-Fattah, A. R. F.,** *Kerntechnik,* 50, 188, 1987.

69. **Buehler, M. F., Lawrence, W. E., and Norton, J. D.,** "Evaluation of Aqueous Na+/Cs+ Separation by Electrodialysis," PNL-9022, Pacific Northwest Laboratory, Richland, Washington, 1993.

70. **Norton, J. D. and Buehler, M. F.,** "Separation of Monovalent Cations by Electrodialysis," *Sep. Sci. Technol.,* 29, 1553, 1994.

71. **Shaw, D. J.,** *Introduction to Colloid and Surface Chemistry,* Butterworths, London, 1980.

72. **Hunter, R. L.,** *Zeta Potential in Colloidal Science,* Academic Press, New York, 1981.

73. **Acar, Y. B., Gale, R. J., and Marks, R.,** "Fundamentals of Removing Contaminants from Soils by Application of Electrical Currents," Presented at the Electrokinetics Workshop, Atlanta, Georgia, January 22–23, 1992. Office of Research and Development, U.S. Department of Energy, Washington, D.C.

74. **Kozak, M. W.,** "Fundamentals of Electrokinetics," Presented at the Electrokinetics Workshop, Atlanta, Georgia, January 22–23, 1992. Office of Research and Development, U.S. Department of Energy, Washington, D.C.

75. **Probstein, R. F. and Hicks, R. E.,** "Removal of Contaminants from Soils by Electric Fields," *Science,* 60, 498, 1993.

76. **Lindgren, E. R., Mattson, E. D., and Kozak, M. W.,** "Electrokinetic Remediation of Contaminated Soils," Presented at Environmental Remediation '91, Pasco, Washington, September 8–11, 1991.

77. **Lindgren, E. R., Mattson, E. D., and Kozak, M. W.,** "Electrokinetic Remediation of Contaminated Soils: An Update," Presented at Waste Management '92, Tucson, Arizona, March 1–5, 1992.

78. **Bibler, J. and Meaker, T. F.,** "A Demonstration of Electrokinetic Migration Technology at the SRS Old TNX Seepage Basin for Mercury Removal," EPRI Workshop — In Situ Electrochemical Soil and Water Remediation, Abstract #22 EPRI, Palo Alto, California, February 28 – Mar. 1, 1994.

79. **Virden, J. W., Surma, J. E., Buehler, M. F., Mattigod, S. V., Lessor, D. L., Peurrung, L. M., Pillay, G., and Gauglitz, P. A.,** "Electrokinetic Remediation of Hanford Soil," PNL-9389, Pacific Northwest Laboratories, Richland, Washington, 1993.

80. **Bostick, W. D., Bush, S. A., Marsh, G. C., Henson, H. M., Box, W. D., and Morgan, I. L.,** "Electroosmosis Decontamination of Concrete," K-TCD-1054, Oak Ridge K-25 Site, Oak Ridge, Tennessee, March, 1993.

81. **Acar, Y. B. and Alshawabkeh, A. N.,** "Principles of Electrokinetic Remediation," *Environ. Sci. Technol.,* 27, 2638, 1993.

82. **Acar, Y. B., Alshawabkeh, A. N., and Gale, R. J.,** "Fundamentals of Extracting Species from Soils by Electrokinetics," *Waste Manage.,* 13, 141, 1993.

83. **EPA Cooperative Agreement No. CR816828-01-0,** "Final Report Phase 1, Risk Reduction Engineering Laboratory," USEPA, Cincinnati, Report No. EK-BR-009-0292, Electrokinetics, Inc., July 10, 1992.

84. **Shvedov, V. P. and Yakushev, M. F.,** "The Use of Electrophoresis, Electrocoagulation, and Electroflotation for the Purification of Radioactive Waters," *Radiokhimiyia,* 12(6), 871, 1970.

85. **Svarovsky, L.,** *Solid–Liquid Separation,* Butterworths, London, 1981.

86. **Clarke, A. N. and Wilson, D. J.,** *Foam Flotation, Theory and Applications*, Marcel Dekker, New York, 1983.
87. **Shvedov, V. P. and Yakushev, M. F.,** "The Use of Nickel Ferrocyanide as a Collector in the Purification of Radioactive Solutions from Cesium-137 and Strontium-90 by the Method of Electroflotation," *Radiokhimiyia,* 12(6), 876, 1970.
88. **Shvedov, V. P. and Yakushev, M. F.,** "Effect of Detergents on the Deactivation of Waste Solutions by Electroflotation," *Radiokhimiyia*, 15(3), 428, 1973.

Chapter 7

Other Separation Processes — Alternative Separation Technologies for Treating Sodium-Containing Acidic Wastes

Richard D. Boardman

I. INTRODUCTION

Chemical solutions containing sodium compounds have historically been used in nuclear fuels-reprocessing plants to clean up extraction solvents, to decontaminate process equipment, to neutralize acidic wastes, and to convert the contaminants in the waste streams into insoluble metal-hydroxide precipitates. The latter practice has resulted in a legacy of tank farm sludges that now must be recovered and immobilized into environmentally acceptable waste forms.

At the Idaho Chemical Processing Plant (ICPP) the uranium extraction process resulted in acidic, high-level radioactive waste (HLW) that contained the metal cations from the dissolved fuel and fuel cladding in solution with the dissolver acids. In addition, decontamination and solvent recovery operations have resulted in the accumulation of low to intermediate radioactive, high sodium concentration, acidic waste (referred to simply as sodium-bearing waste or SBW). Both the liquid HLW and SBW are stored separately in corrosion-resistant stainless steel tanks.

The current practice at the ICPP is to convert the liquid HLW and SBW into an intermediate granular solid in a fluidized-bed calciner operating at 500°C. However, due to the high concentration of alkaline salts (primarily sodium nitrate) in the SBW, a chemical diluent (such as aluminum nitrate) must be added during calcination to prevent the calcine from agglomerating. The diluent increases the volume of the calcine produced, resulting in higher costs for immobilization and disposal of the calcine. Consequently, alternative treatment technologies are being considered to minimize the volume of immobilized waste that will be generated from the SBW.

The focus of this chapter is on emerging technologies that have shown potential for treating the acidic SBW. More widely documented technologies, such as the TRUEX extraction option, are not discussed in great detail because they are adequately presented in other chapters of this text. Although the alternative technologies discussed in this chapter are applicable to the acidic ICPP SBW, they may also be applicable to other sodium-containing wastes such as alkaline tank farm supernatant resulting from neutralizing the liquids with NaOH.

Much of the information and data given in this chapter were obtained from experimenters at the ICPP. The program to identify and develop alternatives for treating the SBW was initiated in 1992; thus, comprehensive details of the work presented herein are currently being prepared for external publication. Additional information on specific research results and technology applications can be obtained by contacting the investigators cited.

Table 1 SBW Composition

Species	Av. conc. (M)	Range (M)
Major Chemical		
Acid (H^+)	1.45	0.43–1.9
Nitrate (NO_3^-)	4.36	2.9–5.8
Aluminum (Al^{3+})	0.55	0.21–0.81
Sodium (Na^+)	1.26	0.78–2.00
Potassium (K^+)	0.15	0.10–0.23
Calcium (Ca^{2+})	0.04	0.00–0.07
Chloride (Cl^-)	0.02	0.008–0.043
Iron ($Fe^{2+,3+}$)	0.03	0.01–0.05
Chromium ($Cr^{2+,3+,6+}$)	0.006	0.002–0.013
Cadmium (Cd^{2+})	0.002	0.000–0.004
Lead ($Pb^{2+,4+}$)	0.001	0.001–0.002
Mercury ($Hg^{1+,2+}$)	0.002	0.001–0.003
Sulfate (SO_4^{2-})	0.04	0.01–0.07
Specific gravity	1.22	1.15–1.26
Radiochemical		
^{90}Sr (mCi/l)	81.0	27–220
^{134}Cs (mCi/l)	2.63	0.03–6.1
^{137}Cs (mCi/l)	94.9	24–270
^{154}Eu (mCi/l)	0.68	0.04–1.8
U (mg/l)	78.2	57–97
Pu (mg/l)	1.04	0.27–1.5
^{241}Am (mg/l)	0.10	0.005–0.018
^{237}Np (mg/l)	1.36	0.62–3.7

II. ICPP SBW COMPOSITION

The SBW is stored at the ICPP in six 1100-m³ stainless steel tanks. The current inventory is approximately 5700 m³. Additional SBW will be generated with ongoing facility and equipment decontamination as wet chemical rinsing and flushing is continued. The chemical and radiochemical composition of the current inventory of SBW is given in Table 1. The range in major and minor constituents reflects the diverse sources from which the wastes in each tank were derived.

From a regulatory viewpoint, the SBW is a mixed (radioactive and hazardous) waste. The concentrations of TRU in the waste are greater than Nuclear Regulatory Commission (NRC) Class C LLW limits. The waste also contains significant quantities of RCRA toxic metals (i.e., Hg, Pb, Cd, and Cr) and it carries several RCRA "listed" waste codes corresponding to past laboratory and decontamination solution disposal practices.

III. ALTERNATIVE TREATMENT TECHNOLOGIES

The alternatives for treating the SBW include (1) waste decontamination; (2) sodium removal; or (3) direct immobilization of the liquid waste, possibly by chemically combining the sodium with constituents in the calcine.

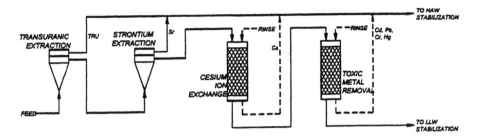

Figure 1 Separation of contaminants by Truex, SREX, and Cs ion exchange.

A. WASTE DECONTAMINATION OPTIONS

Waste decontamination follows the traditional approach to radioactive waste treatment; that is, separation, isolation, and/or recovery of the trace contaminants. Separation of the transuranium elements (TRU or tranuranics) will generally reduce the SBW to a NRC Class C LLW designation. Subsequently, either most of the Cs or most of the Sr must be removed in order to meet NRC Class B limits. In order to achieve NRC Class A limits, nearly all of the Cs and the Sr must be separated from the bulk waste.

1. TRU, Sr, and Cs Separation

Several technologies have been developed for separating actinides, fission products, and toxic metals from neutral and basic solutions; however, only a few technologies have been developed for treating highly acidic solutions. Following a critical review of alternative technologies, TRUEX,[1] SREX,[2] and ion exchange were identified as the leading technologies for extracting TRU, Sr, and Cs, respectively, from the SBW (see Figure 1). Extensive investigations of the TRUEX process on actual and synthetic SBW have demonstrated the success of this technology for treating the SBW (see earlier chapters in this text). Parametric studies with synthetic and actual SBW also show that SREX is a viable process for acidic wastes with high alkali and alkaline-earth metal concentrations, such as contained in the SBW.[3]

Several materials have been identified for extracting cesium from the acidic SBW. Inorganic ion-exchange materials are attractive because they tend to be resistant to attack by nitric acid, resistant to degradation by radiolysis, have good selectivity and capacity for cesium or strontium, and are generally compatible with glass formulations. Following a literary screening of potential sorbents for cesium separation (see, for instance, the extensive testing of Cross and Hooper[4]), approximately ten different ion-exchange materials were tested for their abilities to remove cesium and/or strontium from simulated SBW.[5,6]

Based on the results of these scoping studies, the agents listed in Table 2 were selected for extensive testing on both synthetic and actual waste at the ICPP. While high distribution coefficients, k_d, were measured for all of the candidate materials, AMP has emerged as a process favorite due to its high selectivity for cesium over a range of acid concentrations (0.1 to 10 M nitric acid). It is possible to achieve greater than 99.9% removal of the cesium from SBW with a multistage or fixed-bed AMP contact process.

2. Neutralization/Precipitation

In the context of waste decontamination, nearly all of the polyvalent cations, and a large amount of the divalent cations, can be rejected from the SBW by neutralizing the excess

Table 2 Test Materials and Suppliers

Test material	Description	Supplier
Phosphotungstic acid (PTA)	White powder with formula $H_3[P(W_3O_{10})_4] \cdot 24H_2O$	Sigma Chemical Co., St. Louis, MO
Copper(II) hexacyanoferrate(II) (CuHCF)	Dark-brown granular powder	Pfaltz & Bauer Inc., Waterbury, CT
Ammonium molybdophosphate (AMP–SiO$_2$)	Yellow silica gel extrudates/granules (AMP ≈ 20 wt%)	The PQ Corporation, R & D Center, Conshohocken, PA
Ammonium molybdophosphate (AMP–polytetrafluoroethylene)	Yellow polytetrafluoroethylene membrane (AMP = 90 wt%)	3M Company, St. Paul, MN
Ammonium molybdophosphate (AMP–Zeocarb)	Black zeolite and carbon matrix (AMP ≈ 58, 67, 77 wt%)	ZEOTECH Corp., Albuquerque, NM
Ammonium molybdophosphate (AMP–PAN)	Yellow polyacrylonitrile resinous beads	Sabiesta and John Czeck Technical University, Prague
Cystalline silica titanates	White powder, formulation disclosure patent pending	Sandia National Lab., Albuquerque, NM; Texas A&M University, College Station, TX

acid with NaOH and raising the pH to 8–10. This results in a sludge containing the TRU, toxic metals, and most of the strontium along with the inert divalent and polyvalent cationic precipitates (see Figure 2). Further separation of Sr and Cs from the supernate would be required to produce an NRC Class B or Class A LLW. In order to reduce the amount of precipitating agent that is used, electrohydrolysis is being considered to recover and recycle the alkali hydroxides. Nitric acid is also recovered by electrohydrolysis and can be used to redissolve the precipitate. Following filtration, the precipitate may require washing to remove residual sodium. Subsequently, the precipitate can be directly vitrified or redissolved and calcined in the ICPP calciner. The neutralization process for the ICPP SBW is similar in chemistry to the neutralization process already implemented at the Savannah River Plant[7] and interim in-tank storage of HLW at Hanford.[8]

Studies have been performed to determine the optimum procedure for the hydroxide precipitation process using a SBW simulant.[9] Some of the technical issues addressed include: selecting suitable neutralizing and flocculation agent(s); determining the best scheme to mix the waste and hydroxides; and determining the optimum method for filtering and washing the precipitates.

A series of screening tests were performed to evaluate whether a single- or a two-stage pH precipitation system is required to remove the nonalkali polyvalent metals. Target pH levels of 6.5, 8, 9.5, 10.5, 11, and 12.5 were tested. Table 3 lists the removal efficiencies for the nonalkali metals analyzed in each test.

The tests demonstrated that most of the aluminum was removed from the SBW simulant between a pH of 6.5 and 9.5. As the pH values increase, the aluminum hydroxide precipitate redissolves and reduces the removal efficiency. Thus, the optimum pH value for single-stage precipitation is between 8 and 9.5. The data also show that with only one pH adjustment, more than 97% (at a pH of 8) and 99% (at pH of 9.5) of the aluminum and other nonalkali metals were removed from the waste solution.

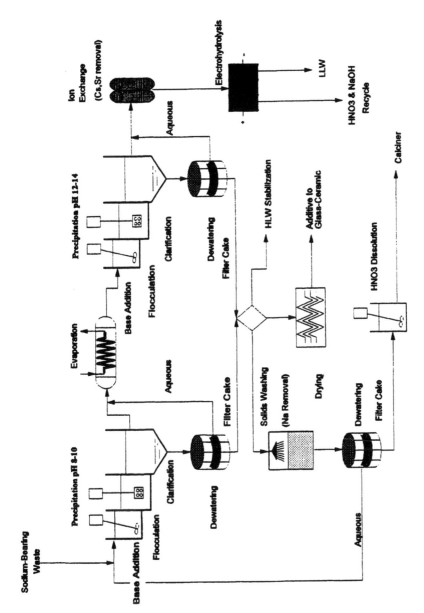

Figure 2   Separation of contaminants by precipitation.

Table 3 Overall Nonalkali Metals Removal Efficiency

1st Precipitation			2nd Precipitation		
Test number	pH	Removal efficiency (%)	Test number	pH	Removal efficiency (%)
1–1	6.5	94.8	1–2	10.5	95.8
2–1	8.0	97.0	2–2	11.0	97.3
3–1	9.5	99.2	3–2	12.5	99.3
4–1	10.5	96.1	4–2	6.5	98.9
5–1	11.0	91.8	5–2	8.0	99.2
6–1	12.5	32.5	6–2	9.5	98.9

Table 4 Composition of Precipitate Derived from Neutralization of SBW

Species	Wt% of species in precipitate			
	Batch mode (pH = 8)	Continuous mode (pH = 8)	Batch mode (pH = 9.5)	Continuous mode (pH = 9.5)
Al	12.0	13.0	10.0	9.60
Cd	0.16	0.16	0.14	0.13
Ca	0.51	0.55	0.79	0.78
Fe	1.0	1.1	0.88	0.84
Pb	0.16	0.17	0.14	0.13
Hg	0.27	0.32	0.30	0.27
K	0.89	0.82	0.70	0.67
Na	15.0	16.0	14.0	14.0

Additional tests were conducted to evaluate the effects of using batch mixing vs. continuous mixing to precipitate the nonalkali metals over the range of optimum pH. The overall removal efficiency of the nonalkali metals is slightly greater at a pH of 9.5 (see Table 4) than at a pH of 8.0. However, the results do not indicate any significant difference between batch and continuous mode processing.

The sludges that result from the precipitation process contain metal oxides or hydroxide precipitates that are colloidal in nature and highly stable. Thus, a chemical coagulant is needed to improve slurry dewatering and filtration efficiency. The selection of the best coagulant and optimum coagulant dosage is controlled by (1) the characteristics of the waste solution in the sludge (i.e., pH, alkalinity, polyvalent cation concentration, and ionic strength) and (2) the characteristics of the sludge particles (i.e., solids concentration, surface charge, and particle size). Commercial coagulants were screened using these criteria and two coagulants were selected for laboratory testing. Various amounts of the coagulant solutions were added to the sludges and the ease of dewatering was evaluated using the specific resistance to filtration test.[10] Cyanamid 844A* was found to be effective in reducing the filtration time and was subsequently used in precipitation tests with radioisotope tracers.

The efficiency of actinide and strontium precipitation were studied by adding radioactive tracers to the SBW simulant. The test procedures for pH adjustment, sludge

---

* Product of American Cyanamid Company, Wayne, NJ.

conditioning, and filtration were similar to those used in the cold simulant tests. All solutions were kept tightly capped when not in use to avoid carbon dioxide absorption from the air. Carbon dioxide can form complexes with the actinide species, increasing their solubility. The effects of aging (up to 4 d) were also evaluated because the characteristics of the sludges may change as they age.

The results of the tracer studies revealed that the actinide precipitation efficiency improves with aging up to 24 h. More than 99.99% of Pu, 99.8% of U, and 80% of Sr were removed. The amount of ^{238}U in all supernate samples was below the detectable limit of 0.09 ppm. With a target solution pH of 9.5, the total activity of the SBW was reduced from 32,000 dps/ml to less than 22 dps/ml gross alpha. There was no significant improvement in Pu and Sr removal with longer aging times. In part, this was attributed to the ability of the coagulant to adsorb plutonium.

B. SODIUM REMOVAL

Selective separation of the sodium from the SBW is a challenge because sodium is the most prevalent cation in the waste (see Table 1) and sodium salts are highly soluble in aqueous solutions. Few, if any, membranes or sorbents are sodium selective. One exception is macrocyclic crown ethers, which can possibly be customized to capture sodium ions. Alternatively, a physical separation of sodium nitrate ($NaNO_3$) precipitate is possible if coprecipitation of the contaminants in solution is not a problem. The saturation limit of $NaNO_3$ is reached with removal of approximately 50% of the water from the SBW. Theoretical calculations show that as much as 75% of the sodium could be precipitated before other constituents begin to coprecipitate.[11] Freeze crystallization and evaporation were identified as possible technologies to concentrate the SBW in order to precipitate $NaNO_3$.

Figure 3 illustrates the general steps of a sodium-separation process. The sodium-depleted waste can then be calcined without the addition of a diluent or directly immobilized by vitrification without concern for a high sodium content in the glass. The sodium

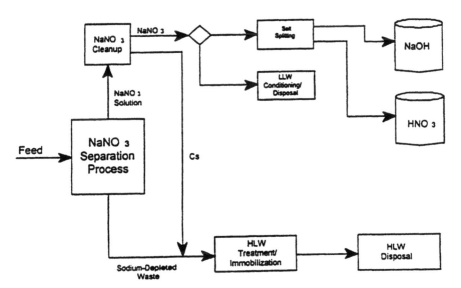

Figure 3 Process steps for separation and recycle of sodium from SBW.

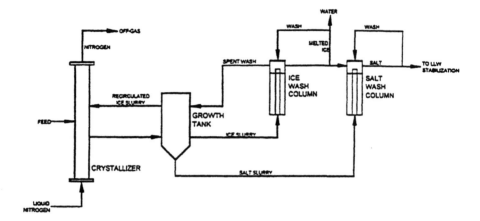

Figure 4 A schematic representation of a direct-contact freeze crystallization pilot plant.

nitrate can either be disposed as low-level waste or split into streams of NaOH and HNO_3 for use as decontamination solutions. Due to the chemical similarities of Cs to Na, Cs cleanup from the low-level waste could possibly be required.

1. Freeze Crystallization

Freeze crystallization is a developing technology that separates solution components by converting one or more to a solid or crystalline phase. This is accomplished by removing heat and reducing the solution temperature below the freezing point of one or more of the solution components. Usually pure-component crystals are formed as a result of the natural rejection of substances that are foreign to the basic crystal structure. The crystal purity is maintained by controlling the temperature in the crystallizer vessel, which in turn controls the size of the crystals. Crystal purity permits operation in a single theoretical stage, as contrasted to the incremental separation and multistage processes often required for a vapor–liquid separation. The energy efficiency of a freeze-separation process results from both this single-stage capability and the lower latent heats associated with solid–liquid phase change. Crystallization latent heats are one half to one tenth those of vaporization.

The process crystallizer is usually one of three types: (1) direct contact, where the solution is injected with an immiscible refrigerant and its evaporation causes heat removal from the solution: (2) indirect contact, where the heat is removed from the solution through a heat transfer surface in contact with an evaporating refrigerant; or (3) triple point, where the solvent is the refrigerant.

The applicability of freeze crystallization to SBW is being investigated by three U.S. companies through research subcontracts.[12] The pilot-plant process illustrated in Figure 4 was first used to test direct-contact cooling/freezing of the SBW with liquid nitrogen. Unfortunately, due to the extremely low freezing point depression that was exhibited by the SBW (–21°C), the tests were not successful. Ice formation around the nitrogen injection nozzle resulted in the buildup of ice, which restricted the flow through the nozzle. Thus, as a minimum, significant development work would be required to implement a full-scale direct-contact freeze crystallization treatment facility for treating the SBW.

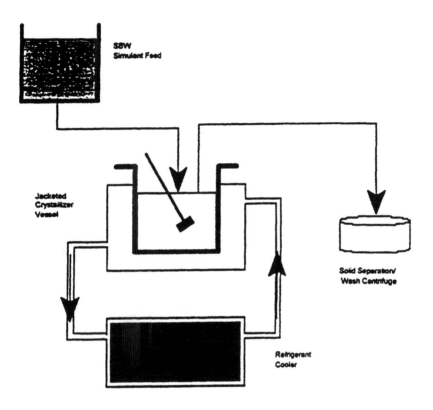

Figure 5 Indirect-contact freeze crystallization bench-scale apparatus.

Indirect-contact freeze crystallization testing was accomplished in the bench-scale equipment illustrated in Figure 5. The jacketed crystallizer vessel, of glass construction, provided the heat transfer, and was the location where ice formation occurred. In the solids separation/wash centrifuge, the ice was separated from the concentrated solution. Samples of the ice melt (containing redissolved salt precipitate) were analyzed for the major constituents in the SBW. Analytical results indicated that with 45% of the solution volume removed, approximately 55% of the original sodium content was removed as sodium nitrate precipitate. A small amount of a phosphate compound (likely sodium phosphate) was also present in the precipitate. Other notable findings include: process conditions favor crystal nucleation over growth; the freezing point depression was −19°C (concurring with the direct-contact testing); and the freezing point was −35°C at the end of the concentration testing.

One problem associated with freeze crystallization is washing the $NaNO_3$ precipitate to remove the contaminants that adhere to the salt during centrifugation. Due to the extremely low freezing point of the ice/salt captured in the centrifuge, washing of the solids resulted in wash water freezing and contaminant "capture" rather than washing. Successful washing was accomplished by partially melting the ice in the centrifuge. The meltwater then served as a wash solution.

2. Evaporation/Concentration/Precipitation

The evaporation/concentration/precipitation process overcomes some of the difficulties exhibited by the freeze crystallization process, albeit at the expense of higher energy demand. Tests were conducted in the laboratory with SBW simulant containing radioactive tracers.[12] The test solutions were incrementally concentrated and allowed to cool overnight. Any precipitate that was formed was removed by vacuum filtering. The filtered solids were washed with a solution of saturated $NaNO_3$ to remove contaminants that may have adhered to the precipitate prior to analysis. The heating, evaporation, cooling, filtering, and sampling was repeated several times, until the remaining liquid volume was approximately 30% of the original volume.

Precipitation of $NaNO_3$ crystals was first observed after the solution was concentrated to approximately 50% of the original volume. Up to 70% of the SBW was evaporated, with sodium being the only cation precipitating out of solution. Approximately two thirds of the sodium was removed from the solution while none of the radioisotopes added to the solution (^{137}Cs, ^{239}Pu, ^{238}Pu, and ^{241}Am) were detected in appreciable amounts in the precipitate.

3. Crown Ethers

Crown ethers have unique properties that allow them to selectively capture cations from multicomponent solutions. The selectivity is attributed to the high binding potential, which is a function of the crown ether cavity size and functional group(s), as well as synergistic effects of the diluent and additives(s) in the system. A crown ether system may exhibit the ability to perform any of the following reactions: (1) selective cation extraction of Na^+, (2) ion-pair extraction ($NaNO_3$), and (3) liquid–liquid ion exchange. These reactions are written as follows for sodium:[13]

$$Na^+_{aq} + CE_{org} = NaCE_{org} \tag{1}$$

$$Na^+_{aq} + NO^-_{3aq} + CE_{org} = NaNO_3CE_{org} \tag{2}$$

$$Na^+_{aq} + HCE_{org} = NaCE_{org} + H^+_{aq} \tag{3}$$

where H is hydrogen and CE is crown ether.

Several successful experiments involving crown ether solvent extraction of radionuclides from acidic systems[14-16] in conjunction with successful demonstration of sodium removal from neutral to basic solutions[17-19] provided the incentive to initiate an investigation of the capability to selectively extract sodium from the acidic SBW.

Following a literature review,[20] and the advisement of some of the scientists cited above, seven crown ethers were selected or custom designed (see Table 5) to study the capability of selectively extracting sodium from the SBW. Three diluents were also tested as shown in the table.

Despite efforts to optimize the extraction system, including the addition of 0.001 M sulfamic acid to the feed to serve as a nitrous acid scavenger, neither effective nor selective extraction of the sodium was observed. It was concluded that alternative crown ether configurations, and possibly alternative diluents, would be required for extraction of sodium from complex mixture, nitric acid solutions.

Table 5 Crown Ethers and Diluents Investigated for Sodium Removal from SBW

Crown ether	Diluents used
Di[4(5)-*tert*-butylbenzo]-16-crown-5	TBP/dodcane, decanol, TP/dodecane/HDDNS
Dibenzo-16-crown-5	TBP/dodcane, decanol, TP/dodecane/HDDNS
3-{*sym*-(Di[4(5)-*tert*-butylbenzo]-16-crown-5-oxy} propane-1-sulfonic acid	TBP/dodcane, decanol
3-{*sym*-(Propyl)Di[4(5)-*tert*-butylbenzo]-16-crown-5} propane-1-sulfonic acid	TBP/dodcane, decanol
Di{4[5]-2′-ethyl-1′-hydroxyhexyl)-benzo}-16-crown-5	TBP/dodcane, decanol
Di[4(5)-*tert*-butylbenzo]-14-crown-4	TBP/dodcane, decanol
Di[4(5)-*tert*-butylbenzo]-15-crown-5	TBP/dodcane, decanol

IV. SUMMARY

Successful laboratory and pilot-plant testing has shown that the ICPP SBW can be treated by several unique alternatives. Although several research issues must still be resolved, the tests completed thus far demonstrate that an industrial-scale process can be developed.

Successful waste decontamination can be accomplished using either the TRUEX/ SREX/Cs ion-exchange process or a conventional precipitation process scheme. Due to the high success of laboratory tests and modeling results, the TRUEX process is emerging as the most viable process for treating the SBW. In the case of precipitation using NaOH as a neutralizing agent, actinide removal efficiencies are very high; however, Cs, and possibly Sr, separation from the supernatant would be required to reach NRC Class A limits. Fortunately, the technologies for removing Cs and Sr from alkaline wastes are well proven. For instance, at the Savannah River Plant in South Carolina, Cs and Sr are routinely removed from an alkaline supernatant.

Sodium removal can be accomplished by concentrating the SBW solution beyond the saturation limit of $NaNO_3$. This can be best accomplished using an evaporation/concentration scheme due to the extremely low freezing point depression exhibited by the SBW. A technology to separate sodium from the SBW using a crown ether liquid–liquid extraction system was not proven successful.

The selection of the best treatment process for treating the SBW is being made by decision makers with the aid of a system analysis tool as discussed later in this text.

REFERENCES

1. **Horwitz, E. P., Kalina, D. G., Diamond, H., Vandergrift, G. F., and Schulz, W. W.,** The TRUEX process — a process for the extraction of the transuranic elements from waste solutions utilizing modified PUREX solvent, *Solv. Extract. Ion Exch.*, 3, 75, 1985.
2. **Horwitz, E. P.,** Combining extractant systems for the simultaneous extraction of transuranic elements and selected fission products, *DOE Report*, ANL/AP – 73631, July 1991.
3. **Wood, D. J., Tanter T. J., and Todd T. A.,** Effect of alkali and alkaline-earth metal ions on the extraction of strontium from acidic nuclear waste solutions by 18-crown-6 derivatives, *Solv. Extract. Ion Exch.*, submitted, 1994.
4. **Cross, J. E. and Hooper, E. W.,** The application of inorganic ion exchangers to the decontamination of radioactive liquid effluents, Chemistry Division, Harwell Laboratory, United Kingdom Atomic Energy Authority, Harwell, Oxfordshire, U.K., 1988.

5. **Marshall, D. W. and Miller, C. J.,** *personal communication,* Lockheed Idaho Technologies Company, Idaho Falls, ID, 1994.

6. **Boardman, R. D., Marshall, D. W., Nenni, J. A., and Pao, J.,** Investigation of alternative technologies for treatment of sodium-bearing aqueous radioactive waste at the Idaho Chemical Processing Plant, *Spectrum '94, Scientific Proceedings,* Atlanta, GA, August 14–18, 1994.

7. U.S. DOE, Final Environmental Impact Statement, defence waste processing facility, Savannah River Plant, DOE/EIS–0082, February 1982.

8. U.S. DOE, Final Environmental Impact Statement, disposal of Hanford defence waste high-level, transuranic and tank waste, Hanford Site, DOE/EIS–0113, December 1987.

9. **Pao, J. and Olson, L. G.,** *personal communication,* Lockheed Idaho Technologies Company, Idaho Falls, ID, 1994.

10. **Dentel, S. K., Abu-Orf, M. M., and Griskowitz, N. J.,** Guidance manual for polymer selection in wastewater treatment plants, Water Environment Research Foundation, Alexandria, VA, 1993.

11. **Heist, J. A.,** Development and demonstration of sodium-bearing radioactive liquid solution treatment by freeze crystallization, *Proposal to Westinghouse Idaho Nuclear Company, Inc. in response to Request for Proposal No. 227732,* FTC Acquisition Corporation, Raleigh, NC, April 1993.

12. **McCray, J. M.,** *personal communication,* Lockheed Idaho Technologies Company, Idaho Falls, ID, 1994.

13. **Inoue, Y. and Gokel, G. W.,** *Cation Binding by Macrocycles,* Marcel Decker, New York, 1990.

14. **McDowell, W. J., Case, G. N., McDonough, J. A., and Bartch, R. A.,** Selective extraction of cesium from acid nitrate solutions with didodecylnaphthalenesulfonic acid synergized with (bis tert-butylbenzo)-21-crown-7, *Anal. Chem.,* 4(2), 217, 1986.

15. **Horwitz, E. P. and Dietz, M. L.,** U.S. Patent 5,100,585, Process for recovery of strontium from acid solutions, March 31, 1992.

16. **Swanson, J. L. and Lumetta, G. J.,** Batch testing of the TRUEX process on some unique Hanford wastes and of the TRUEX and SrEX processes on commercial fuel reprocessing high-level wastes, *DOE Report,* PNL-Sa–19681, October 1991.

17. **Bartch, R. A., Walkowiak, W., and Robison, T. W.,** Selectivity in stripping of alkali-metal cations from crown ether carboxylate complexes, *Sep. Sci. Technol.,* 27(7), 989, 1992.

18. **Inoue, Y., Wada, K., Liu, Y., Ouchi, M., Tai, A., and Hakausi, T.,** Molecular design of crown ethers. 6. Substitution effect in 16-crown-5, *J. Org. Chem.,* 54, 5268, 1989.

19. **Mohite, B. S. and Khopkar, S. M.,** Solvent extraction of sodium by crown ethers, *Anal. Lett.,* 19(15–16), 1603, 1986.

20. **Wendt, K. M.,** *personal communication,* Lockheed Idaho Technologies Company, Idaho Falls, ID, 1994.

CASE STUDIES OF PROCESS SELECTION

Chapter 8.1

Wastes at the West Valley Nuclear Site, New York

Dean E. Kurath and Lane A. Bray

I. INTRODUCTION

The West Valley Demonstration Project Act, passed in 1980, authorized the U.S. Department of Energy (DOE) to conduct a nuclear waste management project at the former commercial nuclear fuel-reprocessing facility at the Western New York Nuclear Service Center near West Valley, New York.[1] Between 1966 and 1972, a total of 640 metric tons (MT) of spent fuel was reprocessed using the plutonium–uranium extraction (PUREX) process. These operations generated 2.1 million liters (528,000 gal) of high-level waste (HLW), which was stored in an underground carbon steel tank (8D-2) inside a cement vault. A smaller quantity (35,000 l) of acidic thorium extraction (THOREX) waste was stored in a stainless steel tank (8D-4).

II. PROJECT OBJECTIVES

One of the West Valley Demonstration Project's main objectives is to solidify the HLW into a form suitable for transportation to, and disposal at a federal repository. A related objective is to minimize the amount of radioactivity disposed of as low-level waste (LLW). Other project goals include minimizing the construction of new facilities by using existing facilities where possible and demonstrating the reuse of decontaminated facilities.

III. WASTE CHARACTERISTICS

Approximately 90% (v/v) of the waste in tank 8D-2 was an alkaline supernatant liquid; the remainder was a metal hydroxide sludge that had settled to the bottom of the tank. The supernatant consisted mainly of sodium nitrate and nitrite salts (see Table 1) and significant amounts of other cations. In 1987, the supernate contained about 7.3 MCi of ^{137}Cs and smaller but significant amounts of other radionuclides, including all of the ^{99}Tc, and trace amounts of ^{90}Sr and various Pu isotopes. The insoluble sludge contains primarily iron, aluminum, manganese, uranium, and nickel as insoluble hydroxides, oxides, and phosphates. A number of significant radionuclides, such as ^{90}Sr and the transuranics, are typically insoluble in an alkaline environment and thus were found primarily in the sludge.

0-8493-4876-5/96/$0.00+$.50
© 1996 by CRC Press, Inc.

Table 1 Alkaline Supernatant Liquid Composition

Chemical Species	Concentration (M)
Na^+	6.4
K^+	0.2
Rb^+	0.00037
Cs^+	0.00134
NO_3^-	3.5
NO_2	2.1
CO_3^{2-}	0.4
SO_4^{2-}	0.2
^{137}Cs (Ci/l)	3.4
pH	10–11
Volume (l)	2,100,000

IV. DESCRIPTION OF SUPERNATANT TREATMENT PROCESS AS IMPLEMENTED

The processing approach involves separating most of the radioactivity from the bulk of the chemical components so that the radioactive fraction can be solidified as high-quality borosilicate waste (Figure 1). After separation of the supernatant from the sludge, an inorganic zeolite (IONSIV IE-96 [c]*) ion-exchange process removes the ^{137}Cs from the supernatant. The sludge is then washed several times with water adjusted to pH 12.5 to remove soluble salts, thereby minimizing the volume of waste that must be vitrified. Since the sludge wash solutions contain traces of plutonium as well as cesium, PNL developed a new ion exchanger (IONSIV TIE-96)* and implemented it at West Valley. The cesium-loaded zeolite will be combined with PUREX-washed sludge and THOREX acid waste (Figure 1) for vitrification. The decontaminated low-level supernatant waste and sludge wash water are combined, concentrated, and solidified as a cement LLW that meets the U.S. Nuclear Regulatory Commission (NRC) 10CFR61 criteria.

The ion-exchange process is implemented by suspending four 3-ft diameter, 10-ft tall ion-exchange columns, connected in series, inside spare tank 8D-1. The supernatant liquid from tank 8D-2 is pumped through a sintered metal cross-flow filter to remove suspended solids and is then routed to a feed tank. The 6 M Na^+ supernate is typically diluted (2 volumes of water:1 volume of waste) and chilled to between 12 and 18°C to improve the cesium loading on the zeolite, thereby reducing the total required mass of exchanger. The columns are typically operated with three in series and one off-line for discharging the loaded zeolite into tank 8D-1 and recharging the column with fresh zeolite. Loading of a column is typically stopped when the first column reaches 80 to 90% breakthrough, at which point the first column is greater than 98% loaded.

V. APPROACH TO PROCESS EVALUATION

Systems engineering was used to select a separation process for treatment of the supernatant.[2,3] This involved developing the process evaluation criteria, identifying processes, collecting the necessary data, and performing the evaluation. The data were collected from experiments and from literature sources.[4]

---

* Registered trademark of UOP, Chicago, IL.

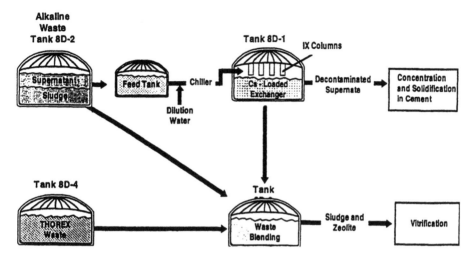

Figure 1 West Valley supernatant treatment process.

Several organizations were involved in the process development and selection process. In 1981, DOE awarded a contract to West Valley Nuclear Services (WVNS) Inc., a Westinghouse Electric Corporation subsidiary, to carry out the West Valley Demonstration Project. Much of the experimental testing was conducted at Battelle, Pacific Northwest Laboratories (PNL); some was performed by Westinghouse Research and Development (R/D) and others. Organizations that consulted on process selection for West Valley include PNL, Westinghouse R/D, Savannah River Laboratory (SRL), and Ebasco Services.

VI. PROCESS REQUIREMENTS

The supernatant decontamination process was selected and developed to meet the following requirements:

- Process performance — The process must be capable of removing 99.9% (decontamination factor [DF] = 1000) of the cesium from the alkaline supernatant liquid and sludge wash solutions.
- Equipment and process complexity (process reliability) — The complexity of the process and associated equipment should be minimized. The goal of this requirement is to minimize project risk, since increasing process complexity generally results in increased risk. The radioactive environment in which the ion-exchange process must be operated requires a simple process because much of the equipment is in a remote environment and is therefore difficult to maintain.
- Impact on the HLW vitrification system — Factors potentially impacting the waste vitrification system include the quality of the waste form, the amount of glass produced, and operation of the melter or the melter feed system.
- Impact on the liquid waste treatment system (LWTS) — Factors potentially impacting the supernatant LWTS include introduction into the system of materials requiring special treatment and additional volume.
- Safety and environmental considerations — The process must be easily controlled and must present no abnormal safety or environmental concerns beyond those present in

existing production-scale radiochemical plants (e.g., limiting the introduction of materials that are hazardous or produce hazardous products). An additional environmental goal was maximizing the radioactivity disposed of in HLW.

VII. ALTERNATIVES CONSIDERED

Processes considered for removing cesium from the supernatant include electrodialysis, hyperfiltration, crystallization, liquid membranes, inorganic or organic ion exchange implemented in a column or batch contracting configuration, and precipitation.[3] Inorganic ion exchangers considered included Durasil, DeVoe/Holbein compositions, natural zeolites, and synthetic zeolites (IE-95/96). The organic exchangers included Duolite CS-100(c)* and ARC-359 (granular phenol-formaldehyde types). Configurations for implementing the inorganic exchangers included batch contacting, single- and multiple-column contacting, and elution with acid. The organic exchangers were evaluated as a regenerable system in which the resins would be eluted with formic acid and the cesium concentrate vitrified with the sludge. Precipitation processes considered included sodium tetraphenyl boron (previously investigated for use at the Savannah River Plant), nickel ferrocyanide, and phosphotungstic acid (PTA). Precipitation with PTA would require acidification of the alkaline waste; the other two would act directly on alkaline waste.

VIII. EVALUATION

Several of the alternatives were eliminated early in the evaluation process because they did not meet the process performance criteria (DF = 1000) or they required extensive development efforts. Crystallization and liquid membranes were ruled out by their limited state of development. Electrodialysis and hyperfiltration did not appear capable of meeting the Cs DF requirements with a feasible design. The cesium nickel ferrocyanide precipitation process was not pursued because of concerns about whether it could produce adequate DFs, its potential to increase the glass volume, and its potential for an explosive reaction with nitrates/nitrites at elevated temperatures (350°C).

The remaining alternatives (inorganic ion exchange, organic ion exchange, and precipitation with tetraphenyl boron or PTA) were evaluated by the following criteria:

- Process performance — All were judged capable of meeting the cesium decontamination requirement of 1000, with the once-through use of zeolite IONSIV IE-96 providing the largest DFs.
- Equipment and process complexity — Once-through use of zeolite was judged to be the least complex and organic ion exchange the most complex of the alternatives, because of its equipment and operational requirements for the elution and regeneration of the resin. The sodium tetraphenylboron precipitation process [$Na(C_6H_5)_4B$] was also relatively complex because of the need to treat the organic-containing precipitate and the gaseous effluents (benzene) generated during processing. The PTA process was judged relatively complex because of the large volume of supernatant that had to be acidified and subsequently reneutralized.

---

* Now owned by Rohm and Haas; material is now different.

- Safety and environmental considerations — The zeolite process presented no significant safety or environmental concerns beyond those normally encountered in radioactive operations. There was concern that the formic acid used to elute organic resins would decompose during vitrification to yield a potentially explosive mixture of H_2 and CO. The PTA process presented limited concern in that acidification and neutralization of the supernate generates heat and reaction gases. The benzene and other organic vapors generated in the NaTPB precipitation process presented a significant safety and environmental concern. Controls to prevent formation of hazardous concentrations and treatment of the process effluent streams to prevent release to the environment would have been necessary.
- Impact on the HLW vitrification system — The once-through zeolite ion exchanger had the potential to impact several aspects of the vitrification process, for example, increased glass volume, decreased glass quality (spinel formation), and transport of the zeolite/slurry feed mixture to the melter. The organic ion-exchange and NaTPB precipitation processes had the potential to reduce metal oxides (Ni, Ru) to metals, which could collect in the bottom of the melter and short-circuit the electrodes. The PTA process presented no negative impact on the vitrification system.
- Impact on the LWTS — The zeolite ion-exchange process was judged to have no impact on the LWTS. Organic ion exchange would have sent the largest volume of effluents for evaporation, and NaTPB would have contributed substantial quantities of organic material requiring treatment at the LWTS. Acidification and neutralization in the PTA process would have increased the mass of LLW requiring disposal by 40%.

The process evaluation was performed by ranking the various alternatives against the criteria in the following priority:

- Safety and environmental considerations
- Equipment and process complexity (process reliability)
- Impact on the disposal functions
- Process performance

The evaluation ranked the processes as follows:

1. Inorganic ion exchange (once-through)
2. PTA precipitation
3. Organic ion exchange (regenerative)
4. Sodium tetraphenyl boron precipitation

The zeolite ion-exchange process (IONSIV IE-96) ranked high because of its relative simplicity, low level of safety and environmental concerns, and low impact on the LWTS. The PTA process ranked second only because of its relative complexity and greater LLW volume. Organic ion exchange received a low ranking because of poor decontamination performance (due to the difficulty in achieving a large decontamination of the resin during elution), relatively high equipment and process complexity (due to the need for regeneration), and impacts on the disposal functions (increased cement volume due to addition of NaOH to adjust pH and increased glass volume). The NaTPB precipitation process received the lowest ranking because of safety and environmental concerns, complex processing requirements, and significant impact on the disposal functions.

IX. RESULTS OF OPERATION

The operation of the West Valley supernatant treatment process was initiated in 1988,[5] and 21 campaigns with IE-96 were completed. The average cesium DF has been about 40,000, but peaks have reached 250,000 (the design goal was 1000). These campaigns used approximately 35,000 kg (dry weight) of the zeolite to remove 5135 KCi of ^{137}Cs from 1,643,957 gal of diluted feed. Approximately 10,394 drums of certifiable class C cement LLW meet the waste form criteria specified in the U.S. NRC 10CFR61. The high DFs resulted in a low average drum contact dose rate (10 mR/h), which allows contact handling of the drums and occasional contact maintenance of the cement plant. The LLW cement is class C because transuranic materials (Pu) are estimated to be present at a level of 21 nCi/g (the class A limit is 10 nCi/g).

On completion of processing the alkaline supernatant solution in the 8D-2 tank, the remaining solution along with the 1.5 ft of hydroxide sludge remaining in the tank was washed three times with water adjusted to a pH of 12.5. This step dissolved soluble salts and washes the interstitial liquid out of the insoluble hydroxide sludge. Preliminary sludge-washing experiments performed at PNL indicated that the transuranic (TRU) content of the cement would increase during sludge washing because transuranic carbonates are somewhat soluble. Additionally, the State of New York wanted the TRU content to be less than 10 nCi/g and meet the class A limit. To increase Pu removal, PNL developed a method to treat the IE-96 zeolite with sodium titanate so that traces of Sr and Pu are also removed.[6] Within 2 years of discovery, this new product (IONSIV TIE-96) was commercialized by the UOP Corporation with close technical transfer support from PNL. West Valley has procured approximately 27,000 kg of the product for processing the sludge wash solutions.

West Valley initiated processing of the first sludge wash solutions in April 1992; the first column contained TIE-96 and processed about 127,000 gallons of solution (pH 12.5). This campaign resulted in a DFs of 64 for plutonium and 38,800 for cesium, and it produced the first LLW cement that meets Class A limits.

X. SUMMARY

Operations at the West Valley Nuclear Site have demonstrated the successful application of a zeolite ion-exchange process that removes cesium from PUREX alkaline supernatant liquid. The performance of this ion-exchange column system (DF = 40,000) has exceeded the requirements (DF = 1000). Plutonium was also separated from sludge wash waters by use of a modified zeolite ion-exchange material developed by PNL and commercialized by UOP. The high DF achieved by this process has minimized the radionuclides in the LLW form, as well as reduced the dose received by the workers.

REFERENCES

1. **McIntosh, T. W., W. W. Bixby, J. E. Krauss, and D. R. Leap.** 1988. "An Overview of Waste Management Systems at the West Valley Demonstration Project." In *Waste Management 88, Vol. 2*, R. G. Post and M. E. Wacks, Eds., pp. 785–790, Arizona Board of Regents, Tucson, AZ.
2. **Bray, L. A., L. K. Holton, Jr., B. M. Wise, J. M. Pope, and D. E. Carl.** 1984 "Selection of a Reference Process for Treatment of the West Valley Alkaline Waste." In *Fuel Reprocessing and Waste Management, Vol 1*, pp. 345–356, ANS International Topical Meeting on Fuel Reprocessing and Waste Management, Jackson, WY.

3. **Carl, D. E. and I. M. Leonard.** 1987. *Selection of the Treatment Method for the West Valley Alkaline Supernatant.* DOE/NE/44139-25, West Valley Nuclear Services, West Valley, NY.

4. **Bray, L. A., L. K. Holton, Jr., T. R. Myers, G. M. Richardson, and B. M. Wise.** 1984. Experimental Data Developed to Support the Selection of a Treatment Process for West Valley Alkaline Supernatant. PNL-4969, Pacific Northwest Laboratory, Richland, WA.

5. **Kurath, D. E., L. A. Bray, W. A. Ross, and D. K. Ploetz.** 1990. "Correlation of Laboratory Testing and Actual Operations for the West Valley Supernatant Treatment System," PNL-SA-16871. Presented at the Fourth International Symposium on Ceramics in Nuclear Waste Management, American Ceramic Society Annual Meeting, April 23–27, 1989, Indianapolis, In *Ceramic Transactions, Nuclear Waste Management III,* Vol. 9, G. B. Mellinger, Ed, pp. 529–538, American Ceramic Society, Westerville, OH.

6. **Bray L. A. and L. L. Burger,** March 1994. U.S. Patent 5,298,166, Method for Aqueous Radioactive Waste Treatment.

Wastes at the Idaho National Engineering Laboratory Chemical Processing Plant, Idaho Falls, Idaho

J. A. Murphy, R. D. Boardman, L. F. Pincock, and N. Christiansen

I. BACKGROUND

This chapter presents a systems analysis approach, focused on wastes at the ICPP, that is being taken to help decision makers to (1) identify optimum radioactive waste treatment and disposal form alternatives; (2) assess trade-offs between various optimization criteria; (3) identify uncertainties in performance parameters; and (4) focus development efforts on options that best satisfy stakeholder concerns. The systems analysis approach evaluates candidate waste treatment processes and disposal options using a consistent set of assumptions and data. Additionally, the potential impacts of existing uncertainties are evaluated, resulting in the recommendation of programs to investigate and resolve those uncertainties.

DOE Order 4700.1 outlines a logical systems engineering approach to problem evaluation. The system analysis approach outlined in this chapter follows a similar twofold approach: (1) the following of a formal systems engineering process; and (2) the application of systems analysis tools and techniques to support informed decision making throughout the process. The formal systems engineering process consists of six principal steps:

- Define the problem and establish the program goal
- Determine the functional (baseline) requirements
- Identify alternatives that meet the functional requirements
- Develop and evaluate the alternatives
- Optimize the alternatives
- Recommend preferred alternatives for selection and implementation

Systems analysis functions in support of systems engineering efforts include:

- Enhancing program integration and coordination by applying systems thinking and organizational learning concepts, tools, and techniques (i.e., Hexagon Idea Modeling, mind maps, decision trees, causal loops, etc.) to clarify and maintain focus on program objectives, to facilitate group interaction, and to support informed selection of optimum strategies and technologies.
- Providing systems modeling capabilities, such as developing flow sheets and computer simulations (models).
- Conducting performance assessments to help establish waste acceptance criteria and estimate final product performance.

In the context of this chapter, the discussion centers on the application of the systems analysis approach to processing (meaning all treatment steps necessary to prepare the waste for final disposition), storage, and disposal of radioactive liquids and calcine at the Idaho Chemical Processing Plant (ICPP). At the time of this writing, the evaluation of candidate treatment processes was still in progress; consequently, all data comparisons presented are for illustrative purposes only. But first, an overview of the ICPP wastes requiring treatment needs to be given.

Historically, two types of radioactive liquid waste have been generated at the ICPP. The uranium extraction process resulted in acidic, high-level radioactive waste (HLW) that contained the metal cations from the dissolved fuel and fuel cladding. A low-activity radioactive acidic waste resulted from cleanup of the extraction solvents, from 2nd/3rd-cycle raffinates, and from equipment decontamination using solutions containing sodium compounds. Because this waste contains a high concentration of sodium nitrate (1 to 2 M), it is referred to as sodium-bearing waste (SBW).

The HLW and SBW are stored separately in stainless-steel tanks prior to solidification in a fluidized-bed calciner operating at 500°C. The calcine produced from HLW primarily consists of various metal oxides and halides, such as Al_2O_3, ZrO_2, and CaF_2, depending on the composition of the HLW. However, a large percentage of the alkali cations in the SBW remain as nitrate salts during calcination. Due to the low melting temperatures of $NaNO_3$ and KNO_3, calcine agglomeration can be a problem in the calciner and solid storage vessels. Consequently, the SBW is typically blended with the HLW during calcination to dilute the molten alkaline-metal salts with metal oxides.

In April 1992, the U.S. Department of Energy (DOE) curtailed fuel reprocessing at the ICPP. Consequently, there is no remaining or new HLW to blend with the existing and projected inventory of SBW. In order to calcine all of the remaining SBW, a nonradioactive diluent, such as aluminum nitrate, would have to be added to avoid calcine agglomeration. This would cause a significant increase in the volume of high-activity calcine produced and could ultimately increase the cost of disposal in a suitable HLW geological repository. Therefore, alternative treatment technologies are being considered to minimize the volume of immobilized waste that will be generated from the SBW.

In previous chapters, some of the candidate options for treating the SBW were presented. The options range from direct immobilization to chemical or physical separation of the radioactive contaminants from the benign constituents in the waste. Likewise, there are technically viable options for treating the calcine, including dissolution in nitric acid and separation/isolation of the radioactive contaminants. The systems analysis approach allows the treatment of both waste types to be jointly considered, resulting in the most cost-effective, stakeholder-acceptable option possible.

II. EVALUATION PROCESS

A. PROGRAM GOAL

Through a series of facilitated meetings, varying perspectives were debated in regard to project milestones and overall desired results before a consensus was reached. As a result, the following program goal was established for the evaluation of technologies for radioactive liquid and calcine waste treatment and disposition: "To support DOE in developing a strategic plan for ICPP radioactive liquid and calcine waste management by gathering, evaluating, optimizing, and presenting performance data (including a recommendation of preferred treatment options and waste disposal forms) for viable candidate processes and technologies."

B. BASELINE REQUIREMENTS

After definition of the program goal, the following items were established as baseline requirements for the technologies evaluation and optimization process.

- Meet all current federal and state regulations, court orders, and agreements between DOE and the State of Idaho, or the EPA.
- Treatment alternatives must be able to ultimately immobilize both radioactive liquid and calcine waste.
- Treatment alternatives must immobilize high-activity waste (HAW) using vitrification (glass or glass-ceramic). Treatment technologies other than vitrification would require an equivalency, determination, or variance.
- Treatment alternatives must stabilize low-activity waste (LAW) to meet concentration-based standards, or seek equivalencies, determinations, or variances.
- Treatment alternatives must be compatible with future radioactive liquid waste generated at the ICPP.
- Treatment alternatives must accommodate on-site storage of immobilized waste.
- New facilities must be designed and built to required radiation-dosage and environmental-release standards.

C. IDENTIFY ALTERNATIVES

Extensive investigations were conducted to identify potential technologies that could be employed at the ICPP to treat and dispose of radioactive liquid and calcine wastes. Several candidate technologies, described in Table 1, were identified and combined into processing systems (flow sheets) capable of handling radioactive liquid and calcine wastes at the ICPP. (While some separation technologies can be employed as pretreatment processes, pretreatment is used in the tables below to indicate those nonseparation technologies necessary for further processing or immobilization. Separations are used to indicate all separation technologies under consideration, whether or not they are used as a pretreatment.)

D. DEVELOP ALTERNATIVES

Mass balance calculations resulting from literary studies and laboratory testing were made for processing systems compatible with ICPP radioactive liquid and calcine wastes. Radioactive waste compositions of current and projected inventories were used as inputs. Mass balances were then used to evaluate candidate technology flow sheet performance relative to ICPP radioactive waste streams and to develop estimates of cost, schedule, throughputs, and radioactive waste volumes. Additionally, a panel of technical experts evaluated the maturity of each candidate technology. Processes that did not fulfill the baseline requirements, that could not be developed in time to meet regulatory requirements, or that were viewed as being technically inferior to other technologies were eliminated from further consideration. Processes that still showed promise following initial development were advanced to the next stage of analysis.

Radioactive liquid and calcine wastes contain many elemental components; separating some components prior to final immobilization results in smaller volume of HAW requiring immobilization and disposal. The following technologies (see Table 2), along with direct vitrification, provide 27 potential combinations of technologies (each with their own immobilization and waste form options) that meet the baseline requirements and support the program goal.

Table 1 Initial Candidate Technologies

Alternatives	Description	Application
	Pretreatment	
Freeze crystallization (FC)	Preparation of a high-sodium solution for calcination by addition of ANN or activated silica to maintain the fluidized bed operation	Calcination of radioactive liquid waste
Calcination	Conversion of a radioactive liquid waste to a granular solid form	Radioactive liquid waste treatment
Dissolution	Conversion of solid radioactive waste to a liquid by chemical reaction with acid	Calcine treatment
	Separations	
Halide volatility	Conversion of an element to a halide gas can then be preferentially separated by trapping the gas at different temperatures	Calcine treatment
Oxidation/reduction	Conversion of elemental valences for use with preferential separation methods, such as precipitation, extraction, etc.	Radioactive liquid and calcine waste treatment
Radionuclide partitioning by liquid-liquid solvent extraction (TRUEX, SrEX, NaEX)	Process using countercurrent aqueous and organic streams to selectively remove components from the aqueous solution; extractants being investigated are CMP and CMPO for transuranics (generally called TRUEX in the report) and crown ethers for both strontium and sodium removal (SrEX and NaEX, respectively)	Radioactive liquid and dissolved calcine waste management
Electrohydrolysis	Process that separates components tusing a chemical transformation driven by electrical current or voltage	LAW from freeze crystallization or precipitation by neutralization
Fractional condensation	Process that uses the differential ability of certain components to leave a gaseous state and form a solid or liquid as temperature is reduced	Product from halide volatility
Freeze crystallization (FC)	Separation of crystallized water and saturated sodium nitrate, which currently limits the efficiency of the new waste calcining facility (NWCF) from radioactive liquid wastes by freezing. This results in a smaller waste volume of HAW than would occur if processed directly into calcine	Radioactive liquid waste treatment only, specifically the sodium nitrate
Ion exchange (IX)	Process that uses an organic resin or inorganic clay to preferentially remove ions in exchange for other ions predeposited on these surfaces	To remove cesium (via CsIX) from waste streams in liquid form
Off-gas treatment	Filtration or chemical process to remove hazardous and/or radioactive contaminants from the process off-gas	Will be used in all processes as needed to control effluents
Precipitation (PPT) by neutralization	Process that uses the solubility properties of certain components to segregate those that do not form a solid	Radioactive liquid and dissolved calcine waste treatment

Table 1 Initial Candidate Technologies *(continued)*

Alternatives	Description	Application
Pyrochemical/ pyrometallurgical	High-temperature processes that convert a solid to a vapor or metallic liquid form	Calcine waste
Selective leaching	Selective dissolution of radioactive solids into liquid by chemical reaction, for example, the dissolution of cesium from calcine with water	Calcine waste
Supported liquid membranes	Membrane filter used to separate target elements from an aqueous solution	Radioactive liquid and dissolved calcine waste treatment
Volatillization/ distillation	Process using the differential ability of certain components to enter a vapor phase upon heating	Calcine waste; part of a pyrochemistry methodology
Immobilization		
Canning	Sealing waste inside a metal container	Final waste forms and potentially calcine
Cementation/grouting	Incorporastion of waste into a matrix of solidified hydrated oxides	LAW immobilization alternative
Glass	Incorporation of waste into a matrix of noncrystalline solidified liquid	Both LAW and HAW immobilization alternatives
Glass-ceramic	Incorporation of waste into a form that contains some glass, but which also contains a large portion of crystalline matrix	Both LAW and HAW immobilization alternatives

E. DECISION TREE

Potential combinations of technologies are arranged according to seven key process decisions, the first four of which are presented in Figure 1. These seven decisions, referred to as nodes, are presently discussed. The first node allows selection of construction and operating expenditures for the HAW immobilization facility during the initial project phase (when the radioactive liquid waste is processed) or during the second phase of the project (when calcine processing occurs). By delaying the HAW immobilization unit until a second phase, the initial capital investment would be delayed; operation costs would be saved on that part of the project, but interim liquid to solid processing and interim storage would be required.

The second decision node deals with the radioactive liquid waste-processing technology selection. Here, there are four possibilities: (1) freeze crystallization, (2) precipitation through neutralization, (3) TRUEX radionuclide partitioning, and (4) nonseparations processing, with each of the three separation technologies producing a LAW stream requiring immobilization. The freeze crystallization technology can be used only with the radioactive liquid waste, while the other alternatives can be used for both liquid and calcine waste.

In the third decision node, the continued use of the existing NWCF is decided. Current inventories of radioactive sodium-bearing and high-level liquid waste cannot be calcined in the NWCF without the addition of either first-cycle raffinates from the reprocessing of spent nuclear fuel (SNF), nonradioactive aluminum nitrate nanohydrate (ANN), or silica. Raffinates from past reprocessing are now depleted, leaving blending with nonradioactive ANN or silica (both of which will significantly increase the calcine waste volume) as the

Table 2 Technologies Under Evaluation

Technology	Technology use	Decision node(s)
	Pretreatment	
ANN or silica addition	Added directly to radioactive liquid waste or to the resulting HAW fraction from FC or PPT	3
Calcination	Conversion of sodium-bearing liquid plus additive to a granular solid form	3 & 4
Dissolution	Used on all calcine wastes	4
	Separations	
Freeze crystallization (FC)	Used on radioactive liquid waste with high sodium concentrations	2
Precipitation (PPT) by neutralization	Used on both radioactive liquid and calcine wastes	2 & 4
Radionuclide partitioning by liquid-liquid solvent extraction (TRUEX, SREX, NaEX)	Used on radioactive liquid and dissolved calcine wastes, to extract transuranics (TRU) and strontium	2 & 4 for TRU; 7 for strontium
Electrohydrolysis	Used on the LAW fraction from PPT separation and potentially, FC	2
Ion exchange (IX)	Used on the LAW fraction from FC, PPT, or TRU removal methods	7
	Immobilization	
Canning	All final HAW forms	Assumed for all HAW forms
Cementation/grouting	Immobilization method for LAW	6
Glass	Immobilization method for LAW or HAW	6 & 7
Glass-ceramic	Immobilization method for LAW or HAW	6 & 7

only methods to continue calcination. The NWCF is presently equipped for ANN calcination; if the decision is made, however, to use silica instead, modifications would be required to the calciner prior to performing calcination. Regardless of which additive is used, the decision to continue calcination varies according to the processing technology selected for the sodium-bearing liquid waste. For freeze crystallization, precipitation, and direct calcination, the calciner could be used to process the HAW stream using either ANN or silica. If, however, a HAW immobilization facility is built for Phase I operations, the calciner is no longer needed. Similarly, if TRUEX is used to process sodium-bearing liquid waste, continued calcine operations would not be cost-effective based on estimates that the resulting volume of concentrated liquid would feed the calciner for only 45 to 60 d after keeping the calciner on standby for a 12-year period.

Selection of technologies for processing calcine waste occurs in decision node 4, resulting in 27 potential technology combinations, as mentioned above. Possible calcine waste-processing options include precipitation (through neutralization), TRUEX radionuclide separation, and direct immobilization of calcine into a HAW waste form.

F. DECISION TREE OF IMMOBILIZATION TECHNOLOGIES

Nodes 5, 6, and 7, shown in Figure 2, represent integrated decisions regarding disposal form and waste classification options for each radioactive waste-processing alternative.

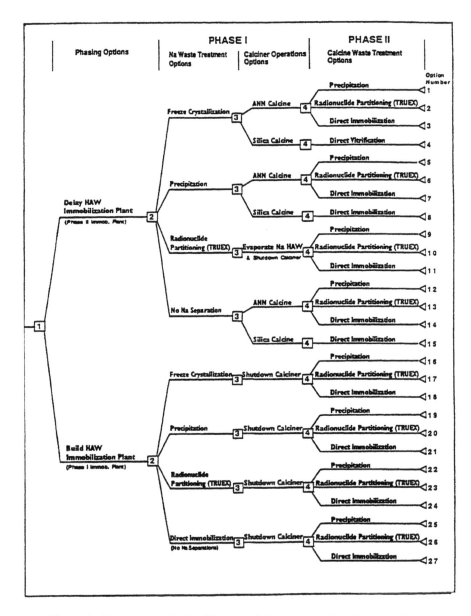

Figure 1 Decision tree (nodes 1 through 4) for waste treatment technologies.

Decision node 5 involves selecting a HAW immobilization technology; HAW immobilization processes available are glass and glass-ceramic. The glass process involves combining the HAW stream with appropriate glass forming additives (typically SiO_2, B_2O_3, and Na_2O) at high temperatures in a glass melter to produce a glass-immobilized waste form. There are three possible feeds to this process: (1) sodium-bearing waste, (2) calcine waste, or (3) the concentrated actinide and fission product stream from the separation process(es). The glass melter would be operated at temperatures between 1050

136

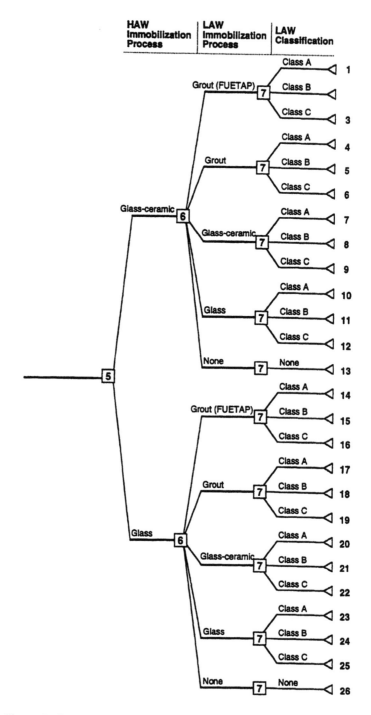

Figure 2 Decision tree (nodes 5 through 7) for waste form and classification.

and 1250°C, producing a molten glass product that would be poured into canisters, sealed, and temporarily stored, pending final disposal. Similarly, the glass-ceramic process involves combining and thoroughly mixing the HAW stream with appropriate additives, sealing the mixture in stainless steel canisters, and processing the canisters at a high temperature and under moderately high pressure. A hot-isostatic press (HIP) is typically used to form the glass-ceramic waste forms at temperatures between 1000 and 1200°C, and at pressures ranging from 15,000 to 20,000 psi.

LAW immobilization process selection between glass, glass-ceramic, normal grout (Portland-type cement), and grout formulated under elevated temperature and pressure grout (FUETAP) is made in decision node 6. If no separations are performed, then LAW immobilization and LAW class are not necessary; it is described in the tree as no LAW. Glass and glass-ceramic processes are analogous to those described above for HAW. Grouting or cementation technology refers to the process of encapsulating or stabilizing the LAW in a material that can be pumped into transportable containers where it is cured and solidified. FUETAP grout requires applying heat and pressure externally to accelerate the curing process. The key to controlling grout consistency is selecting the most appropriate mixing process and controlling the mixing of ingredients. The process of mixing the grout can be divided into three categories: batch mixing, in-line mixing, and in-container mixing.

The final decision node is used to select desired radioactivity levels for LAW according to waste class; currently, the NRC classification guide is used to distinguish the possibilities (i.e., Class A, B, or C). If a Class A or B waste is selected, then strontium and cesium removal via SREX and IX, respectively, is required in addition to all separations technologies.

III. OPTIMIZE AND EVALUATE ALTERNATIVES

Within each of the 27 processing combinations are additional variables that may affect the performance of any of the processing options. Additional variables include waste immobilization techniques for high- and low-level waste forms, interim storage, and calciner operations. A computerized model was developed to assist in the evaluation and analysis of alternatives by identifying development activities needed to support technology evaluation, prioritizing development efforts by identifying essential information needed prior to selecting an optimum process, and establishing relationships between available data and process selection criteria. Each of the 27 candidate processes was configured using a uniform basis for each cost element; this ensured an objective comparison of waste treatment alternatives by presenting each alternative in its most cost-effective configuration.

A. TECHNOLOGY EVALUATION AND ANALYSIS METHODOLOGY (TEAM) MODEL

The TEAM model computes the present value (in 10/94 dollars) for life-cycle costs and cost components of user-specified treatment options for radioactive liquid and calcine waste. These costs include development; design, construction, and start-up; operations and maintenance; interim storage; disposal; and decontamination and decommissioning (D&D) costs associated with radioactive liquid and calcine waste treatment. Although the primary focus is on cost analyses of the various waste treatment options, other issues, such as the timing of expenditures and environmental impacts, are expected to be important factors in the selection and implementation of waste treatment technologies. Consequently, the TEAM model also computes the cost–time profile of expenditures and the volumes of each waste type produced.

Determining the relative costs for each option is complicated by the interactions among the different cost components and by uncertainties in key parameters. For example, because each option generates both HAW and LAW, which have substantially different long-term disposal costs, options that appear relatively inexpensive in regard to short-term expenditures may be overshadowed by high disposal costs if they generate even a modest increase in the amount of resulting HAW. The TEAM model automatically selects the most cost-effective HAW and LAW immobilization processes and LAW waste classes, and includes a user's option to change the selected immobilization technology to explore other options where cost is not a primary driver or criterion.

B. STRATEGY FOR EVALUATION

To identify the most economical treatment option, the TEAM model was first used to evaluate the 27 alternative processes against a number of valuative criteria (e.g., life-cycle costs, life-cycle cost components, waste volumes between alternatives, and cost–time profiles) using base case parameters with user variation. Because the relative importance of valuative criteria differ among individual stakeholders and decision makers, results of the TEAM model can be displayed from various viewpoints to better help decision makers ultimately identify one or two preferred treatment and waste form alternatives. No one alternative is likely to satisfy all stakeholder concerns; therefore, alternatives are based on choices or trade-offs between competing values. This approach makes trade-offs clearly visible while allowing individual values to be applied to the interpretation of technical data. Identified trade-offs are discussed in greater detail as evaluation results are displayed. The TEAM model was also used to investigate key sensitivities to help prioritize development work needed prior to further evaluation and final selection.

Over the next few months, the model will continue to be used in facilitated meetings with key decision makers at DOE-ID and HQ to discuss potential changes to options and get immediate feedback on the results of those changes. Additionally, the model will be used by scientists and engineers to compare optimization scenarios of favorable options. This information will be useful in identifying the alternatives with the greatest potential for meeting the radioactive liquid and calcine waste-processing demands. These evaluations will continue to identify changes that may increase the favorability of a given option, resulting in the best informed decision for the ICPP waste management.

C. INVESTIGATION OF UNCERTAINTY

As mentioned above, the evaluation is complicated by uncertainties in many of the performance parameters (e.g., efficiencies and costs of various candidate processes) that could have a major impact on the technology evaluation and selection process. To identify the parameters for which further investigation will be most valuable, a sensitivity analysis was performed, using the TEAM model, to assess the range of parameter uncertainty and to determine if varying certain parameters had a significant impact on the ranking of options. First, the range of uncertainty was assessed for each parameter of interest, and the high and low ends of that range were defined. The model was then run setting the specified parameter at each end of the established range. If setting the parameter at either one end of the range or the other makes no significant difference in the choice of options, then further investigation of that particular parameter is not useful at this time, since it

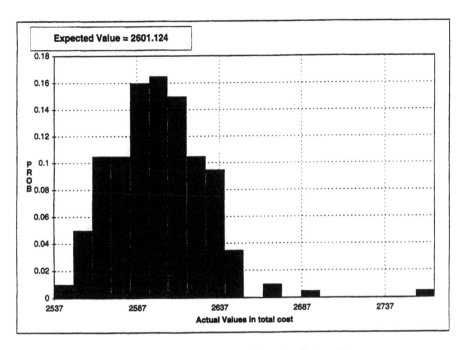

Figure 3 Distribution of total cost for Option 13.

would not affect the outcome of the overall evaluation. If, however, variations in parameter settings do affect the choice of options, then it may be important to investigate the parameter further.

The separation factors, densities, and weight loadings in the proposed waste immobilization facility (WIF) could vary from the estimates in the TEAM model even though estimates are based on the best data and calculations available. A subjective probability distribution was assigned to each separation factor, density, and weight loading in the model and a Monte Carlo simulation was run for Options 10, 13, 6, 4, 23, and 27. This type of simulation provides a probability distribution of the output (total cost), which gives some idea as to what the uncertainty of the output is, based on the uncertainty of the data. The distribution of the total cost for Option 13 is shown below in Figure 3.

The expected value of the total cost for Option 13 is $2.601 billion. The range of uncertainty, using 10th and 90th percentiles, is $68 million (from $2.567 billion to $2.635 billion). This represents a 2.6% uncertainty in the total cost due only to the uncertainties in separation factors, density, and weight loadings for Option 13. Option 10 has an uncertainty range of 3.1% due to uncertainties in the same three variables; the distribution of total cost for Option 10 was similar to Option 13. Several other options were run using the Monte Carlo simulation, yielding an uncertainty range of about 2 to 4% and similar bell-shaped output distributions. We can conclude that this small range in uncertainty will not effect the decision of one option over another where differences in total costs are more than $70 million. To determine which of three groups of variables stated above (i.e., separation factors, densities, and weight loadings) has the most significant impact on total cost, the total costs for all 27 options were calculated with the variables set at the 10 and 90 percentiles for their probability distribution. This resulted in an average change in total

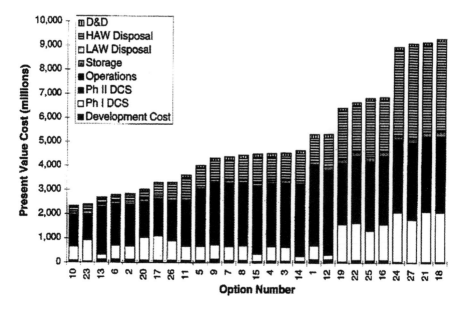

Figure 4 Life-cycle costs.

cost of 5.8% for separation factors only, 1.1% for densities only, and 15.4% for weight loadings only. These results show that weight loadings have a potential for high payoff if investigated further.

In the TEAM model, costs are compared using net present value. One uncertainty is the discount rate being used. The model uses a default discount rate of 2.8%, the current 30-year rate from the Office of Management and Budget. Two runs were completed, one with a discount rate of 0% and the other with a discount rate of 4%. With the discount rate varying from 0 to 4%, the top options remained unchanged from default settings when sorted by total cost. From this analysis it can be concluded that variations in the discount rate have no significant impact on the ranking of options.

Another uncertainty is future disposal costs. To determine if future disposal costs could affect the ranking of options, the model was run using a value of zero for all disposal costs. With future disposal costs set at zero, the top options remained unchanged when sorted by total cost. The ranking of options also remained unchanged when the disposal costs were doubled. From this analysis it can be concluded that variations in future disposal costs have no significant impact on the ranking of options (Figure 4). The sensitivity analysis at this point does not address uncertainties in cost estimates for the different technologies. The cost outputs from the TEAM model should be considered a relative cost analysis, because there is more confidence in the cost difference between options than the total absolute cost of a single option. More confidence in absolute costs can only be obtained through further development of both technology and cost data.

D. STAKEHOLDER INVOLVEMENT

Evaluation, recommendation, and selection of treatment and immobilization technologies for dispositioning ICPP radioactive liquid and calcine waste should include the consideration of stakeholder input to better identify and establish the relative importance of decision criteria. The basic premise is that technical experts can rank the performance of various alternatives against given parameters; however, the identification of all parameters important to stakeholders and the relative importance of each parameter is a value choice rather than a purely technical decision. For this reason, input (i.e., the identification of additional parameters and relative values) should be solicited from various stakeholders prior to making a final technology recommendation to the DOE.

Proposed stakeholder involvement activities to further support the technology evaluation and recommendation include (1) targeting special interest groups — DOE, the State of Idaho, local governments, Chambers of Commerce from local cities, INEL employees, Shoshone-Bannock Tribal members, the Oil Chemical and Atomic Workers (OCAW) labor union, environmentally concerned citizens, and the general public — via the Site-Specific Advisory Board (SSAB); (2) soliciting input on the adequacy of established criteria, including additions to the criteria if necessary; and (3) gaining stakeholder input on the relative importance of evaluation criteria as it applies to the decision-making process.

IV. EVALUATION DATA

Evaluations were performed for all 27 waste treatment options. This section provides an overview of optimization criteria, representative of diverse decision-maker concerns, and presents technology evaluation data as a function of those criteria; data for each of the 27 waste treatment alternatives are displayed graphically. Cost data are subdivided into the following life-cycle expenditure components: development costs; Phase I construction costs; Phase II construction costs; operations costs; interim storage costs; LAW disposal costs; HAW disposal costs; and D&D costs. These components, discounted to the start of FY-95, are added together to give total life-cycle expenditures for each waste treatment option.

Criteria discussed include the following:

- Cost
  - Life-cycle costs
  - Life-cycle cost minus HAW disposal costs
  - Development plus Phase I DCS (design, construction, and start-up) costs
  - Phase I DCS vs. total construction
  - Cost–time profile
    Peak annual cost
    Average annual cost
  - 5-year costs
- Waste volumes before immobilization
- Waste volumes after immobilization

Additional criteria not presented herein include tank farm capacity, time required to complete final immobilization, LAW radioactivity levels and resulting waste form classification, and technical maturity. Criteria were determined by the project technical staff and are based on past experience in the DOE and U.S. government process for selecting and funding new technology programs. Stakeholder input will be solicited to broaden these criteria and attempt to include concerns of the public into the decision making process.

As stated previously, the options were developed, evaluated, and compared in their most cost-effective configurations, according to the following default parameters:

- Phase I title design starting in 1998
- Phase II title design starting in 2005
- Phase I operating time of 7 years
- Phase II operating time of 30 years
- Calciner operations - two campaigns except Options 12 to 14, which assumed five
- High-level liquid waste evaporator operation assumed for all options
- New tank farm employed by 2000, if necessary for a given option

These conditions were selected primarily to allow for a consistent evaluation of all the options. Each parameter can be changed, however, to reflect possible variations in schedule and facility availability, to explore areas of technical uncertainty, and to optimize preferred waste treatment options. Further optimization will occur in the future on selected options.

The data presented in the following sections are relative values based on available information. The graphic representations of radioactive waste treatment options are consistent evaluations comparisons and should not be considered as the actual cost required to implement a given option.

A. LIFE-CYCLE COSTS
Life-cycle costs are measured in present value. Figure 4 shows the 27 processing options sorted according to the life-cycle costs, with the least expensive options on the left of the graph.

The life-cycle costs shown here provide the decision maker with an understanding of the overall relative cost for each option and for the range of life-cycle expenditures ($2.3 billion to greater than $9 billion) for all 27 options. The top third performers of this criterion are Options 10, 23, 13, 6, 2, 20, 17, 26, and 11.

B. PHASE I DCS COSTS VS. TOTAL DCS COSTS
Some options have a small initial investment and still yield a low total cost, such as Options 13 and 10. Another way to look at the trade-off between early investment and total cost is to compare the Phase I DCS costs to the total DCS costs of a given option. Figure 5 shows a scatter plot comparing Phase I DCS costs, on the vertical axis, and total DCS costs, on the horizontal axis.

The most favorable options are in the lower left corner of the graph. Option 10 has the lowest total DCS cost, but Option 13 has a lower Phase I DCS cost. Therefore, a trade-off exists between Options 10 and 13. Option 13 costs about $150 million more than Option 10 in total DCS cost, but costs nearly $350 million less in Phase I DCS cost. Option 10 builds the processing facility up front, thus causing higher Phase I DCS costs. Option 13 delays building the processing facility until Phase II and builds a tank farm during Phase I. The selection of optimum alternatives under this type of scenario will depend entirely on stakeholder and decision-maker values.

C. COST–TIME PROFILE
The timing of expenditures is an important aspect of budgeting for the DOE. If a process has large surges in funding (see Figure 6), priorities within the DOE complex must change to meet increased funding needs. Many DOE programs are short of funds; therefore, expenditures should be close to flat for the lifetime of a given process or the likelihood of success is diminished. Figure 6 shows a cost–time profile for Option 14 where the ICPP

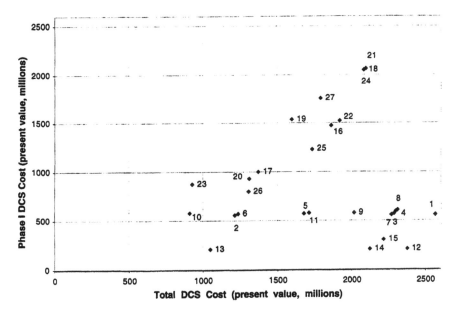

Figure 5   Phase I DCS vs. total DCS costs.

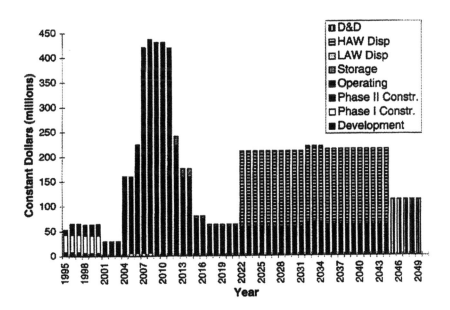

Figure 6   Cost-time profile for Option 14.

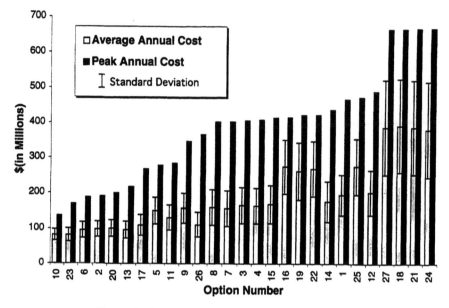

Figure 7 Comparative evaluation of cost-time profiles.

continues to calcine radioactive liquid waste, then builds and operates a glass-ceramic plant to immobilize the calcine. The profile displays the different expenditures that take place throughout the life cycle of an option. In this profile we see a small construction period in the first 5 years, as a new tank farm is built, and a later, much larger spike when a glass-ceramic plant is built. The most important information concerning a cost–time profile is the size of the peaks, the average annual cost of an option, and how much that average varies from year to year.

Figure 7 displays a comparative evaluation of all 27 options sorted by the funding spike on the cost–time profile. Also displayed is the average annual funding for each option along with the standard deviation on the average value.

The range of peak annual costs is between $137 million and $671 million. The annual average cost ranges between $82 and $392 million. The standard deviation is shown as an error bar on the average bar and varies by option as shown. In the criterion of lowest annual peak, Options 10, 23, 6, 2, 20, 13, 17, 5, and 11 are in the top third. In the criterion of lowest average annual cost, Options 23, 10, 6, 13, 2, 20, 17, 26, and 11 rank in the top third.

D. FIVE-YEAR COSTS

With the United States government budget tightening and requested budgets exceeding the amount of funding available, the next few years are critical to any implementation plan. By adding the first five bars on the cost–time profile, the present value of the 5-year costs can be estimated.

Figure 8 shows all 27 options with corresponding 5-year costs (y axis) as compared to corresponding life-cycle costs (x axis). This graph is displayed to show that it is possible to achieve low life-cycle costs and low 5-year costs. These options appear in the lower left corner of the graph. The 5 year costs range from $240 to $500 million. Options in the top third, when ranked by low 5 year costs, are Options 3, 10, 11, 4, 5, 7, 2, 8, and 6.

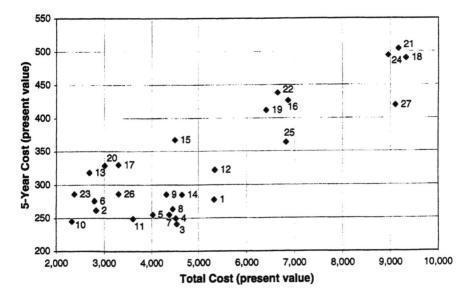

Figure 8   Five-year vs. life-cycle costs.

E. WASTE QUANTITIES BEFORE IMMOBILIZATION

The total volume of waste requiring immobilization is directly influenced by the volume of final waste form, size of the immobilization facilities, and ultimately cost. Some of the options increase the volume of waste requiring immobilization more than others. The radioactive liquid waste can be calcined with the use of nonradioactive chemicals or segregated directly, and both waste types can be segregated into HAW and LAW by radionuclide-partitioning technologies. These processes will add material requiring final immobilization. The waste quantities in Figure 9 display the dry weight equivalent (oxide form) with the water and other volatile components (such as nitric acid) removed; the data is sorted by quantity of HAW.

Under the criterion of minimize the quantity of HAW requiring long-term geological isolation, Options 17, 20, 23, 26, 6, 10, 2, 13, and 16 are in the top third. Concerning total waste quantities requiring immobilization (both HAW and LAW), the top third performers are Options 27, 25, 18, 16, 26, 21, 19, 17, and 5.

V. EVALUATION RESULTS

A. TREATMENT TECHNOLOGIES

Table 3 subdivides the 27 waste treatment options into three groups of nine, according to their performance per the identified criteria. Options are displayed in order of their performance to facilitate the identification of consistently top-performing options. By evaluating the alternatives against the criteria and splitting them into thirds, it becomes apparent which options do not require further development.

Using Table 3, option performance can be measured by observing the cumulative number of times an option appears in the top third. For purposes of evaluation, each time an option appeared in the top third for any of the eight criteria presented, it received one point. There is a distinct break between those options that performed favorably,

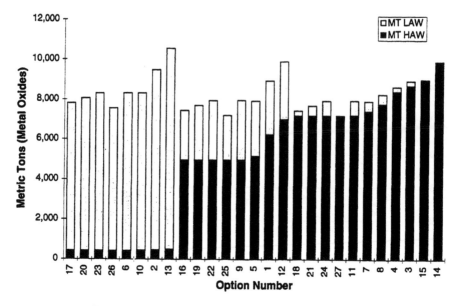

Figure 9   HAW and LAW quantities before immobilization.

Table 3   Summary of Criteria Performance

Criteria	Top third	Middle third	Bottom third
Life-cycle costs	10,23,13,6,2,20,17,26,11	5,9,7,8,15,4,3,14,1	12,19,22,25,16,24,27,21,18
Life-cycle costs minus HAW disposal costs	10,23,13,6,20,2,26,11,17	5,8,9,4,15,14,7,3,1	12,19,25,22,16,27,24,21,18
Development + Phase I construction	14,13,12,15,3,10,11,5,7	4,2,8,6,1,9,26,23,20	17,25,16,19,22,27,18,24,21
Cost–time profile peak annual cost	10,23,6,2,20,13,17,5,11	9,26,8,7,3,4,15,16,19	22,14,1,25,12,27,18,21,24
Cost–time profile average annual cost	23,10,6,13,2,20,17,26,11	5,7,9,8,4,3,15,14,1	12,19,22,16,25,24,27,21,18
Five-year costs	3,10,11,4,5,7,2,8,6	1,14,9,23,26,13,12,20,17	21,24,11,7,8,4,3,15,14
Minimal HAW quantity	17,20,23,26,6,10,2,13,16	1,14,9,23,26,13,12,20,17	24,27,11,7,8,4,3,15,14
Minimal total waste quantity	27,25,18,16,26,21,19,17,5	7,24,22,11,9,20,8,23,6	10,4,1,3,15,2,12,14,13

accumulating four points or more, and those that did not. Table 4 summarizes the accumulated points for each of the 27 waste treatment options. Option 10, with seven points accumulated, employs a phased approach using radionuclide-partitioning technology for both the radioactive liquid and calcine waste treatment phases.

Table 5 shows the results of a direct comparison of criteria of the top-performing options (those with six points or more) against Option 10 and gives either a "more favorable" or "less favorable" percentage for that option on each criterion; Option 10, having received the most points, was selected as the baseline for the direct comparison. Options showing more favorable performance for a given criteria, when compared to

Table 4 Summary of Accumulated Option Points

Option numbers	Points accumulated
None	8
10	7
2,6,11,13,17	6
20,23,26	5
5	4
None	3
3,7,25	2
4,8,12,14,15,16,18,19,21,27	1
1,9,22,24	0

Option 10, represent trade-offs that may require further analysis. Value assessments will be used to rank these trade-offs and to identify the most favorable option under the given circumstance.

There are only five cases where other options had more favorable performance indicators than Option 10. Four of these, HAW quantity before immobilization for Options 6 and 17 and total waste quantity before immobilization for Options 11 and 17, show a resulting difference of 6% or less. Because the percentage of difference is insignificant, these trade-offs are not investigated further. One criteria that is significantly more favorable is the development plus Phase I construction costs for Option 13. This option delays building any new processing facilities until Phase II and is displayed clearly by examining the cost–time profile peak costs, where Option 13 is 95% higher than Option 10. The resulting peak shows that the Phase II construction cost for Option 10 is not avoided, just delayed and compounded, resulting in the need for a tank farm, a 30% higher 5-year cost, NWCF operation for an additional five campaigns, the filling of a new calcined solids storage facility, and an increase in the size of the new processing facility. These factors combine to generate higher life-cycle costs and waste quantities for Option 13. However, this trade-off of delayed investment costs may be important if reducing costs for the next 13 years is more important to decision makers than 5-year cost reduction. Stakeholder value assessments will give a clearer understanding of the importance of this trade-off to various stakeholders. It is important to note that the technology utilized in Options 10 and 13 is exactly the same. Therefore, technology selection based on performance alone (not applying value sets) would result in the selection of radionuclide partitioning. The proper phasing of the technologies, however, will require additional interaction with the DOE and the State of Idaho.

Technology maturity is an important consideration even though a methodology for numerically assigning a value has not been agreed upon. The available technologies that can be employed during Phase I (processing the radioactive liquid waste in the 2008 through 2015 time period) is somewhat limited. For example, the only direct immobilization method available during the 2008 Phase I time period, without resolution of difficult technology issues, is continued calcination. One issue that must be evaluated prior to attempting operation of a glass plant directly on zirconium calcine wastes containing fluoride is corrosion control of the melter system. Due to the technology issues associated with direct vitrification of the calcine, several options in the bottom half of the decision tree in Figure 1 are unavailable for title design until 2001. On the other hand, radionuclide partitioning, grouting, FC, and PPT technologies are sufficiently technically mature to be considered for Phase I construction. The phased approach, as described in

Table 5 Direct Comparison of Options to Option 10

Criteria	Baseline: Option 10 (TRUEX, TRUEX)	Option 2 (FC, Truex)	Option 6 (PPT, TRUEX)	Option 11 (TRUEX Direct Immob.)	Option 13 (Calcination, TRUEX)	Option 17 (FC, TRUEX)
Total life-cycle costs ($ millions)	2,329	2,837 22% worse	2,801 20% worse	3,610 55% worse	2,692 22% worse	3,302 42% worse
Life-cycle minus HAW disposal costs ($ millions)	2,135	2,634 23% worse	2,611 22% worse	2,788 31% worse	2,478 23% worse	2,888 35% worse
Development plus Phase I construction costs ($ millions)	656	678 3% worse	704 7% worse	663 1% worse	329 50% better	1,085 65% worse
Cost-time profile peak annual cost ($ millions)	137	191 39% worse	187 36% worse	283 107% worse	217 58% worse	267 95% worse
Cost-time profile average cost ($ millions)	82	98 20% worse	96 17% worse	140 70% worse	96 17% worse	110 34% worse
Five-year cost ($ millions)	245	262 7% worse	276 13% worse	249 2% worse	318 30% worse	329 34% worse
HAW quantity before immob. (MT)	449	465 4% worse	434 3% better	7,235 1511% worse	497 11% worse	428 5% better
Total waste quantity before immob. (MT)	8,331	9,495 14% worse	8,329 Same	7,972 4% better	10,562 27% worse	7,972 4% better

the top half of the decision tree, could begin Phase I title design as early as 1998 for any of the specified options.

In light of these analyses, Option 10, which uses radionuclide partitioning via TRUEX as the processing technology, performed best in almost all the criteria. Additionally, the process is adequately mature to support title design in the 1998 time period and does not require a HAW immobilization facility in the first phase of operations.

B. IMMOBILIZATION TECHNOLOGIES

To help identify the differences between immobilization technologies, a comparative evaluation of the different methods available is shown in Table 6. All comparisons were performed using Option 10 — the most cost-effective method — as the baseline, shown in the first row.

In the first two rows of this evaluation, trade-offs can be observed between HAW volume and costs. Total HAW volume can be reduced by 77% of the glass HAW volume by selecting glass-ceramic. This selection would result in: (1) an increase in life-cycle costs of 28%, $656 million in 10/94 dollars, (2) an increase of the cost–time profile peak of 101%, and (3) an increase in the average annual expenditure of 20%. To consider glass-ceramic a viable technology, value assessments of stakeholders would need to show that this reduction in volume was worth the additional cost. Based on this technical evaluation and these trade-offs, glass appears to be the preferred HAW immobilization technology.

The bottom three rows in Table 6 show the alternative methods for LAW immobilization and their trade-offs. The first of these three rows compares the FUETAP grout method on the base case (top row) with a Portland-type grout method. The evaluation results show little benefit (two criteria at 1%) to the Portland-type grout method when compared to FUETAP; note the LAW 208% volume increase. Completing development of the FUETAP grouting technology in the near future will verify whether or not FUETAP is superior to Portland-type grout technology. Additionally, glass does not appear favorable when compared to FUETAP for LAW immobilization. LAW glass requires an increase in life-cycle cost of 95% ($2,204 million in present-value dollars), 81% more LAW volume than FUETAP, and results in 115% larger funding spikes on the cost–time profile. Glass-ceramic shows similar trade-offs as it did in the HAW immobilization discussion above. In the LAW case for glass-ceramic, the life-cycle cost difference was 57% ($1,331 million in present-value dollars), the cost–time profile peak is increased by 146%, and the average cost increases by 43%; however, the LAW volume decreases by 48%. Stakeholder value assessments will give the data required to determine whether or not volume reduction is worth the extra expenditures. It is believed that the stakeholder values will not place the minimizing LAW volume criterion so high in importance that the glass or glass-ceramic methodology would be favored over the FUETAP grout technology.

Additionally, processes that require a glass immobilization facility in Phase I will employ the COGEMA glass process. Both the stir-melter and glass-ceramic technology have technical issues that must be resolved prior to committing funds toward design and construction of these processes. The COGEMA process has low throughput that requires the size and cost of the facility to be substantially larger if the technology treatment option employs direct immobilization or precipitation of the calcine waste. Current HAW immobilization technology development favors phasing of the HAW immobilization process or removing the fluoride components prior to immobilization. If the waste is segregated such that the fluoride-containing components are placed in the LAW fraction, the same technological development issues remain that will require several years to

Table 6 Comparative Evaluation of Immobilization Technologies

Option description	Total cost ($ million)	5-year cost ($ million)	HAW volume (m³)	LAW volume (m³)	Peak cost ($ million/yr)	Average cost ($ million/yr)
Baseline: Option 10 glass-HAW and FUETAP grout-LAW	2,329	245	945	8,714	137	82
Option 10 with glass-ceramic-HAW and FUETAP grout-LAW	2,985 28% worse	249 2% worse	221 77% better	8,714 Same	276 101% worse	98 20% worse
Option 10 with glass-ceramic-HAW and Portland grout-LAW	2,385 2% worse	243 1% better	945 Same	26,814 208% worse	135 1% better	85 4% worse
Option 10 with glass-HAW and glass-LAW	4,533 95% worse	416 70% worse	945 Same	15,730 81% worse	294 115% worse	145 96% worse
Option 10 with glass-HAW and glass-ceramic-LAW	3,660 57% worse	465 90% worse	945 Same	4,493 48% better	337 146% better	106 143% worse

resolve. Thus, if glass or glass-ceramic were selected for use in Phase I LAW immobilization, a project delay would result until remaining technological issues were resolved. This would likely delay a Phase I project long enough to require a new tank farm.

Due to the performance factors measured, technology maturity levels, and a preferred phased approach with LAW immobilization in Phase I and a HAW immobilization in Phase II, FUETAP grout technology for LAW and glass (vitrification) for the HAW are the preferred technologies for the immobilization of radioactive liquid and calcine waste.

VI. SUMMARY

In this case study, a systems analysis approach has been used to develop criteria for evaluating alternative waste treatment options for radioactive liquid and calcine waste at the ICPP. This approach helps identify those options that meet the overall goals and requirements of a program through calculation and display of measurable quantities using a technology evaluation and analysis methodology (TEAM) model. In view of the complexities and interaction among treatment options for both wastes, the need to meet legal drivers, and stakeholder concerns, a systems analysis approach is not only valuable, but it is also essential for successful development, selection, and implementation of technologies to meet program goals.

The TEAM model provides a tool to jointly convolve stakeholder and decision-maker values related to life-cycle costs, short-term costs, final waste volumes, implementation schedule, etc. Once these values have been established, the options can be ranked to determine the most favorable treatment strategy. The TEAM model also provides a tool for evaluating the relative importance of model parameters (e.g., waste loadings in the immobilized waste). This in turn gives guidance to waste form research and development personnel.

The systems analysis approach follows the guidelines established by DOE for complex problem evaluation. The case study presented in this chapter is serving as a successful benchmark for evaluation of other projects being undertaken at the Idaho National Engineering Laboratory. The composition of the TEAM model will vary to fit each unique case; however, the systematic approach, evaluation methodology, and assessment of stakeholders and decision-makers values will be similar.

Section II

Treatment of Solid Wastes

This section covers the separation of radionuclides from solid wastes and residues. These wastes and residues can be divided into materials that have radionuclides either only on the surface, spread throughout the material in discreet phases, or intimately mixed throughout the material. Treatment techniques for each of these classes of wastes are discussed in the section.

Solid wastes produced during nuclear chemical processing, such as salts, oxides, or metals, are contaminated throughout with radionuclides. These wastes can either be dissolved in an aqueous system and the radionuclides separated as discussed in Section I or pyrochemical processes can be employed. Chapters 9 and 10 of this section discuss the pyrochemical processes that are being used or can be used to separate radionuclides from waste salts, metals, and oxides.

Process equipment and materials of construction are generally only contaminated on the surface. Techniques that selectively remove either the surface radionuclide or the base material are discussed in Chapter 11. Both aqueous and nonaqueous surface treatment are discussed in this chapter.

Wastes that contain the radionuclides in discreet phases may be processed using physical separation techniques. Physical separation techniques employ material properties such as density, magnetic susceptibility, or hydrophobicity. Techniques that use a physical property of a material to effect separations are discussed in Chapter 12.

Mark C. Bronson

Chapter 9

Pyrochemical Treatment of Metals and Oxides

Mark C. Bronson and Charles C. McPheeters

I. INTRODUCTION

Nuclear industry wastes have traditionally been treated by aqueous processing techniques to remove actinides or fission products. Aqueous processing systems have proven to be very effective at achieving high separation efficiencies but in general generate large amounts of low-level nuclear waste. For some wastes, pyrochemical processing techniques can achieve the desired separation while minimizing the generation of additional waste. Pyrochemical processing techniques for separating radionuclides from waste oxides and metals can be divided into the following three areas: metal decontamination, oxide residue treatment, and spent nuclear fuel treatment. Each of these areas are discussed in this chapter.

II. METAL DECONTAMINATION

There are two types of radionuclide contamination of metal. In the first type, the radionuclides are only on the surface. The separation of radionuclides from metal surfaces will be covered in Chapter 11. In the second type, the radionuclides are subsurface in pores, cracks, along grain boundaries, or mixed throughout the metal matrix. Two pyrochemical technologies that have been used to remove subsurface contamination from metals are melt slagging and carbonyl refining. Both of these technologies are discussed in this section.

A. MELT SLAGGING

Melt slagging or melt refining separates radionuclides from more noble metals, e.g., copper, nickel, or steel, by selectively oxidizing the radionuclides into a flux or slag. The oxides and halides of the actinides and fission products are thermodynamically more stable than most base metal oxides and halides. Therefore, the actinides and fission products can be selectively oxidized and fluxed away from most molten metals. Reactive metals such as zirconium and aluminum have similar thermodynamic stabilities as the actinides and fission products, and, therefore, less efficient separations are achieved using this technology.

The melt-slagging operation is relatively simple. The contaminated metal is placed into a crucible with fluxing agents and melted. The radionuclides react with fluxing agents or injected gases to form compounds that are soluble in the flux. The radionuclides report to the slag phase and are separated from the molten metal. Typical fluxes are mixtures of metal oxides and silicates similar to those used to form slags in the production of primary metals. However, fluxes of carbonates, halides, and other compounds have also been examined or proposed.

A review of the available literature on melt slagging for removal of uranium from common metals was performed by E. W. Mautz et al. of National Lead in 1975.[1] Results

Table 1 National Lead Company — Review of
Decontaminating Uranium-Contaminated Scrap Metals

Material	Product ingot U content (ppm)			Feed U content (ppm)
	Av.	Low	High	
Common steel	0.4	0.00	3.5	Up to 2000
Stainless steel	0.6	0.01	3.2	~71
Ni-bearing steel	0.5	0.02	2.4	~77
Nickel	1.3	0.9	1.6	Up to 6050
Monel (Ni alloy)	0.5	0.01	4.0	—
Copper	0.4	0.01	2.5	Up to 3360
Brass	0.4	0.01	2.5	—
Yellow brass	2.1	0.30	3.2	—
Bronze	0.3	0.04	1.2	—
Aluminum	200	3	1400	Up to 6500

of this review, provided in Table 1, show that melt slagging can provide a high degree of uranium separation from copper and various steels. Some removal from aluminum can be achieved, but the aluminum product may still contain significant uranium levels (1400 ppm). The tests were performed on a range of sample sizes, from laboratory-scale to large-scale melts of several tons each.

Laboratory-scale investigations at Argonne National Laboratory,[2,3] Oak Ridge National Laboratory,[4,5] and Hitachi Energy Research Laboratory[6] have shown a high degree of decontamination of uranium, plutonium, and americium from contaminated metals. Generally, actinide concentrations around 1 ppm can be obtained using melt slagging technology for iron alloys, copper, nickel, lead, and zinc. Much higher levels are obtained for aluminum and aluminum alloys.

At Siemplekamp Giesserei in Krefeld, Germany, over 3500 Mg of contaminated metal scrap from nuclear plants has been treated by melt slagging.[7,8] Tests have resulted in over 98% of the radioactivity removed from copper wire on average and between 90 and 95% removed from brass. In one test on copper that used a new crucible to keep the copper clean, activity levels in the copper after melting were below the limit of detection.

Although melt slagging has been shown to be very effective at separating radioactive materials from certain metals, the ultimate use and disposition of the "cleaned" metals is still in question. In the United States there is currently a nondiminumus rule, which can be stated as follows: solid materials that have been in contact with radionuclides are always considered contaminated even if contamination cannot be detected. Therefore, metals that have been decontaminated are limited to use within nuclear facilities as waste storage containers or shielding blocks.

B. CARBONYL REFINING

Carbonyl refining is based on the formation of gaseous metal carbonyls. Nickel, iron, and cobalt form carbonyls. Iron pentacarbonyl, $Fe(CO)_5$, is volatile but the rate of reaction is slow. Cobalt can form tetracarbonyl, $Co_2(CO)_8$, and tricarbonyl, $Co_4(CO)_{12}$, but both of these compounds have low volatility. Nickel has been purified commercially using carbonyl technology since 1902.[9] Nickel carbonyl is formed by reacting nickel metal and carbon monoxide gas at atmospheric pressure and a temperature of 50°C. The equation of this reaction is as follows:

$$Ni + 4CO(g) \Rightarrow Ni(CO)_4(g) \qquad (1)$$

Impurities in the nickel and radionuclides are left behind in the solid phase and the nickel carbonyl is transported through the gaseous phase, resulting in a separation. Reaction 1 is reversible with increased temperature. The nickel carbonyl molecule decomposes to nickel metal and carbon monoxide at temperatures in the range of 150 to 300°C.

There are large quantities of nickel metal used in gaseous diffusion and gas centrifuge plants to isotopically enrich uranium hexafluoride. This nickel is contaminated with uranium compounds, and carbonyl technology is a proposed method to separate the nickel from the uranium compounds.

III. OXIDE RESIDUE TREATMENT

Separation of radionuclides from oxide residues has been almost exclusively performed using acid dissolution and aqueous separations technology. With increased emphasis on waste and residue reduction, pyrochemical technologies have been examined for removing plutonium from incinerator ash[10] and separating nonradioactive constituents from waste calcine.[11]

A. SEPARATING PU FROM INCINERATOR ASH

An examination of pyrochemically separating plutonium from intractable transuranic (TRU) residues stored at DOE facilities was performed at the Lawrence Livermore National Laboratory and the Argonne National Laboratory.[12] The residues were considered intractable because actinide separation using traditional aqueous process technologies was problematic. A residue that was examined in demonstration tests was incinerator ash and leached incinerator ash, known as ash heel.

The proposed pyrochemical process consisted of the following operations and is shown schematically in Figure 1. First, the feed material is calcined to remove carbonaceous material and moisture. Then the oxide residue is reduced using calcium as the reductant and $CaCl_2$–CaF_2 as the solvent salt in order to solvate the product CaO. A zinc–calcium alloy is added to solvate the product metals in order to keep the metal phase molten. The waste salt from the direct oxide reduction is recycled by using molten-salt electrolysis to recover the calcium metal and release the oxygen as CO. The plutonium is removed from the zinc alloy by electrorefining. During the electrorefining, the americium reports to the electrorefining salt. This salt is periodically removed and the americium is scrubbed out by calcium reduction. The depleted zinc alloy (containing tramp elements) is scrubbed to oxidize any remaining plutonium and then retorted to remove zinc, which is recycled.

Most the of the unit operations were demonstrated on a small scale. Demonstration of the reduction operation resulted in up to 99.5% of the plutonium extracted into the calcium–zinc alloy. However, demonstration of the complete process was not performed.

B. SEPARATING WASTE CALCINE

The feasibility of several pyrochemical approaches for separating nonradioactive constituents from Idaho Chemical Processing Plant (ICPP) high-level waste (HLW) calcine were examined at the Lawrence Livermore National Laboratory and the Idaho National Engineering Laboratory.[13] The calcine contains less than 1% fission products and actinides

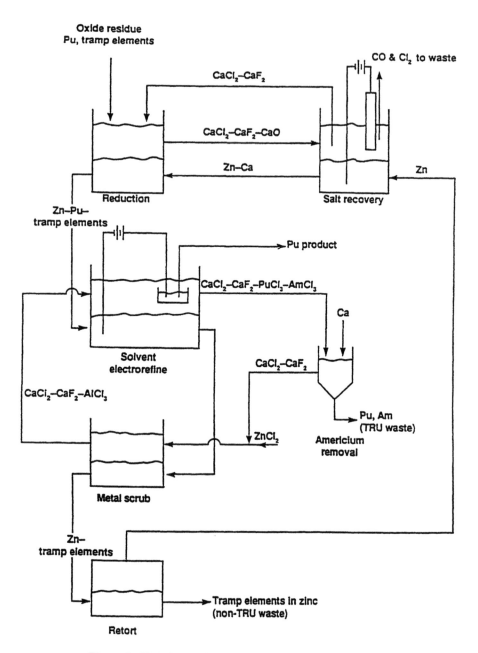

Figure 1 Plutonium residue recovery for intractable residues.

(elements that make the calcine HLW). Separation of the fission products and actinides into a HLW and the inert materials into a low-level waste (LLW) may significantly decrease disposal costs. Currently LLW disposal costs are orders of magnitude less than estimated HLW disposal costs. Both aqueous and pyrochemical processing methods have

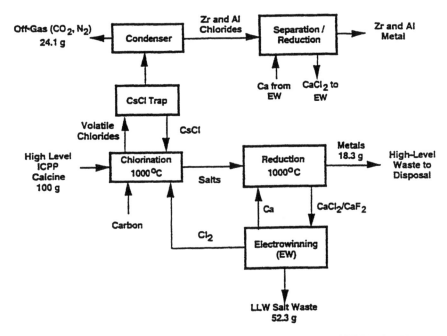

Figure 2 Baseline pyrochemical flow sheet for treating zirconia ICPP HLW calcine.

been proposed to remove the HLW components from the calcine. In proposed aqueous processes, actinides and fission products are separated from the nonradioactive constituents by dissolution followed by hydroxide precipitation, ion exchange, or solvent extraction. In the proposed pyrochemical processes, the actinides and fission products are separated from the nonradioactive constituents by chloride volatilization and molten-salt reduction.

The removal of aluminum by carbochlorination of the Al_2O_3 to form gaseous $AlCl_3$ was successfully demonstrated with alumina-based calcine. The examination of pyrochemical processes to remove nonradioactive constituents, primarily aluminum, zirconium, calcium, and fluorine, from zirconia calcine showed that fluorine removal was required in order to chlorinate the calcine. Fluorine removal from simulated ICPP zirconia calcine was demonstrated by both a $CaCl_2$ wash method and a sulfuric acid treatment method. The $CaCl_2$ wash method uses lower-melting $CaCl_2$ to solvate the high-melting CaF_2. The sulfuric acid treatment method reacts the CaF_2 with H_2SO_4, forming HF and $CaSO_4$. X-ray fluorescence analysis indicated that greater than 97% of the fluorine can be removed by either method. Regeneration of $CaCl_2$ from the salt and oxide residues of the $CaCl_2$ wash experiments was also demonstrated by an aqueous dissolution method. From the salt residue, 85% or more of the $CaCl_2$ was recovered and contained little, if any, CaF_2 or CaO. From the oxide residue, close to 100% of the $CaCl_2$ was recovered. A proposed flow sheet is shown in Figure 2.

IV. SPENT NUCLEAR FUEL TREATMENT

Recovery of the actinide elements, particularly the transuranic (TRU) elements, from spent nuclear reactor fuels offers several benefits for future energy resource conservation and waste management. The TRU elements (primarily neptunium, plutonium, americium, and curium) are toxic, radioactive materials that constitute the primary long-term hazards in the spent fuels. Disposal of spent fuels as HLW in a geological repository presents unique challenges for waste qualification. The Environmental Protection Agency (EPA) has established limits on the quantities of TRU elements and fission products that can be released to the environment over the first 10,000 years of isolation in the repository.[14] If the concentration of TRU elements can be reduced by a factor of 1000 (99.9% removal), then essentially all of the residual TRU elements can be released over the 10,000 year period, and the waste form will still meet the EPA limits. Obviously, allowing essentially complete release is much more desirable than having to prove substantially complete containment.

The potential benefits in developing the technology for recovery of TRU elements from spent oxide reactor fuels and their recycle to a metal-fueled reactor, such as the Integral Fast Reactor (IFR),[15] include facilitating disposal of spent oxide fuels in compliance with EPA regulations, preserving the TRU elements in these fuels as an energy resource for the future, and reducing the long-term heat burden in the geological repository (due to radiation from the TRU elements). The pyrochemical processes that have been developed for recovery of TRU elements from spent oxide fuels are discussed in this section.

A. EBR-II PYROCHEMICAL PROCESSING

The Experimental Breeder Reactor Number Two (EBR-II) was designed as a breeder reactor, and it was operated during the 1960s in conjunction with a fuel cycle facility to demonstrate the breeding concept.[16] During that demonstration, approximately 35,000 fuel pins were processed, and the fuel was recycled to the EBR-II core for irradiation. The pyrochemical fuel cycle used in that demonstration is illustrated schematically in Figure 3.[17] The metal fuel was melted and equilibrated in a zirconia crucible. The volatile fission products were released from the fuel and either retained in a fume trap or released. The rare-earth elements, as well as barium, strontium, and tellurium, reacted with the crucible and the noble metal fission products were carried along with the fuel to the reactor to be recycled. The process was later improved to separate the noble metal fission products from the fuel and to recover the 5 to 10% of the fuel that was trapped in the crucible "skulls". Since the fuel in the skulls was either oxides or a mixture of metals and complex oxides, it was necessary to develop a process to reduce the skull oxides to metals and to separate the uranium and TRU elements from the fission products.

The process thus developed is shown as a schematic flow sheet in Figure 4. The crucible skulls were first oxidized completely to allow removal of the heavy-metal components from the crucibles. Some of the noble metal fission products associated with the skulls did not oxidize, but remained as metal particles mixed with the oxides. These noble metal fission products were extracted by liquid zinc in a liquid-metal/molten-salt system. Some of the weaker oxides were reduced to metals by the zinc and dissolved in the zinc. The heavy-metal oxides remained in the molten-salt phase as the zinc and salt phases were separated. The salt phase was placed in contact with a Zn–Mg alloy, and the magnesium reduced the oxides to metal, producing a solid MgO precipitate in the salt phase. The uranium and TRU elements were dissolved in the Zn–Mg alloy until the uranium became saturated (about 0.2 wt%) at which point it precipitated as a separate phase. At the time this process was developed and operated, no attempt was made to recover the minor actinides (Np, Am, and Cm), which were considered a waste product.

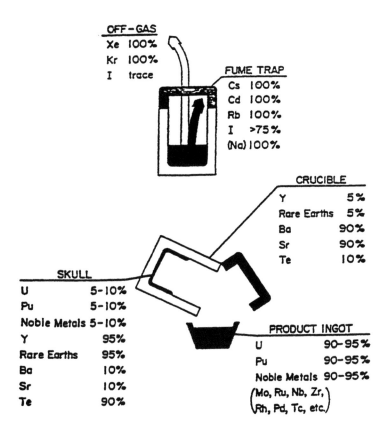

Figure 3 Distribution of elements in the EBR-II fuel cycle process.

The Zn–Mg alloy and molten salt were transferred away from the uranium precipitate and discarded as waste. To recover the uranium precipitate, a high-zinc Zn–Mg alloy was added to dissolve the uranium, and the alloy was transported to a retort crucible. The Zn–Mg alloy was evaporated, leaving the uranium behind as a consolidated ingot.

B. PYROCHEMICAL PROCESSING OF LWR FUEL

The requirements of the pyrochemical process for recovery of TRU elements from oxide spent fuel are as follows: (1) reduce the TRU element oxides to metals, (2) separate the TRU elements from the bulk uranium fuel, (3) recover at least 99.9% of the TRU element mass for feeding into the IFR fuel cycle, and (4) reduce the waste volume as much as possible below the volume required for direct disposal of the oxide spent fuel. In addition, the process must be economical and have a throughput of about 2000 MT/year.

The processes that were developed in the 1960s and 1970s[16-18] for recycle of the EBR-II fuel and recovery of fuel values from the crucible skulls were adapted to meet the requirements for recovery of TRU elements from spent oxide reactor fuels. The generalized concept for this process is shown as a schematic flow sheet in Figure 5. The process steps are briefly described below.

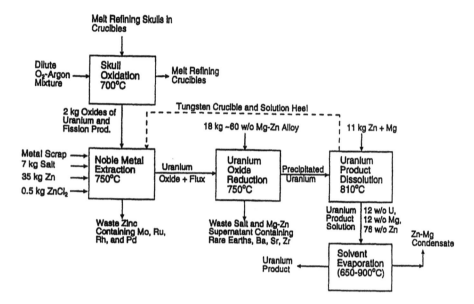

Figure 4 Simplified liquid-metal process for reclamation of fuel from melt refining skulls.

1. Disassembly and Decladding

The first step is dismantling the fuel assemblies and separating the cladding (Zircaloy or stainless steel, depending on the reactor type) from the oxide fuel. An alternative is to chop the fuel assemblies into small pieces and feed the chopped material into the process. This approach is feasible because the quantities of Zircaloy cladding and stainless steel assembly hardware are 235 and 57 kg per metric ton of initial heavy metal (kg/MTIHM), respectively. These quantities of zirconium and iron could be handled in the process and are expected to follow the noble metal fission product waste stream. Most of the fission gases would be released in the decladding step, and the remainder in the oxide reduction step discussed below.

2. Oxide Reduction

The second step is reduction of the oxides to metals. Depending on the concept being considered, this reduction is accomplished at elevated temperature by reaction of the oxide with either calcium or lithium in a molten-salt solution. The reduction reactions are described by the equation

$$MO_2 + 2xR \Rightarrow 2R_xO + M \tag{2}$$

where M represents the actinide elements, R represents the calcium or lithium metal reductant, and x is one for the calcium case and two for the lithium case. All of the reductant metal oxide and some of the reductant metal must be in solution in the molten-salt phase so that the reduction reaction goes to completion and rapid reaction kinetics is assured. Therefore, enough salt must be used to ensure that these concentrations remain below saturation. The reaction is thought to be heterogeneous at the liquid-salt/solid-oxide interface; therefore, vigorous agitation is required to enhance the reaction rate. In

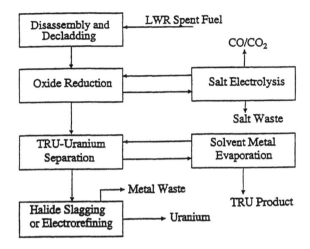

Figure 5 Generalized flow sheet for recovery of actinides from LWR spent fuel.

the calcium case, calcium is added in a Cu–Mg–Ca alloy, and the uranium, TRU metals, rare-earth metals, and noble metal fission products enter into solution in this alloy during reduction. Because most of the mass of the spent fuel is uranium, the Cu–Mg alloy quickly becomes saturated with uranium, and the uranium metal precipitates as a separate solid phase. In the lithium case, the lithium is added as the pure metal, and all of the reduced metallic fuel components are precipitated as solid particles.

The spent fuel fission products are distributed between the salt and metal precipitate in the reduction step, according to the relative stabilities of their oxides and chlorides. Some of the fission products are present as metal particles or weak oxides dispersed in the spent fuel. This so-called noble metal group includes all of the noble metal and transition metal fission products whose oxides and chlorides are less stable than the reductant-metal oxides and chlorides. They are easily reduced by lithium or calcium and precipitate with the actinide elements in the reduction step. Some of the fission products form stable chlorides that are soluble in the salt phase. This group of fission products includes cesium, strontium, barium, tellurium, and possibly europium and samarium. This group also includes fission product iodine that will likely form alkali-metal iodide. The rare-earth oxides are reduced by calcium. They follow the actinide metals but are not reduced by lithium. Instead, they remain in the salt phase as oxides, chlorides, or oxychlorides. At the end of the reduction step, the salt phase is transferred to the electrolysis step, and the metal phase is transferred to the uranium–TRU separation step.

3. Salt Electrolysis

The salt electrolysis step involves electrochemical reduction to regenerate the reductant metal and to allow recycle of the salt. The oxygen is evolved in the form of CO and CO_2 gases. To minimize waste volume, as much of the salt is recycled as possible. However, a small portion of the salt (~1%) must be discarded as waste to control the level of the heat-producing fission products, primarily cesium and strontium. The electrochemical reactions that occur at the cathode are given below.

$$Ca^{2+} + 2e^- \Rightarrow Ca° \qquad (3)$$

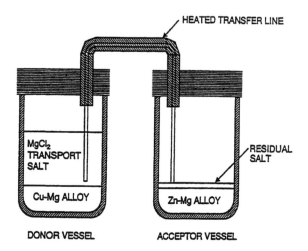

Figure 6 Salt transport conceptual arrangement with salt in donor vessel.

or

$$2Li^+ + 2e^- \Rightarrow 2Li° \tag{4}$$

The electrochemical reaction that occurs at the anode is given below.

$$O^{2-} + C \Rightarrow CO + 2e^- \tag{5}$$

In the calcium system, the cathode is liquid Cu–Mg alloy in which the calcium is soluble. As the calcium metal is produced at the cathode it enters into solution. This Cu–Mg–Ca solution then becomes the reductant alloy for the next reduction step. In the lithium system, the cathode is a steel rod, on which lithium metal collects and rises to the top of the molten salt. The lithium pool then becomes the active cathode for the remainder of the electrolysis. In both cases, the anode is graphite, and the anodic reaction is assisted by reaction of the product oxygen with the graphite to form carbon monoxide and/or carbon dioxide, which can be filtered and vented to the atmosphere. It is well known that graphite anodes form carbon dust during electrolysis in molten salts. This dust can be controlled by use of a porous ceramic separator between the anode and cathode.[19] It is important that this carbon is not recycled with the molten salt because it could react with the actinide elements to form very stable carbides that would be difficult to recover. Alternative electrodes, such as inert, oxygen-evolving anodes may have some advantages in avoiding this potential carbon problem.

4. TRU–Uranium Separation

The differences among various process options lie primarily in the methods used to separate the TRU metals from the uranium metal. The separation method used in the calcium system is a salt extraction and transport process, illustrated in Figure 6. The Cu–Mg alloy that contains the TRU elements in solution (donor) is equilibrated with molten $MgCl_2$ salt. Because the activity coefficients of the TRU elements in the Cu–Mg alloy are fairly large, their mass distribution favors the salt phase at equilibrium. The TRU-laden salt is transferred through a heated pipe to a crucible containing a Zn–Mg alloy (acceptor). The activity coefficients of the TRU elements in the Zn–Mg alloy are low, so their mass distribution favors the Zn–Mg alloy at equilibrium. Therefore, the TRU elements are stripped out of the $MgCl_2$ salt phase into the Zn–Mg alloy. With repeated cycles of equilibration and salt transfers between the donor and acceptor, essentially all of the TRU content is transported to the Zn–Mg alloy. A small quantity of uranium and essentially all of the rare-earth fission products are transported along with the TRU elements by the same extraction and transport mechanism. The stripped Cu–Mg alloy is recycled to the electrolysis step, and the uranium precipitate is transferred to the final cleanup step. The Zn–Mg alloy, along with its TRU burden, is transferred to the solvent-metal evaporation step.

5. Solvent-Metal Evaporation
The TRU product is consolidated by evaporation of the volatile solvent Zn–Mg alloy and melting of the TRU product ingot. This product contains about equal masses of the TRU elements (Pu, Np, Am, and Cm), uranium, and rare-earth fission products. The TRU product is suitable as feed material for the IFR electrorefiner, where it is further purified and used in the fabrication of new fuel elements.

6. Uranium Purification
Final cleanup of the uranium is necessary to achieve at least 99.9% recovery of the TRU element mass. Experiments have shown that about 15 to 20 % of the neptunium coprecipitates with uranium from solution in the Cu–Mg alloy during the reduction reaction. Two methods have been found for recovering the neptunium: halide slagging and electrorefining. Halide slagging is a chlorination step that is performed during uranium melting and casting. It results in molten salt slag, such as calcium chloride containing a chlorinating agent such as uranium trichloride. As this salt slag equilibrates with the molten uranium, the neptunium (and other residual TRU elements that might be present) is preferentially chlorinated and extracted into the salt phase (the TRU trichlorides are more stable than uranium trichloride). The melt is then cooled, and the frozen salt is separated from the uranium ingot and returned to the reduction step to recover the TRU elements.

In the case of lithium reduction, the uranium–TRU separation is accomplished by electrorefining.[20] The metal precipitate is placed in an electrorefiner where the uranium is transported to a solid steel cathode as a pure product. The TRU elements are anodically dissolved along with the uranium; however, they cannot be deposited on the solid cathode under ordinary operating conditions. The TRU trichlorides are more stable that uranium trichloride, so when uranium trichloride is present in the salt phase, it will react with any TRU metals that might deposit on the cathode, producing uranium metal and TRU trichloride in the salt. Thus, the TRU trichlorides accumulate in the electrolyte salt. As pure uranium is deposited and removed from the electrorefiner in sequential batches, the TRU content of the electrolyte builds up to appreciable levels. When the TRU:U ratio reaches values greater than two, the TRU may be extracted with a liquid cadmium cathode. The cadmium is distilled and the TRU metals are melted, resulting in the desired TRU product for feeding to the IFR fuel cycle.

7. Waste Treatment
One of the requirements of the processes for recovery of TRU elements from spent oxide fuels is minimization of radioactive waste volumes. Four types of wastes are generated by these processes: (1) mineral wastes produced from the spent salts, (2) metal wastes that

include fuel assembly hardware, cladding, and noble metal fission products, (3) fission gases such as krypton, xenon, and tritium, and (4) process wastes such as broken equipment and miscellaneous hardware.

The spent salt removed from the reduction/electrolysis cycle will be ion-exchanged and occluded on zeolite.[21] This salt-occluded zeolite will be consolidated into a leach-resistant waste form by one of two methods: (1) glass bonded to form a monolithic waste form or (2) thermal conversion of the zeolite structure to sodalite, a naturally occurring mineral that contains chlorides.

The metal wastes will be consolidated by melting the stainless steel or Zircaloy cladding to form a matrix material. The noble metal and transition metal fission products will be incorporated into the matrix along with any discarded steel-processing equipment. Some of the process waste, such as broken equipment or miscellaneous wastes, may be LLW and will be disposed of separately.

The fission gases, such as krypton, xenon, and tritium, will be released to the inert gas (argon) atmosphere of the process cell. The gas purification system will remove and concentrate the tritium in a small quantity of water. A cryogenic distillation system, downstream of the purification system, will separate and concentrate the krypton and xenon and compress them for long-term storage in tanks for decay and eventual release to the environment.

8. Development Status

All of the process steps have been shown to be chemically feasible, for both the calcium system and the lithium system, by means of small-scale laboratory experiments.[22] The lithium reduction step has been demonstrated in an engineering-scale system at a batch size of 4.4 kg, and equipment is under construction that will allow completing the demonstration at batch sizes up to 20 kg. The electrorefining process has been operated at a 10-kg scale for over 6 years, and it is a proven technology.

REFERENCES

1. **Mautz, E. W., Briggs, G. G., Shaw, W. E., and Cavendish, J. H.,** "Uranium Decontamination of Common Metals By Smelting — A Review," National Lead Company of Ohio, NLCO-1113 Special, Cincinnati, OH, February 5, 1975.
2. **Gerding, T. J., Seitz, M. G., and Steinder, M. J.,** "Salvage of Plutonium and Americium Contaminated Metals," Paper 1096, 70th AIChE Annual Meeting, Session 109, New York, November 13–17, 1977.
3. **Seitz, M. G., Gerding, T. J., and Steinder, M. J.,** "Decontamination of Metals Containing Plutonium and Americium," Argonne National Laboratory, ANL-78-13, Argonne, IL, June 1979.
4. **Heshmatpour, B. and Copeland, G. L.,** "The Effects of Slag Composition and Process Variables on Decontamination of Metallic Wastes by Melt Refining," Oak Ridge National Laboratory, ORNL/TM-7501, Oak Ridge, TN, January 1981.
5. **Copeland, G. L., Heshmatpour, B., and Heestand, R. L.,** "Melting Metal Waste for Volume Reduction and Decontamination," Oak Ridge National Laboratory, ORNL/TM-7625, Oak Ridge, TN, March 1981.
6. **Uda Tatsuhiko, Iba Hajime, and Tsuchiya Hiroyuki,** "Decontamination of Uranium Contaminated Mild Steel by Melt Refining," *Nucl. Technol.*, Vol. 73, 109–115, 1986.
7. **Sappok, M.,** "Recycling of Metallic Material form the Dismantling of Nuclear Plants," *Kerntechnik*, (6), 376–378, 1991.
8. **Hebrant, P.,** "Leading Piping a New Lease on Life," *Nucl. Eng. Int.*, January 1989, pp. 48–49.

9. **Queneau, P., Ed.,** *The Winning of Nickel,* D. Van Nostrand Company, Inc., Princeton, NJ, 1966, 374.

10. **Pittenger, L. C., Alire, R. M., Coops, M. S., Landrum, J. H., Priest, R. E., Thompson, D. S., Gregg, D. W., Burris, L., Pierce, R. D., Morrissey, L. J., and Mulcahey, T. P.,** "Plutonium Residue Recovery (PuRR) Project Quarterly Progress Report April–June 1988," Report UCID-21542, University of California, Lawrence Livermore National Laboratory, Livermore, CA, October 24, 1988.

11. **Riley, D. and Bronson, M.,** "Continuous Oxidation/Reduction System (CORS)", *Actinide Processing Methods and Materials,* The Minerals, Metals, and Materials Society, Warrendale, PA, 1994, 217–224.

12. **Gregg, D. W., Hickman, R. G., Landrum, J. H., Johnson, G. K., Johnson, I., Mulcahey, T. P., Pierce, R. D., and Poa, D. S.,** "Plutonium Residue Recovery (PuRR) Project Quarterly Progress Report April–June 1989," Report UCID-21542-89-2, University of California, Lawrence Livermore National Laboratory, Livermore, CA, October 20, 1989.

13. **Bronson, M. C., Ebbinghaus, B. B., and Riley, D. C.,** "Pyrochemical Processing of Idaho Chemical Processing Plant High Level Waste Calcine," *Light Metals 1995,* The Minerals, Metals, and Materials Society, Warrendale, PA, 1995.

14. Title 40 Code of Federal Regulations, Part 191, EPA Regulations for Disposal of High-Level Nuclear Waste, Appendix A.

15. **Chang, Y. I. and Till, C. E.,** "Actinide Recycle Potential in the Integral Fast Reactor (IFR) Fuel Cycle," Special Publication, LMR: A Decade of LMR Progress, Winter ANS Meeting, Washington, D.C., November 11–16, 1990.

16. **Pierce, R. D. and Burris, L., Jr.,** "Pyroprocessing of Reactor Fuel," Argonne National Laboratory, Reactor Technology, Selected Reviews, TID 8540, 1964.

17. **Lawroski, S. and Burris, L., Jr.,** "Processing of Reactor Fuel Materials by Pyrometallurgical Methods," Atomic Energy Review, Vol. 2, No. 3, IAEA, Vienna, Austria, 1964.

18. **Steunenberg, R. K., Pierce, R. D., and Burris, L., Jr.,** "Pyrometallurgical and Pyrochemical Fuel Processing Methods," Progress in Nuclear Energy Series III Process Chemistry, Vol. 4, Pergamon Press, Oxford and New York, 1969.

19. **Poa, D. S., Pierce, R. D., Mulcahey, T. P., and Johnson, G. K.,** "Electrowinning Process with Electrode Compartment to Avoid Contamination of Electrolyte," U.S. Patent No. 5,225,051, July 1993.

20. **Pierce, R. D., Johnson, T. R., McPheeters, C. C., and Laidler, J. J.,** "Progress in the Pyrochemical Processing of Spent Nuclear Fuels," *J. Metals,* 45 (2), 42–44, February 1993.

21. **Lewis, M. A., Laidler, J. J., and Fischer, D. F.,** "Immobilization of High-Level Chloride Salt Waste," Argonne National Laboratory, Unpublished work 1993.

22. **McPheeters, C. C., Pierce, R. D., Poa, D. S., and Maiya, P. S.,** "Pyrochemical Methods for Actinide Recovery from LWR Spent Fuel," Proceedings of the International Conference on Future Nuclear Systems: Emerging Fuel Cycles and Waste Disposal Options, Global '93, Seattle, WA, Vol. 2, pp. 1094–1101, September 12–17, 1993.

Chapter 10

Pyrochemical Treatment of Salts

Karen E. Dodson

I. INTRODUCTION

Waste and residue salts are generated in the production of uranium and plutonium metal and in plutonium metal purification. Magnesium fluoride salts that contain up to 5% of the feed uranium are produced during the production of depleted uranium metal. Currently these salts are disposed of directly as low-level waste (LLW). In the production of highly enriched uranium or plutonium metal, calcium fluoride salts are produced. These salts have relatively high melting points, and, therefore, are treated by aqueous means to separate the actinides from the waste salts. The plutonium pyrochemical processes, direct oxide reduction (DOR), molten-salt extraction (MSE), and electrorefining (ER), use lower-melting chloride-based salts. Waste and residue salts from these processes can be treated by either pyrochemical or aqueous technologies to separate the plutonium from the waste salt. This chapter covers the pyrochemical processing of plutonium residue salts.

DOR is used for converting plutonium oxide to metal, MSE for extracting the radioactive decay product of americium from plutonium metal, and ER is used for final purification of plutonium metal. Unlike aqueous processes, these pyrochemical operations are conducted in furnaces at temperatures high enough to achieve a molten mix of plutonium metal, and chloride salts that are used as a solvent, reactant, or electrolyte. Upon completion of the pyrochemical processes, the products are cooled, and the bulk metal and salt freeze out as two distinct phases because of the large difference in their densities; however, small quantities of residual metal will be dispersed through the salt cake. The plutonium metal is separated from the solid residue salt, and the residue salt must be further processed to meet strict waste acceptance criteria (WAC) at the Waste Isolation Pilot Plant (WIPP) in Carlsbad, New Mexico before it can be discarded.

Presently, no more than 200 g-eq. fissle [239]Pu per 55-gal waste disposal drum is allowed per the WAC. A typical MSE or ER salt cake can contain hundreds of grams of plutonium; therefore, several 55-gal drums would be required for disposal of this single cake. Without removing the radioactive actinides from the salts, a significant impact to the WIPP would be realized, possibly exceeding the WIPP capacity. Exceeding the WIPP capacity would significantly impact the country because of the huge costs to build another WIPP-like facility. Therefore, most of the actinides must be scrubbed from salt residues to minimize the total waste volume stored at the WIPP. The actinide level in the salts must be decreased to a level that would allow 55-gal waste drums to be completely filled.

Aqueous processing is an accepted method for processing the residue salts to recover the actinides; however, aqueous treatment of this salt to meet disposal criteria can generate up to 3 kg of grouted waste per kilogram of plutonium feed. Additionally, special precautions must be made to account for the problems associated with chloride corrosion in an aqueous system, and complete aqueous processing of a single salt cake may take several days, while pyrochemically processing a salt cake would take less than one day.

Pyrochemically processing residue salts eliminates the waste generated during aqueous processing and the associated chloride corrosion problem.

II. PYROCHEMICAL RESIDUE SALTS

A. DILUENT SALTS

To decrease side reactions between plutonium (Pu) and the diluent salt, highly stable chloride salts such as calcium chloride ($CaCl_2$), potassium chloride (KCl), sodium chloride (NaCl), or a mixture of these is normally used as the diluent for the DOR, MSE, and ER pyrochemical processes. At process completion, the bulk of the residue is the diluent salt mixed with compounds of actinide oxides, actinide chlorides, and residual metal flakes and shot. The MSE and ER residue salts contain a substantial amount of plutonium, and the MSE residue salt also contains a relatively high quantity of americium (Am). DOR residue salts contain small quantities of plutonium and, depending upon the environmental discard limit for plutonium, may not require further processing for plutonium recovery.

B. MSE RESIDUE SALT

Weapons- and reactor-grade plutonium metal contain a mixture of isotopes: [238]Pu, [239]Pu, [240]Pu, [241]Pu, and [242]Pu. The decay of [241]Pu results in the formation of its daughter product, [241]Am, which is an alpha emitter with a half-life of 13.2 years. Americium-241 further decays to its daughter product [237]Np, a 60-keV gamma emitter with a half-life of 458 years.[1] This radiation emission presents a health hazard to those working in both the energy- and weapons-related fields. In addition, the impurities in the plutonium metal alloy change both its nuclear and mechanical properties. Americium can be periodically removed from plutonium by the MSE process. MSE separates americium from plutonium metal by a pyrochemical oxidation reaction followed by molten-salt solvent extraction.

In MSE, plutonium metal is contacted with a molten-salt mixture containing an americium oxidant such as magnesium chloride ($MgCl_2$), or more recently plutonium chloride ($PuCl_3$) added as either $PuCl_3$ or dicesium hexachloroplutonate (Cs_2PuCl_6). The Cs_2PuCl_6 decomposes to $PuCl_3$ and cesium chloride (CsCl) as the temperature of the process is increased. The $MgCl_2$ or $PuCl_3$ selectively oxidizes Am^0 to a chloride that is extracted into the molten-salt phase. Some of the plutonium metal feed is also entrapped in the salt phase as metal shot or flakes. This plutonium metal and that in unreacted $PuCl_3$, as well as the extracted americium, must be recovered before the salt residue can be discarded or recycled.

It has been determined by X-ray diffraction that the MSE residue salts generated from runs with $PuCl_3$ as the americium oxidant and $CaCl_2$ as the solvent not only contain the solvent $CaCl_2$, but also $PuCl_3$, plutonium oxychloride (PuOCl), americium chloride ($AmCl_3$), americium oxychloride (AmOCl), and plutonium dioxide (PuO_2).[2] The quantity of plutonium and americium in the residue salt can be as high as 30 wt% but is dependent on the amount of oxidant added as feed to the MSE process, the MSE operating conditions (e.g., operating temperature, stirring, etc.), the concentration of americium in the MSE feed metal, and the total salt-to-metal ratio of the MSE process.

Pyrochemical processing is normally done under an atmosphere that contains little or no oxygen; therefore, the supply of oxygen to the process is generally from an oxide layer on the plutonium feed metal or from the reactor crucible. For MSE operations, a tantalum crucible with an oxide surface coating is used. The oxide coating is necessary to prevent the plutonium metal from wetting the tantalum, allowing the metal to be removed from the crucible after cool-down. The oxides react with $PuCl_3$ and the actinide metals to produce oxychloride compounds.

C. ER RESIDUE SALT

In plutonium ER, pure plutonium metal is produced from impure feed metal by ionic transport of the plutonium through a molten-salt. Impure plutonium metal is cast into an anode that is placed in a magnesium oxide (MgO) cup within a MgO ER cell. The anode cup is surrounded by an equimolar NaCl–KCl or a $CaCl_2$ molten-salt bath. To provide plutonium ions initially in the molten-salt electrolyte, $PuCl_3$ or $MgCl_2$ is added. The $MgCl_2$ reacts with plutonium metal to produce ionic plutonium ($PuCl_3$). After the salt and anode are heated to a molten state, a positive potential is applied to the liquid plutonium anode. Plutonium oxidizes at the molten metal–salt interfaces to form Pu^{+3}. The Pu^{+3} ions are transported through the molten salt to the negatively charged cathode where the plutonium is reduced back to metal. Metal droplets of plutonium are formed on the cathode that drop through the salt and form a pool of purified metal.

The plutonium metal in the cast anode has already been processed through MSE to remove the americium; therefore, ER residue salts have little americium and mainly consist of the solvent salt, $PuCl_3$, PuOCl, and Pu^0.[3] The Pu^0 concentration in the residue salt can be as high as 20 wt% and is in a variety of particle sizes. The solvent salt has traditionally been an equimolar NaCl–KCl eutectic mix that would be scrubbed with magnesium (Mg). More recently, $CaCl_2$ has been examined as the solvent salt that would be scrubbed with calcium (Ca). The metal product produced from the Ca scrub of $CaCl_2$ salt can be recycled back through ER because it is almost pure plutonium. The cleaned salt can be recycled or stored as TRU waste.

D. DOR RESIDUE SALTS

The DOR process converts feed PuO_2 to Pu metal through Ca reduction. The solvent of eutectic $CaCl_2$–26 mol% calcium fluoride (CaF_2) or pure $CaCl_2$ is used for the pyrochemical process. The quantities of actinides, predominantly present as unreacted PuO_2 and uncoalesced metal, in the DOR residue salts are most dependent on the amount of Ca fed to the process, the salt-to-oxide ratio, and the mixing geometry. The actinide concentration in the residue salt is normally low and would not be recovered if the salt meets the environmental discard limit for plutonium, and if the actinide concentration in the salts was low enough to allow the complete filling of a waste drum. However, additional processing of the salt to convert residual Ca to a nonreactive form such as calcium oxide (CaO) may be required before disposal at the WIPP.

E. URANIUM SALTS

The pyrochemical salt scrub process can also be used to recover the remains of depleted uranium present in magnesium fluoride salt produced during the reduction of uranium fluoride to metal. A scrub alloy metal could be used to reduce uranium fluoride and uranium oxide compounds to uranium metal. The cleaned salt could be discarded as LLW and the uranium metal recovered.

III. PYROCHEMICAL SALT SCRUB PROCESS

In the pyrochemical salt scrub process shown in Figure 1, an actinide chloride reductant is used to reduce the actinide chloride salts and PuO_2 to metal. Historically, stationary, resistance-heated furnaces have been used for melting the reactants. Tantalum flat or

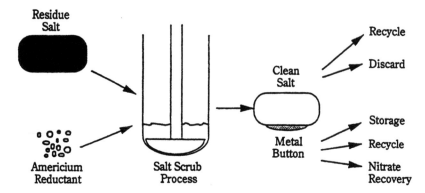

Figure 1 Pyrochemical salt scrub process.

opposed-pitch dual propeller paddles are used to stir as much as 6 kg of molten salt at a rate of several hundred rpm, depending on the type of stirrer used. Either MgO or alumina (Al$_2$O$_3$) crucibles are used as the reactor vessel. The ceramic can be easily broken away from the tall salt cakes and metal product after the run is completed.

The process steps include adding the residue salt to the ceramic crucible, loading the crucible into the stationary furnace, heating the crucible at a rate no greater than 200°C/h, and stirring the molten salt. The reductant can be added to the crucible with the residue salt or can be added incrementally after the salt is molten. When Ca is used as the actinide reductant, an exothermic reaction will occur that can generate enough heat to thermally shock and crack the ceramic crucible. However, by adding the Ca incrementally, large temperature spikes and consequent crucible failure can be avoided. After all the Ca is added, and after the molten mix has been stirred for several minutes, the run is concluded, and the crucible is allowed to cool.

A metal button is produced during the salt scrub process that is easily separated from the "cleaned" salt. The metal button can be stored, converted to an oxide for storage, or further refined using aqueous processing to separate the plutonium from the americium, if plutonium recovery is required. Americium is an intense emitter of gamma radiation. In metal form, this radiation is attenuated; therefore, radiation exposure to personnel is minimized when the MSE salt scrub button is stored as a metal rather than converted to an oxide and stored. The cleaned salt can be stored as transuranic (TRU) waste or could be recycled back through MSE or other pyrochemical operations.

A. REDUCTION TECHNIQUES
1. Calcium Reduction
When Ca is used as the actinide reductant, a plutonium–americium–calcium metal button is produced by the following, thermodynamically favorable reactions:

$$2PuCl_3 + 3Ca = 2Pu + 3CaCl_2 \tag{1}$$

$$2AmCl_3 + 3Ca = 2Am + 3CaCl_2 \tag{2}$$

$$2PuOCl + 3Ca = 2CaO + 2Pu + CaCl_2 \tag{3}$$

$$2AmOCl + 3Ca = 2CaO + 2Am + CaCl_2 \tag{4}$$

$$PuO_2 + 2Ca = Pu + 2CaO \tag{5}$$

Because the melting temperature of Ca metal is 839°C, the salt scrub operating temperature should be above 850°C to ensure the Ca is molten. Both MgO and Al_2O_3 crucibles react with Ca at this operating temperature. However, MgO is more stable than Al_2O_3 as shown by the following reactions:[4]

$$MgO + Ca = CaO + Mg \quad \Delta G_R^{1200K} = -9.497 \text{ kcal/mol} \tag{6}$$

and

$$Al_2O_3 + 3Ca = 3CaO + 2Al \quad \Delta G_R^{1200K} = -59.111 \text{ kcal/mol} \tag{7}$$

Salt scrub tests have been conducted at Lawrence Livermore National Laboratory (LLNL), in conjunction with the EG&G Rocky Flats Plant (RFP), to determine the applicability of a Ca–gallium (Ga) pellet for scrubbing plutonium metal from spent salts.[5] The Ca was used to reduce the Pu^{+3} to Pu^0 and the Ga to alloy with the Pu^0 to form a gallium–plutonium alloy. A $CaCl_2$–$PuCl_3$–CsCl salt mix was used to simulate residue salts generated during MSE operations with $CaCl_2$–Cs_2PuCl_6 (Cs_2PuCl_6 decomposes to $PuCl_3$ and CsCl at MSE operating temperatures). Tests were conducted with both MgO and Al_2O_3 crucibles. It was found that the actinides would not reduce to metal when Al_2O_3 crucibles were used, but more than 99% of the actinides were recovered in metal form, under the same test conditions, when MgO crucibles were used. When alumina crucibles are used, the salt becomes saturated with CaO from the Al_2O_3 reduction shown in Equation 7 and the Ca solubility is decreased as explained below. With decreased Ca solubility, the reduction of the actinide salts to metal is decreased; therefore, MgO crucibles should be used when Ca is used as the reductant.

The $CaCl_2$-based salts are particularly amenable to Ca metal scrubbing. At 850°C, approximately 2.8 mol% Ca is soluble in $CaCl_2$, and the solubility is increased to approximately 3.5 mol% at 900°C.[6] However, as the CaO concentration of the system is increased, as would be the case for Equations 3 to 5, the Ca solubility in $CaCl_2$ decreases. Perry and Shaw,[6] and Axler and DePoorter[7] have demonstrated this as shown by the data in Tables 1 and 2.

The addition of noncontaminated $CaCl_2$ diluent to the salt scrub process may be required if the CaO production is expected to be high. LLNL has found this to be the case for residue salts from the MSE of plutonium metal that had been in storage for more than 20 years and was coated with an oxide layer. The plutonium oxide from the aged material would have been transferred to the salt and would have been converted to CaO by the reaction shown in Equation 5. Even though an excess of calcium was added to the residue salts, the reduction of all the actinide compounds to metal was not possible until additional $CaCl_2$ was added to dilute the CaO and increase the Ca solubility. However, as shown in Tables 1 and 2, when the temperature of the salt mix is increased by 50°C, the CaO solubility increases from about 3.4 to 6.0 mol% while the Ca solubility is maintained at approximately 1.8 mol%. Thus operating at higher temperatures would decrease the impact of CaO on the Ca solubility, and consequently on the "scrub" efficiency.

LLNL has successfully processed more than 34 kg of MSE and ER residue salts consisting of either pure $CaCl_2$ or 74 mol% $CaCl_2$–KCl eutectic as the solvent salt. Seven kilograms of actinides were recovered in metal form by the calcium reduction process.

Table 1 Ca–CaO–CaCl$_2$ Solubility Data for 850°C

Ca (mol%)	CaO (mol%)	CaCl$_2$ (mol%)
2.8	0.0	97.2
2.7	0.0	97.3
2.3	0.5	97.2
2.4	0.7	96.9
2.0	1.2	96.8
2.0	1.1	96.9
1.8	3.4	94.8
1.8	3.4	94.8

Table 2 Ca–CaO–CaCl$_2$ Solubility Data for 900°C

Ca (mol%)	CaO (mol%)	CaCl$_2$ (mol%)
3.7	0.1	96.2
3.2	1.4	95.4
3.2	1.4	95.4
2.9	4.0	93.1
2.7	4.6	92.7
1.9	5.9	92.2
1.9	6.0	92.1

Americium concentrations in the metal buttons ranged from 8 to 22 wt%, and all the salt met TRU waste discard limits. Scrub efficiencies ranged from 91.2 to 99.9 wt% actinides recovered. The lower efficiency runs occurred when there was less than 50 g of actinides initially present in the residue salt. With small quantities of actinides, a molten-metal pool is not produced that provides a mechanism to coalesce metal droplets. Starting the process with a pool of metal, or adding a metal that will alloy with plutonium (such as aluminum, gallium, or zinc) would improve the actinide recovery of the residue salts with lower actinide concentration.

Los Alamos National Laboratory (LANL) has used a pyroredox process to recycle spent anodes from the ER process.[8] The anode heels are produced after most of the plutonium has been transferred from the anode to the cathode, and the ER process has been concluded because the impure anode begins to solidify or because the back EMF within the cell reaches a predetermined value. In the pyroredox process, the anode heel is reacted with a 31 mol% zinc chloride (ZnCl$_2$)–KCl eutectic salt. Impurities like americium, uranium, and aluminum will be oxidized, along with the plutonium, to form chloride salts, and the Zn^{+2} will be reduced to metal and will combine with other metal impurities like gallium, tantalum, iron, and tungsten. Calcium reduction has been used to recover greater than 98% of the plutonium from the salt mix. After the Ca reduction, a gravity separation process was used to separate the plutonium from the other metals. After the separation, the plutonium was pure enough to recycle back through ER.

LANL has also used Ca reduction to recover the actinides from equimolar NaCl–KCl salts used in the ER of plutonium metal. Christensen and Mullins[9] found that up to 95.6% of the plutonium in the salts could be recovered in metal form after a single-stage reduction. However, an excess of Ca is required to compensate for the consumption of some of the Ca in its reaction with the less stable NaCl to produce sodium (Na) metal and the more stable $CaCl_2$ salt.

Giebel and Wing[10] have demonstrated plutonium recovery from DOR residue salts by using Ca, clean $CaCl_2$, and plutonium metal fines and pieces. The Ca was to reduce any residual plutonium oxide to metal, the $CaCl_2$ was to dilute the CaO and increase the Ca solubility, and the plutonium metal was added to provide a pool for the consolidation of the plutonium in the DOR residue salt. As much as 89% of the plutonium in the residue salts could be recovered by this method; however, the method is not as efficient as the zinc–magnesium alloy scrub described in Section III.B.

2. Magnesium Reduction

Magnesium reductant is normally used for MSE salts containing substantial quantities of NaCl because the Na^{+1} will be reduced by Ca to produce pyrophoric and volatile Na^0 metal. A MSE process using an equimolar NaCl–KCl salt mix, with $MgCl_2$ as the americium reductant, was used at the RFP for approximately two decades. There are tons of the NaCl–KCl salt mix presently in storage at RFP. This salt would be processed by adding excess Mg to react with the actinide compounds to produce an actinide metal button. The Mg would be oxidized to produce $MgCl_2$ salt and the actinides would be reduced by the following equations:

$$2PuCl_3 + 3Mg = 2Pu + 3MgCl_2 \tag{8}$$

$$2AmCl_3 + 3Mg = 2Am + 3MgCl_2 \tag{9}$$

$$2PuOCl + 3Mg = 2MgO + 2Pu + MgCl_2 \tag{10}$$

$$2AmOCl + 3Mg = 2MgO + 2Am + MgCl_2 \tag{11}$$

$$PuO_2 + 2Mg = Pu + 2MgO \tag{12}$$

The Gibbs free energy for the reaction between Mg and the actinide chlorides is close to zero. To force the reaction, aluminum (Al) metal is added to the process to form a plutonium–aluminum intermetallic, thus lowering the plutonium activity[11] and shifting the reaction to the right as described by Le Chatelier's principle: *When a stress is brought to bear on a system at equilibrium, the system tends to change so as to relieve the stress.*[12] By producing the strongly bonded intermetallic, the reaction proceeds between the excess Mg and actinide chlorides to relieve the stress brought on by the change in concentration of the reactants. The reaction between Al and Pu is shown by the following equation:

$$Pu + xAl = PuAl_x \tag{13}$$

The quantity of Al initially added to the salt scrub process is approximately equal to the quantity of Pu + Am in the residue salt.

An aluminum–magnesium–plutonium–americium alloy button is produced by the Mg–Al salt scrub process; however, the homogeneity of the metal button is dependent on the quantity of magnesium fed to the salt scrub process. Cusick et al.[13] found that even though the plutonium loading in the alloy button increased as the Mg above the stoichiometric requirement was increased from 20 to 50%, the metal consolidation and uniformity decreased. When Mg additions exceeded 150% of the stoichiometric requirement, button consolidation was poor.

During a 6-month period from September 1983 to February 1984, RFP produced 127 aluminum–magnesium–plutonium–americium metal buttons containing 37.4 kg of plutonium and 1.9 kg of americium (approximately 30 wt% actinides), and a discardable salt.[14] At that time, the RFP discard limit was 0.0645 g plutonium per gram salt cake and the salts produced during this campaign contained an average 0.0079 g plutonium per gram of salt cake. Approximately 33 kg of the plutonium was recovered from 113 of the scrub alloy buttons after aqueous processing at the Savannah River Plant.

B. ZINC–MAGNESIUM ALLOY

During the development of DOR for the Special Isotope Separation program at LLNL, a discard limit of 1 g Pu per DOR salt cake was established. Approximately one half of the DOR salt cakes contained more than this limit and thus required further processing to recover the actinides. Those salt cakes weighed from 5 to 6 kg and contained from about 1.3 to 51.1 g actinides. To recover the actinides, a zinc (Zn)–30 wt% percent Mg alloy was used as a PuO_2 reductant and metal scavenger.[15] Magnesium reduced the PuO_2 to metal as shown in Equation 12 and the metal readily alloyed with the Zn. Less than 1/2 g of plutonium remained in the salt after scrubbing them with the Zn–Mg alloy. The Zn–Mg alloy could be used in consecutive runs until the plutonium loading was about 50 wt%. The zinc–magnesium–plutonium metal product could be stored as is, or it could be aqueously processed to separate out the plutonium. Other scrub alloys could be effectively used for recovering actinides from low concentrated salts, for example, a calcium–zinc or calcium–aluminum alloy could be used.

IV. ADVANCED CONCEPTS

A. CALCIUM TITRATION

In initial salt scrub tests at LLNL, a calculated quantity of Ca was added to each batch run. The Ca addition was calculated with the assumption that all the actinides were chloride compounds (Equations 1 and 2). An additional 10 percent Ca was sometimes added to the stoichiometric quantity to ensure reaction completion. However, it is difficult to accurately determine the quantity of actinides in residue salts. MSE residue salts are concentrated in americium that generates high radiation fields that the analytical laboratory at LLNL is not equipped to handle. Additionally, MSE and ER salts are inhomogeneous and extracting a sample representative of the entire salt cake was not possible; therefore, the salts were analyzed nondestructively by calorimetry and isotopic measurements. Actinide values generated by this method are often in error due to metal shot being dispersed throughout the salt (i.e., inhomogeneity of the actinides).[16]

The calculated quantity of Ca was added incrementally to prevent a large temperature spike, and possible thermal shock of the crucible, caused by the exothermic reactions between the Ca and actinide compounds. The Ca was added in approximately 10-g increments with an approximately 10 to 15°C temperature increase with the first addition, as shown in Figure 2. The temperature was allowed to stabilize before another Ca addition was made. This process was continued until all the predetermined quantity of Ca was added. With each 10-g Ca addition, the increase in temperature was less until finally there was no increase. A decrease in temperature with the last excess Ca additions was often observed because the Ca was cold (room temperature) compared to the molten-salt.

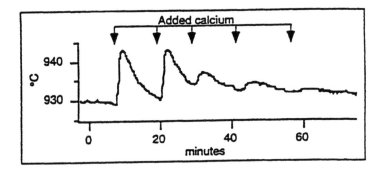

Figure 2 Temperature profile of salt scrub process.

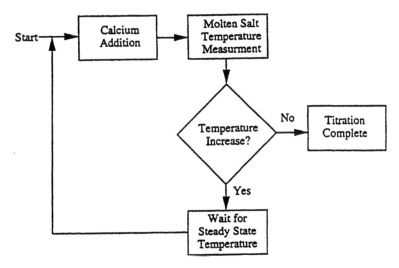

Figure 3 Ca titration flow diagram.

More than 10% excess Ca was being added to each run because some of the plutonium in the salt was already metal shot (i.e., not all the plutonium and americium were present as chlorides). This procedure increased the residue stream, excessively contaminated the product metal, and left unreacted Ca metal that had to be separated from the salt before it could be discarded. It was determined that a more accurate quantity of Ca could be added by monitoring the temperature change of the exothermic reaction.

To eliminate the need for manual Ca addition, an automated Ca feeder and associated control system were designed, built, and tested.[17] The flow diagram shown in Figure 3 was developed where the endpoint for the calcium titration system would be determined by the temperature of the mixture. After the first Ca addition, the temperature of the molten-salt mix is measured. If there is an increase in temperature, then the system waits for the temperature to reach steady state before another Ca addition is made. If there is no temperature increase, then it is assumed that the reactions have been completed, the actinides have been converted to metal, and the run is concluded.

By monitoring the temperature and regulating the addition of Ca, the temperature spikes are controlled, thus eliminating thermal shock to the crucible. Additionally, excess

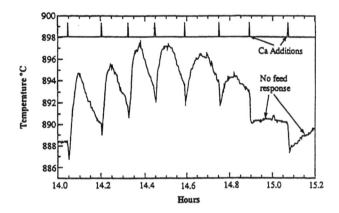

Figure 4 Temperature profile generated by the Ca titration control system.

Ca is not required to assure reaction completion. A reduction in radiation exposure to the operator is realized because the addition of Ca to the furnace is automated, and assay of the residue salts before processing is not required.

Pyrochemical salt scrub tests with the Ca titration system have resulted in automated operations that were comparable to manual operations. Temperature curves showing the automated Ca additions are shown in Figure 4. The top curve of the plot shows when Ca was added at each pulse of the line. Immediately after the Ca addition, the temperature decreases because of the Ca cooling the salt, but then the temperature quickly starts increasing. The change in temperature decreases with each Ca addition until finally there is no increase. For this particular run, the titrator was commanded to continue running until there were two Ca additions with no corresponding increase in temperature. The slow increase in temperature at the second "no feed response" is due to the furnace controller detecting a drop in temperature of the crucible, increasing the power to the heater coils, and slowly heating the salt back up to the set point.

Actinide recovery with the Ca titration process has ranged from 98.32 to 99.87%, and the quantity remaining in the "cleaned" salts ranged from 0.32 to 3.15 g. A 55-gal waste drum will hold 28 1-gal cans standing end on end. If one assumes an average actinide content of 1.5 g per salt cake, and that a 1-gal can would hold no more than two cakes, a 55-gal drum could easily be filled without exceeding the 200 g-eq. fissile ^{239}Pu TRU waste discard limit.

B. ADVANCED REACTORS

A reactor for continuous metallothermic reduction of uranium tetrafluoride (UF_4) has been designed, built, and tested at the Martin Marietta Energy Systems Y-12 plant. The reactor, shown in Figure 5, is made of alumina or graphite and has overflows for both metal and salt.[18] The reactor is initially loaded with a uranium (U)–iron (Fe) alloy and a diluent salt of either NaCl or $CaCl_2$. When the alloy and salt are molten, a blended powder of UF_4, Mg, diluent salt, and Fe is fed through the top of the reactor. Because of the differences in density, three distinct phases are formed. The top layer contains the Mg reductant, the middle layer is the salt, and the heavy bottom layer is the U–Fe alloy. As the UF_4 passes through the Mg, the uranium is reduced to metal and the Mg is oxidized to MgF_2. The heavy U metal falls to the bottom of the reactor where the alloying with Fe

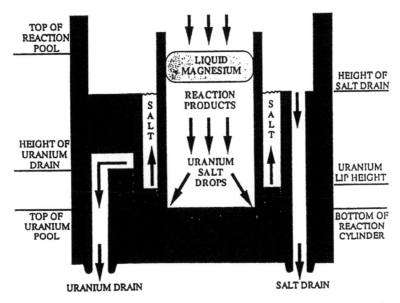

Figure 5 Schematic representation of uranium metallothermic reaction vessel.

is completed. As the metal pool at the bottom of the reactor fills, the excess metal overflows into a collection pot below the reactor. As the salt volume increases, the excess salt is squeezed past the bottom edge of the weir and then out the salt overflow into a graphite container below the reactor. When the salt squeezes between the bottom of the reactor weir and the top of the U–Fe alloy, it is mechanically cleaned of uranium particles. The waste salt contains between 5 and 86 ppm U. Salt containing less than 50 ppm U may be eligible for disposal in sanitary landfills as established in several Nuclear Regulatory Commission and state exemptions.

For plutonium residue salts, the plutonium concentration would have to be decreased to about 1 ppm to dispose of the cleaned salts as LLW rather than TRU waste. A reactor similar to the continuous metallothermic reduction for UF_4 may achieve this low Pu concentration. A total actinide concentration of approximately 1.6 eq. ppm fissile ^{239}Pu (approximately 0.25 ppm weapons-grade Pu) is equal to 100 nCi/g of waste, the LLW limit. By disposing of the plutonium-contaminated salts as LLW, tremendous savings would be realized. Disposal costs for a 55-gal drum filled with TRU waste is several orders of magnitude greater than those for a 55-gal drum filled with LLW.

LLNL has designed an advanced reactor, shown in Figure 6, that will be tested for almost total plutonium recovery from pyrochemical residue salts. The reactor will be constructed from carburized tantalum, so, unlike ceramic crucibles, it can be used repeatedly without thermal shock or breakage. (Carburized tantalum has been shown to be highly stable in the presence of molten Pu.) The reactor would be initially loaded with Pu metal and the diluent salt. When the reactor contents are molten, an initial charge of reductant would be added that would float on top of the salt. If a Pu-alloying agent is required, it could be added initially or after the reactor was at operating temperature. A mixture of residue salts, reductant, and alloying agent would be fed through the top of the reactor. The actinide compounds would be reduced to metal as they passed through the reductant, and the metal would fall to the bottom of the reactor. As the metal volume

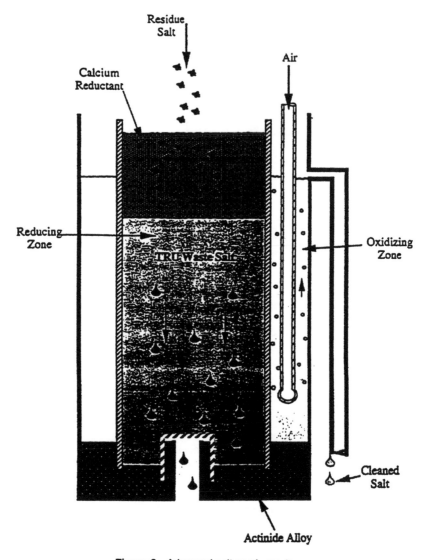

Residue
Salt

Air

Calcium
Reductant

Reducing
Zone

Oxidizing
Zone

Cleaned
Salt

Actinide Alloy

Figure 6 Advanced salt scrub reactor.

increased, it would overflow into a container at the bottom of the reactor. The metal could then be stored in the catch can or oxidized for storage, or further treated to separate the plutonium from the americium. As the salt volume increased, the salt would be pushed between the bottom of the reactor weir and plutonium metal where any plutonium particles would be mechanically removed. The salt would then be contacted with air or chlorine to ensure complete passivation of any reactive metals. The salt would then overflow into a catch can under the reactor, and the can could be used as the disposal container for the salt. The filled catch cans could be retrieved and replaced by temporarily stopping the reactor feeding. Immediate continuation of the process could then be made

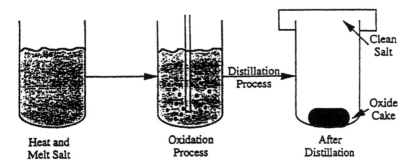

Figure 7 Salt distillation process.

by simply restarting the reactor feeding. A nondestructive assay of the catch cans and their contents could be made to determine the actinide concentrations followed by disposal, storage, or further processing.

Even if the product salt cannot be treated as LLW, there are still some advantages to the advanced reactor concept. It could be operated continuously, thus substantially decreasing process cycle time. The MSE pyrochemical salt scrub process is an operation with substantial dose exposure to the operator because of the [241]Am and its [237]Np daughter product. The exposure could be substantially reduced by eliminating the run breakout traditionally associated with pyrochemical salt scrub runs. During traditional product breakouts, the ceramic crucible must be first broken and removed, followed by cleaving the metal button from the salt, and finally packaging of the metal, pieces of salt cake, and ceramic fragments. With this reactor, breaking of the crucible, and packaging of the products would be eliminated. Additionally, a second unit operation to pacify any residual, reactive metals would be eliminated.

C. SALT DISTILLATION

LANL has developed a salt distillation process for recovering the actinides from residue salts.[19] As shown in Figure 7, the actinide compounds and metal, and any reactive metals like calcium or magnesium, are first oxidized by bubbling oxygen through the molten, residue salt to produce oxide and oxychloride compounds. The highly stable diluent salts, that is, NaCl, KCl, and $CaCl_2$, still remain as chloride salts after the oxidation process. A high-temperature distillation is then used to distill the higher vapor pressure salts away from the nonvolatile, oxygenated compounds. Oxygenated compounds left in the crucible are in powder form and can be easily removed from the reactor and stored, or if necessary, further processed to recover the plutonium.

Theoretical calculations performed by Garcia et al.[19] have shown that the actinide concentration in the distilled salts could be decreased below the approximately 1 ppm LLW discard limit by several orders of magnitude. Even if the salts contain more than the 1 ppm actinide LLW limit, the actinide concentration would certainly be low enough to completely fill 55-gal TRU waste drums without exceeding WIPP discard limits.

As shown in Figure 8, the vaporization of $PuCl_3$ will occur at a temperature lower than that of $CaCl_2$ and close to that of NaCl and KCl, thus the need for oxidizing the $PuCl_3$ to PuO_2 before the distillation process. Distillation of KCl, NaCl, or a mixture of the two salts from the actinide oxides can be accomplished at temperatures above 900°C. Distillation operations for the separation of $CaCl_2$ from PuO_2 must occur at operating temperatures above 1100°C.

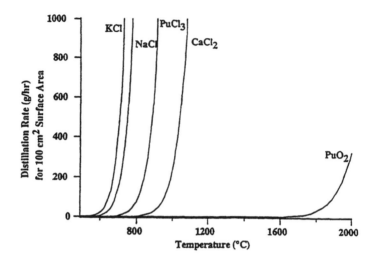

Figure 8 Distillation rates as functions of temperature.

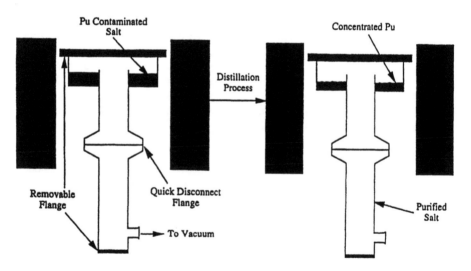

Figure 9 Advanced condenser concept.

LANL has processed more than 19 kg of NaCl–KCl residue salt containing plutonium and americium through the distillation process. More than 99.9% of the actinides were recovered in powder form.[20] Even though some of the product salt from batch tests contained a few hundred parts per million Pu, there have been several tests where the salt contained less than 10 ppm Pu, and even more promising, some product salt batches contained less than 1 ppm total actinides.

LANL has been working on the design of an advanced condenser as shown in Figure 9.[20] The condenser can be used at the higher operating temperatures required for the distillation of $CaCl_2$ salts. Residue salts, pretreated to produce actinide oxides, will be loaded into the reactor through the top flange. During the distillation process, a vacuum will be used to pull the salt vapors down into the cooler area of the reactor where they will condense on the reactor walls. Separation of the lower portion of the reactor, where the "clean" salts collect, from the top portion of the reactor will be accomplished with a quick-disconnect flange.

REFERENCES

1. Bureau of Radiological Health, *Radiological Health Handbook*, U.S. Department of Health, Education, and Welfare, Rockville, MD, 1970.
2. Dodson, K. E., "Molten Salt Extraction (MSE) of Americium from Plutonium Metal in $CaCl_2$-KCl-$PuCl_3$ and $CaCl_2$-$PuCl_3$ Salt Systems," M.S. thesis, UCRL/LR-110870, Lawrence Livermore National Laboratory, 38, June 1992.
3. Dodson, K. E., McAvoy, D. P., and Holck, D. M., "Electrorefining Plutonium Metal with a Calcium Dichloride - Pu^{+3} based salt at Lawrence Livermore National Laboratory," Presented at the Twelfth Annual Pyrochemical Workshop, Gatlinburg, TN, October 28, 1992.
4. (U.S.) Bureau of Mines, *Thermodynamic Data for Mineral Technology*, Albany, OR, 1985.
5. Fisher, D. C., "Applicability of the Calcium/Gallium and Aluminum/Magnesium Salt Scrub Processes at the Rocky Flats Plant," RFP Report RTT-94-001, May 16, 1994.
6. Perry, G. S. and Shaw, S. J., "Solubility of Calcium in CaO-$CaCl_2$," Atomic Weapons Establishment, AWE Report 0 16/90, 1990.
7. Axler, K. M. and Depoorter, G. L., "Solubility Studies of the Ca-CaO-$CaCl_2$ System," *Materials Science Forum*, Vol. 73–75, 19, 1991.
8. McNeese, J. A., Fife, K. W., and Williams, J. D., "Recent Development in Pyrochemistry at Los Alamos," LANL Report LA-UR-84-3233, 1984.
9. Christensen, D. C. and Mullins, L. J., "Salt Stripping, a Pyrochemical Approach to the Recovery of Plutonium Electrorefining Salt Residues," LANL Report LA-9464-MS, October 1982.
10. Giebel, R. E. and Wing, R. O., "Cleanup of Plutonium Oxide Reduction Black Salts," RFP Report RFP-3939, December 17, 1986.
11. Cusick, M. J., Sherwood, W. G., and Fitzpatrick, R. F., "Plutonium and Americium Recovery from Spent Molten-Salt-Extraction Salts with Aluminum-Magnesium Alloys," RFP Report RFP-3614, April 23, 1994.
12. Keenan, C. W., Wood, J. H., and Kleinfelter, D. C., *General College Chemistry*, Harper & Row, New York, 1976, 259.
13. Cusick, M. J., Fitzpatrick, R. F., Boyle, M. F., and Ishmael, R. T., Jr., "Salt Cleanup Development," reported in Chemistry Research and Development Annual Progress Report October 1, 1982 to September 30, 1983, Frend John Miner, Ed., RFP Report RFP-3654, September 24, 1984.
14. Cusick, M. J., Fitzpatrick, R. F., Fray, R. E., and Ishmael, R. T., Jr., "Salt Cleanup Development," reported in Chemistry Research and Development Annual Progress Report October 1, 1983 to September 30, 1984, K. V. Gilbert, Ed., RFP Report RFP-3825, June 13, 1986.
15. Crawford, T. W., "Plutonium Recovery from Direct Oxide Reduction Salt Residues at Lawrence Livermore National Laboratory," Presented at the Eighth Annual Pyrochemical Workshop, Livermore, CA, November 8, 1988.
16. Prindle, A. L., Gunnink, R., Nagle, R., and Ruhter, W. D., "Gamma-Ray Spectrometry Analysis for LLNL's Material Accountability and SIS Programs," LLNL Report LLNL/UCAR 10062-88, 77, 1988.
17. Dodson, K. E., Johnson, G. W., McAvoy, D. P., and Holck, D. M., "Pyrochemical Salt Scrub," in *Actinide Processing Methods and Materials*, B. Mishra and W. A. Averill, Eds., The Minerals, Metals & Materials Society, Warrendale, PA, 1994, 225.
18. DeMint, A. L., Maxey, A. W., and Horton, J. A., "Demonstration of Pilot-Scale Production of Uranium AVLIS Feedstock by Continuous Reduction of Uranium Tetrafluoride," MMES Y-12 Report, Y/DV-1323, September 30, 1993.

19. **Garcia, E., Dole, V. R., Griego, W. J., Lovato, J. J., McNeese, J. A., and Axler, K.,** "Distillation Separation of Actinides from Waste Salts," from Nuclear Materials Technology Division — 1993 Division Review, LANL Report LALP-93-92, 1994.

20. **Garcia, E., Dole, V. R., McNeese, J. A., and Lovato, J. J.,** "Distillation Separation of Chloride Salts from Plutonium," Presented at the Eighteenth Annual Actinide Separations Conference, Durango, CO, May 23–26, 1994.

Chapter 11

Surface Treatments

Mark C. Bronson

I. INTRODUCTION

This chapter provides an overview of technologies presented at various symposia and conferences[1-4] on techniques used to remove radionuclides from the surface of contaminated materials. The primary objectives for decontaminating materials that have been used in nuclear facilities are to reduce the cost of disposal, minimize personnel radiation exposure, reduce risk of environmental contamination, and salvage equipment or materials for possible future use. In recent years, the lack of geologic disposal sites for radioactive wastes has promoted the need to separate radioactive contaminants from nonradioactive materials, thus allowing reuse or disposal of the nonradioactive materials in commercial landfills.

Technologies that remove surface contaminants can be divided into chemical removal technologies and physical removal technologies. Both of these technologies generally remove not only the surface contaminants, but also a portion of the base material in order to remove contaminants from surface defects such as cracks and pits. Many surface decontamination processes use a combination of both chemical and physical removal technologies.

Chemical surface removal technologies generally oxidize the surface materials to form soluble or gaseous species. Physical surface removal technologies use processes such as washing, machining, ablation, or peelable coatings. Both chemical removal technologies and physical removal technologies are discussed in the following sections.

II. CHEMICAL REMOVAL TECHNOLOGIES

Chemical removal technologies remove radioactive materials from surfaces by converting the radioactive contaminants and some of the base material into soluble or gaseous species. Once in a soluble or gaseous form, the contaminants can be "washed away" from the surface of the material. Typical chemical removal technologies include leaching, electropolishing, and fluoride volatility.

A. LEACHING

Leaching or chemical decontamination utilizes concentrated or dilute solvents to dissolve either the contamination film covering the base material, or the base material. Many types of chemical agents are used in chemical decontamination processes.[1] Oxidizing agents are used to oxidize elements to soluble species, such as the conversion of chromium to the soluble +6 valence state. Reducing agents are used to reduce elements to soluble species, such as the conversion of nickel ferrite to the soluble +3 valence state. Complexing agents are used to prevent the formation of sludge from the precipitation of dissolved salts. In addition, inhibitors are used to reduce base material corrosion and surfactants are used as wetting agents. Table 1 gives a list of typical chemical agents and applications.

Table 1 Typical Chemical Agents and Applications

Name	Formula	Application
Alkaline permanganate	NaOH and $KMnO_4$	Pretreatment; oxidize Cr corrosion film
Ammonium citrate — dibasic	$(NH_4)_2HC_6H_5O_7$	Remove corrosion film
Citrox	$H_2C_2O_4$, $(NH_4)_2HC_6H_5O_7$, $Fe(NO_3)_3\cdot9H_2O$, and $(C_2H_5NH)_2Cs$	Reducing agent; dissolution of carbon and 400 series stainless steels
Hydrochloric acid	HCl	Stainless steel, carbon steel, chrom-moly steel, and copper
Nitric acid	HNO_3	Dissolves actinide oxides and fission products
Oxalate peroxide	$Na_2C_2O_4$, $H_2C_2O_4$, H_2O_2	Dissolution of uranium oxide
Oxalic acid	$H_2C_2O_4$	Remove iron oxides and complex fission products
Phosphoric acid	H_3PO_4	Decontaminate carbon steels
Sulfamic acid	NH_2SO_3H	Decontaminate carbon steels
Sulfuric acid	H_2SO_4	Remove organic deposits
Superoxidants	Ag(II), Co(III), or Ce(IV)	Dissolves actinide oxides

Once the radionuclides have been removed from the surface, the surface is washed with a clean (containing no radionuclides) solution and the base material is decontaminated. The leach solution is then processed by conventional aqueous processing techniques.

B. ELECTROPOLISHING

Electropolishing is an electrochemical process used industrially to produce a smooth surface on almost any metal.[5,6] Electropolishing removes the microscopic high points on the surface of a metal. In doing so, contaminants that are adherent on the surface are also removed. Electropolishing is performed by immersing the contaminated item into a electrolytic cell, making the contaminated item anodic, and passing current, resulting in anodic dissolution of the surface material. Electropolishing can decontaminate metal in a few minutes with minimal metal removal and produces smooth surfaces that can easily be rinsed of contaminated electrolyte.

Electropolishing is performed by immersing the contaminated metal into a tank, or by filling the interior of a contaminated component, such as a pipe or tank, with electrolyte, or by an electroswabbing technique in which current is passed through electrolyte absorbed on Dacron felt. Studies at the Pacific Northwest Laboratory[5] have demonstrated, using electropolishing, the complete removal of plutonium oxide contamination, at levels of 1 million disintegrations per min · 100 cm^2, from stainless steel components. The time required for this operation is less than 10 min. Electropolishing has also been demonstrated in the decontamination of plutonium from large components (>15 ft^2). The inside of contaminated piping has been decontaminated by electropolishing using pipe end caps, filling the pipe with electrolyte, and applying a current between the pipe and a center electrode.

Typical operational conditions for electropolishing using phosphoric acid electrolytes are as follows: solution temperatures between 40 and 80°C, phosphoric acid concentrations of 40 to 80%, electrode potentials of 8 to 12 VDC, and current densities of 500 to

2700 A/m^2. Generally, 5 to 50 μm of material is removed from surfaces that are not heavily corroded or pitted.

Electropolishing is limited to conductive materials and requires a pretreatment operation to remove paint or other nonconducting surface material. If an item is to be reused, complex electrode shapes may be required to avoid preferential metal removal at sharp edges. Components made from several metals or alloys may experience preferential attack in a bath immersion system. One major disadvantage of electropolishing for decontamination is that it is labor-intensive, which can result in increase radiation dose to personnel.

C. FLUORIDE VOLATILITY

The use of fluoride volatility to process and separate plutonium and uranium from spent reactor fuels was examined in the 1960s by Jonke.[7] Fluoride volatility is based on the fact that the higher fluorides of uranium and plutonium (UF_6 and PuF_6) have very high vapor pressures at near-ambient temperatures. Thus, reacting surface contaminants with fluorine gas can result in removal of the actinides from the surface. However, elevated temperatures are required in order for the reaction to proceed at a reasonable rate. Although applicable as a possible surface treatment, the highly corrosive nature of hot fluorine gas has not made this technique very attractive. A modification to this technique uses dioxygen difluoride as the fluorination agent.

The use of dioxygen difluoride and krypton difluoride for the recovery of plutonium from lean residues has been investigated at Los Alamos National Laboratory[8] and proposed as a surface treatment option. Dioxygen difluoride can be formed by the following reaction but only has a half-life of a few seconds at room temperature.

$$F_2(g) + O_2(g) = FOOF(g) \qquad (1)$$

The dioxygen difluoride molecule reacts rapidly with plutonium and uranium compounds forming uranium and plutonium fluorides. The key to using dioxygen difluoride for removing actinides from surfaces is the manufacture and transport of the molecule to the surface before it decomposes.

III. PHYSICAL REMOVAL TECHNOLOGIES

Physical removal technologies do not change the chemical state of the materials, but remove materials from surfaces using mechanical means. The following are typical physical removal technologies:

Abrasion
Dry-ice blasting
Grit or shot blasting
High-pressure Freon®, water, or steam
Laser ablation
Mechanical techniques
Scrubbing
Strippable coatings
Ultrasonics
Vibratory finishing

Physical removal technologies are applicable to a variety of base materials, such as metal, rubber, plastic, glass, wood, and concrete. In addition, contaminated surface materials such as paint, oil, grease, dirt, tape, and corrosion layers can be removed effectively. Typically, physical removal technologies are required prior to chemical removal technologies to ensure that the surface material is reactive to chemicals. An example is the removal of paint from a surface by vibratory finishing prior to electropolishing. A review of decontamination techniques was performed by R. P. Allen of Battelle, Pacific Northwest Laboratories, in 1985.[9] This section provides summary information from this review and discusses additional techniques.

A. ABRASIVE CLEANING

Abrasive cleaning includes several different decontamination techniques. The abrasive may be propelled against the surface by a fluid, rubbed against the surface, or used in a mechanical system to grind or hone the surface. Abrasive cleaning techniques include scrubbing, sanding, dry-ice blasting, grit blasting, and shot blasting.

Scrubbing is the most widely and frequently used surface decontamination technique. Contaminated surfaces are wiped or scrubbed by hand or with power equipment. Loose contamination on smooth surfaces may be removed by simple wiping. Embedded contamination, on the other hand, may require an abrasive pad or cleaning agent. Abrasive cleaning is applicable to a wide range of materials including metal, wood, rubber, and plastic.

Secondary waste from abrasive cleaning can be significantly reduced by using dry ice or water ice as the abrasives. In dry-ice systems, solid carbon dioxide particles are propelled by dry compressed air on to the surface of a contaminated material. The carbon dioxide particles shatter upon impact with the surface and flash into gas. The carbon dioxide gas transports the contaminated particulates to high-efficiency particulate air (HEPA) filters. Commercial units are available and have repeatedly demonstrated contamination removal to levels that allow materials to be released within radiation control areas (RCA).[10] Items that are wrapped in plastic and tape can be directly fed to a dry-ice decontamination station. The impacting dry-ice particles will easily remove the plastic and tape.

B. HIGH PRESSURE CLEANING

High-pressure fluids are commonly used to remove contamination from surfaces. Several fluids have been used, with the majority of systems using high-pressure water or Freon.[9] Water sprays operating at pressures up to 10,000 psi are used to remove loose contamination. Sprays operating at pressures over 10,000 psi have demonstrated removal of not only loose contamination, but also base metal. High-pressure water can be used to decontaminate large and small components with simple and irregular geometries. High-pressure water decontamination can be enhanced by the addition of abrasives or chemical agents. Disadvantages of high-pressure water sprays include possible redistribution of contamination, possible production of large volumes of liquid radioactive waste, and possible formation of airborne contamination.

High-pressure Freon cleaning systems operate at pressures up to 3000 psi and are limited to removal of smearable contamination. These systems operate in a recycle mode with the Freon being filtered and reused. Freon is significantly better than water at removing contamination associated with grease and oil. With current environmental concerns and restrictions regarding CFCs, the use of Freon in future decontamination applications is doubtful. However, alternate solvents such as super critical carbon dioxide are being investigated.

C. LASER ABLATION

Laser ablation can be used to remove both a contaminated surface film, such as paint or base material. The application of laser ablation to decontaminating nuclear wastes is in the early stages of development and demonstration. Laser ablation consists of focusing a powerful laser beam on a contaminated surface and ablation of the material from the surface to the surrounding atmosphere. The contaminated particulates in the gas phase can then be transported and collected using HEPA filtration. Efforts to demonstrate this technology are being conducted at Ames Laboratory and Westinghouse Hanford Co.

D. MECHANICAL TECHNIQUES

Mechanical techniques to remove radionuclides from the surface of materials are generally limited to simple geometries such as lead bricks or the inside of piping. For soft materials like lead, a planing or milling operation can remove surface contamination in addition to a large amount of base metal; however, this technique is very labor-intensive. Machining operations have mainly been used for separating high-value materials like plutonium and highly enriched uranium in nuclear weapons.

For piping, a variety of mechanical techniques have been demonstrated to remove contamination.[9] These techniques include rotating brushes, cutters, and scrapers and devices propelled or forced through the pipes by fluids. These devices are currently used commercially in the oil and gas industry. These technologies can be used to decontaminate surfaces inside pipes *in situ* and are capable of even cleaning pipe elbows.

E. STRIPPABLE COATING

Strippable coatings are generally polymeric materials that can be applied to a surface as a liquid and then cured to form a flexible film that entraps the surface contaminants and can be pulled away from the surface. Strippable coatings are used to remove contamination from large areas such as floors, walls, and the surface of glove boxes. The coating is applied by standard painting techniques, such as airless spraying, brushing, or roller methods. Loose contamination that presents an airborne release concern can readily be controlled by using strippable coatings. The major disadvantages with strippable coating techniques are that only loose contamination can be removed and the application and removal process is labor-intensive.

F. VIBRATORY FINISHING

Vibratory finishing is another commercial process that has been demonstrated as a decontamination technique.[11] This technique uses a bed of rapidly vibrating ceramic or metal media to hone the surface and remove contaminants. Vibratory finishing has been used to remove contamination to below TRU levels, less than 0.2 nCi/g, from a wide range of materials including metals, plastic, glass, and rubber. The remaining contamination is primarily imbedded in the surface and is not smearable. Vibratory finishing can be used to remove paint or other surface coatings that would inhibit decontamination by other methods. Using this technology to decontaminate soft metals such as lead and copper can be problematic. The loose contamination is reduced, but the fixed contamination levels remain high. The major disadvantage with vibratory finishing is that commercial units can only process items that are about 8 to 12 in. Larger items require disassembly or sectioning.

G. ULTRASONICS

Ultrasonic vibrations are used commercially to clean a variety of items. However, most commercial systems are designed for small items such as laboratory ware or watch parts. Ultrasonic cleaners use high-frequency mechanical vibrations transmitted through a fluid, resulting in cavitation of the fluid at the surface of the contaminated item. Ultrasonics have been demonstrated to be effective for decontamination, especially when used in conjunction with chemical agents.[9] The limitation of this technique is that items to be decontaminated are limited to the size of tank employed.

REFERENCES

1. **Manion, W.J. and LaGuardia, T.S.,** *Decommissioning Handbook,* Nuclear Energy Services, Inc., Danbury, CT, November 1980.
2. *Proceedings of the ANS Meeting on the Treatment and Handling of Radioactive Waste,* Battelle Press, Columbus/Richland, 1982.
3. **Mickelson, S., Ed.,** *Proceedings of the 1982 International Decommissioning Symposium,* Technical Information Center, U.S. Department of Energy, Washington, D.C., 1982.
4. **Tarcza, G., Ed.,** *Proceedings of the 1987 International Decommissioning Symposium,* USDOE CONF-871018, National Technical Information Service, Springfield, VA, 1987.
5. **Allen, R.P., Arrowsmith, H.W., Charlot, L.A., and Hooper, J.L.,** "Electropolishing as a Decontamination Process: Progress and Applications," Battelle, Pacific Northwest Laboratories, PNL-SA-6858, April 1978.
6. **Allen, R.P.,** "Electropolishing Applications in the Nuclear Industry," *Electrochemical Engineering Applications,* AIChE Symposium Series, R.E. White, R.F. Savinell, and A. Schneider, Eds., American Institute of Chemical Engineers, New York, (254): 156–160, 1983.
7. **Jonke, A.A.,** "Reprocessing of Nuclear Reactor Fuels by Processes Based on Volatilization, Fractional Distillation and Selective Absorption," *At. Energy Rev.,* 3, 1–60, 1965.
8. **Campbell, G.M., Foropoulos, J., Kennedy, R.C., Dye, B.A., and Behrens, R.G.,** "Extraction of Plutonium from Lean Residues by Room-Temperature Fluoride Volatility," *Emerging Technologies in Hazardous Waste Management,* D. W. Tedder and F. B. Pohland, Eds, American Chemical Society, Washington, D.C., 1990.
9. **Allen, R.P.,** "Nonchemical Decontamination Techniques," *Nucl. News,* 112–116, June 1985.
10. "CO_2-pellet cleaning process cuts costs," *Nucl. Eng. Int.,* 53–54, February 1992.
11. **Allen, R.P., Fetrow, L.K., and McCoy, M.W.,** "An Integrated Decontamination System for Surface-Contaminated TRU Waste," *Proceedings of the ANS Meeting on the Treatment and Handling of Radioactive Waste,* Battelle Press, Columbus/Richland, 104–108, 1982.

Chapter 12

Physical Treatment Techniques

Laura A. Worl

I. INTRODUCTION

Physical separation processes rely on the physical properties of materials and have demonstrated a strong potential for treating radioactive solid wastes. In practice, the majority of radioactive waste is treated by traditional chemical dissolution processes. The physical separation of mixed particulate solids entails selective partitioning and/or concentration of certain solids from a mixture. Radionuclides often exhibit significantly different physical properties and surface chemistry from that of a host material. For example, density, magnetic susceptibility, surface wettability, and electrical charge of actinides and certain fission products are markedly different from that of graphite, sand, quartz, or clay. These differences in physical properties offer a basis for a separation. The physical separation processes that are being developed for solid–solid separations for radionuclide applications are discussed below.

Most physical separation processes have been extensively utilized in the mining industry on a large-scale basis.[1,2] A variety of unit operations exist from the mining industry that are applicable to certain particle size ranges. A convenient guide is presented in Figure 1.[3] Application of these processes for radionuclide separation have been most extensively demonstrated in soil remediation.[4] Soil remediation has been a growing area because of the demand for remediating contaminated sites across the U.S. Department of Energy complex. There are also many forms of solid residues that are contaminated with radionuclides that exist in the nuclear industry. Chemical processing and nuclear reactor residues that have been treated using physical separation techniques include graphite casting molds, insulating sands, pyrochemical salts, incinerator ash, and decladding sludges. Physical separation of solid radioactive waste residues results from the need to address metal-contaminated residues that originate from inefficient processing technologies.

The benefits of concentrating radionuclides in waste residues by physical separation methods are straightforward. One of the major benefits is that a physical separation process concentrates actinides, heavy metals, and most fission products, while generating minimal quantities of secondary wastes. The technologies only partition the existing waste volume. A high contaminant concentration of one component, A, and a low contaminant concentration of a second component, B, are generated. If the separation efficiency is high, component B can be discarded and component A, with reduced volume, can be further processed or packaged for disposal. The ability to concentrate the actinides from extraneous materials before further processing yields more efficient recovery or treatment operations. For example, concentrating the contaminants reduces the volume of chemical reagents (acids) necessary for subsequent operations. Increased concentration of contaminants in dissolved feeds allows a more efficient ion exchange or solvent extraction unit operation. Because less extraneous material is leached and dissolved, the salt load on subsequent waste treatment operations is reduced. In addition, the majority of these techniques have already been demonstrated commercially on a large-scale basis.

0-8493-4876-5/96/$0.00+$.50
© 1996 by CRC Press, Inc.

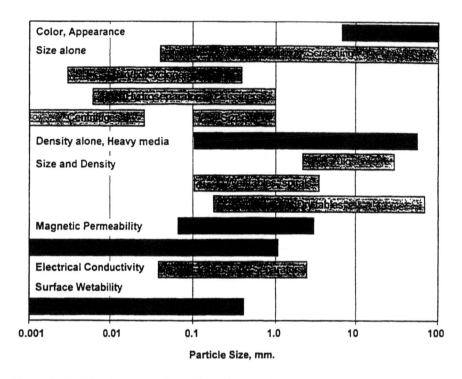

Figure 1 Particle size ranges for solid–solid separations and the physical attribute of the material that controls the separation technique.

II. MAGNETIC SEPARATION

Magnetic separation is a physical separation process that segregates materials on the basis of magnetic susceptibility.[5,6] When a paramagnetic particle encounters a nonuniform magnetic field, the particle moves in the direction in which the field gradient increases. Diamagnetic particles react in the opposite sense. When the field gradient is of sufficiently high intensity, paramagnetic particles can be physically captured and separated from extraneous nonmagnetic material.

Because all actinide compounds are paramagnetic (Table 1), magnetic separation of actinide-containing mixtures is feasible. Magnetic separation of plutonium and chemical-process residues has been demonstrated on an open-gradient magnetic separator.[7,8] The advent of reliable superconducting magnets also makes magnetic separation of weakly paramagnetic species attractive.

Numerous magnetic separation methods exist.[5,6] The methods currently being developed for radioactive waste-processing applications are dry-powder magnetic separators and high-gradient magnetic separators.

Two types of dry-powder separators have successfully demonstrated the applicability for magnetic separation in the treatment of radionuclides. Open-gradient magnetic separation (OGMS) is based on the formation of low-intensity magnetic field gradients where particles are deflected by the magnetic gradient according to their magnetic susceptibility. The available magnetic field is up to 20,000 G. In general, the sample is fed onto a vibrating tray where flow rate along the tray is determined by a combination of slope and

Table 1 Volume Magnetic Susceptibility
of Selected Compounds and Elements

Compound/element	Susceptibility $\times 10^6$ (SI)
FeO	7178.0
Fe_2O_3	1479.0
UO_2	1204.0
Cr_2O_3	844.0
NiO	740.0
Am	707.0
Pu	636.0
U	411.0
PuO_2	384.0
CuO	242.0
RuO_2	107.0
UO_3	41.0
CaO	−1.0
ZrO_2	−7.8
MgO	−11.0
$CaCl_2$	−13.0
NaCl	−14.0
CaF_2	−14.0
SiO_2	−14.0
MgF_2	−14.0
Graphite	−14.0
Al_2O_3	−18.0

vibration amplitude. The magnetic force that is exerted on the particles is in a direction perpendicular to the particle flow and field direction. Diamagnetic particles experience a force that pushes them toward a low-intensity field, while the opposite is true of paramagnetic particles. At the lower end of the tray, a divider separates the particles into two fractions.

A second dry-powder separator is based on a permanent magnetic rare-earth roll. A diagram of a roll separator is shown in Figure 2. In practice, a dry powder is delivered onto a thin belt that moves over rollers. The front roller is a rare-earth permanent magnet that has a field intensity of about 8000 to 9000 G. Ferromagnetic and sufficiently paramagnetic particles are attracted to the permanent magnet and adhere to the belt in the regions of the magnet. As the belt moves away from the magnet, the ferromagnetic and paramagnetic particles disengage from the belt and are collected in a catch pan. Diamagnetic and nonmagnetic particles pass over the magnetic roller relatively unaffected and are collected in a different catch pan. The operation results in a separation. This method is especially effective on a large particle size range of 90 to 850 μm.

The high-gradient magnetic separation (HGMS) method is used to separate magnetic components from solids, liquids, or gases. A diagram of the method is shown in Figure 3. Most commonly, the solid is slurried with water and passed through a magnetized volume. High magnetic field gradients are produced in the magnetized volume by a ferromagnetic matrix material such as steel wool, steel balls, or nickel foam. Ferromagnetic and paramagnetic particles are extracted from the slurry by the ferromagnetic matrix, while the diamagnetic fraction passes through the magnetized volume. The

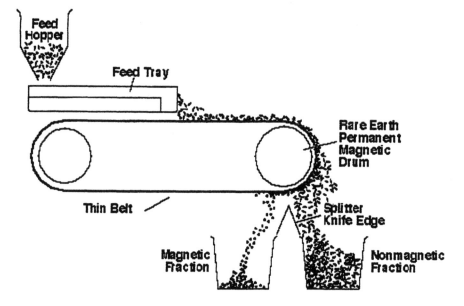

Figure 2 Diagram of a roll magnetic separator.

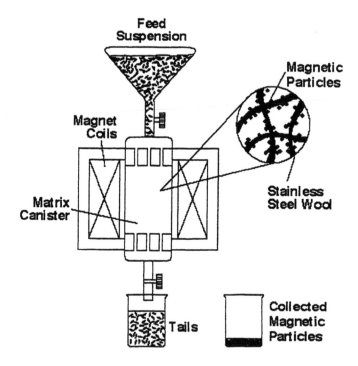

Figure 3 Simplified HGMS diagram.

magnetic fraction is flushed from the matrix later when the magnetic field is reduced to zero or the matrix is removed from the magnetized volume. HGMS effectively addresses ~0.3- to 90-μm sized particles and is complementary to a dry-powder separator.

Conventional electromagnets are limited by the saturation of iron to a magnetic field strength of about 20,000 G. Much higher fields are now routinely available with today's superconducting magnet technology. Higher fields offer the possibility of a broader range of HGMS applications.

A. DRY POWDER MAGNETIC SEPARATION

A variety of tests have been conducted on uranium- and plutonium-contaminated waste materials. Plutonium-contaminated graphite powder, bomb-reduction sand, sand, slag, and crucible (SS&C), electorefining (ER) salt, direct oxide reduction (DOR) salt, incinerator ash, uranium-contaminated magnesium fluoride slag, and contaminated glass shards from molding scraps have been tested by dry-powder magnetic separation. The best results were obtained from residues where the paramagnetic particles are discrete or can be made discrete from the host materials.

Oak Ridge National Laboratory (ORNL) performed bench scale magnetic separation tests using a Frantz Model L-1 open-gradient magnetic separator.[8,9] Several tests were done on uranium-contaminated MgF_2 residues that contained ~2% uranium. In the best results, 95% of the uranium was concentrated in 6% of the initial bulk residue. The particle size range for the high performance was limited to 75 to 100 μm. A less efficient separation was observed from materials with a broader particle size range. Other materials that were tested included uranium-contaminated soil and pond sludge samples from ORNL and Westinghouse Hanford Company (WHC) in Richland. The magnetic separation tests with these materials showed small separation efficiencies. Yet, several magnetic separation parameters remained to be investigated.

Los Alamos National Laboratory (LANL) has tested two forms of dry-powder separators for plutonium-contaminated residues.[7,10] OGMS work was done using a Frantz Model LB-1 separator on an experimental scale. Test residues included SS&C, bomb reduction sand, graphite, ER salt, DOR salt and incinerator ash. Residues were crushed and particle sized prior to a magnetic separation test. Experiments on graphite, SS&C, and bomb reduction residues resulted in a plutonium-rich fraction and a plutonium-lean fraction. The plutonium content of the lean fraction was reduced from about 2% in the feed to 0.1 to 0.2%. The best performance was observed when the fine particles (less than 45 to 90 μm) were removed. For these cases, about 80 to 90% of the plutonium was concentrated in 15 to 30% of the initial bulk material. The plutonium concentrations in the plutonium-lean fractions were low enough to be discarded without further processing. OGMS of pyrochemical salts were less favorable. While a plutonium-lean fraction was obtained, the final plutonium concentrations were too high to be discarded. The incinerator ash samples were not successfully processed by OGMS to give a plutonium-rich and -lean fraction. Yet, the particle size of the ash material was very small, suggesting that the HGMS method would be more applicable.

The OGMS method demonstrated the applicability of magnetic separation for actinide-contaminated residues. Yet, for larger scale applications, the technique is limited because of a low processing rate. To improve processing rates for dry powders, a rare-earth roll separator has been tested at LANL.[11] Several lots of graphite powder, bomb reduction sand, and SS&C residues have been processed with variable results. The best results were obtained with graphite where particles were sized greater than 125 μm. In these cases, ~85% of the plutonium was concentrated into ~4% of the bulk material. The plutonium content of the lean fraction was sufficiently low in plutonium so that it could be discarded without further processing. The roll separator results were not as favorable with MgO or SS&C residues. In the best case for MgO, 68% of the plutonium was concentrated in 44% of the bulk material. For SS&C, the results were inconclusive with no discrete separation. Yet, magnetic separation on these residues has been successfully demonstrated with OGMS[7] as discussed above. For a roll separator, performance can be increased by an increased magnetic field intensity. In magnetic roll separators, the magnetic field intensity acting on the feed particles is inversely proportional to the thickness of the belt. Very thin (0.005 in.) graphite-impregnated kevlar belts and stainless steel belts are available for this improvement. A second approach is processing without a belt; in this case the powders are fed directly onto the permanent magnetic roll. Further tests are planned on chemical-processing residues on an International Process Systems, Inc. roll separator with increased magnetic field intensity.

Argonne National Laboratory researchers in Idaho have utilized a rare-earth roll separator on uranium wastes produced from the Integral Fast Reactor project.[12] A major concern is fuel recovery for reprocessing purposes. Contaminated reactor mold scraps result from the demolding process of the fuel from the quartz molds. The purpose is to recover uranium and minimize actinide loss during the demolding process. Preliminary experiments were done on four different particle sized fractions. Material less than 300 μm showed no separation with the roll separator. The best magnetic separations were observed with particles less than 4 mm and greater than 300 μm. The results indicated that it is possible to recover more than 95% of the uranium while reducing the bulk volume by 80%. It was also observed that as the particle size range decreased, the grade of uranium also decreased. It was concluded that dry-powder magnetic separation is a suitable technology for fuel reprocessing of waste materials.

B. HIGH-GRADIENT MAGNETIC SEPARATION

Currently, high-gradient magnetic separators are used extensively in the clay-processing industry.[13] A highly desirable white clay can be produced for paper, coating, and rubber industries. High-purity white kaolin clay is obtained with HGMS by removing trace amounts of colored paramagnetic particles. HGMS has also been used for treatment of hazardous industrial waste streams originating from mineral beneficiation, coal desulfurization, and water purification.[5] HGMS has been applied extensively to particle/liquid separations[14,15] and a lesser extent to particle/gas separations.[16] For radionuclide applications, HGMS has been studied for the extraction of radioisotope particles from liquid nuclear waste streams. In these cases two approaches are taken: (1) the contaminants are extracted by the nature of their paramagnetic forces[17] or (2) the contaminants are coprecipitated with highly magnetic ferric hydroxides or magnetite, in which case the combined particles are extracted.[18,19] For solid/solid separations in the treatment of solid nuclear wastes, HGMS has been applied to actinide-contaminated sludges, scrap, chemical-processing residues, and soils. These applications are discussed below.

Soil remediation applications are being developed at LANL,[20,21] in a cooperative research and development agreement with Lockheed Environmental Systems and Technologies, to extract and concentrate the actinides from contaminated soil sites. A comprehensive series of HGMS experiments with nonradioactive surrogates have been completed, and work has progressed to tests with radioactive material. The results to date are very promising. The first tests with a spiked clay type soil showed decontamination from about 1500 to less than 4 pCi/g with an applied field of 2 T. Figure 4 shows these results at

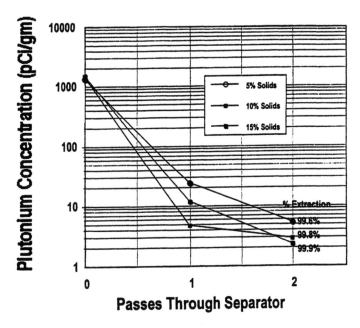

Figure 4 HGMS Extraction of plutonium oxide from soils. External magnetic field strength of 2.0 T; 2- to 5-μm PuO₂ particles.

varied solids concentrations and where there is more than three orders of magnitude decontamination. High-magnetic field HGMS tests have recently been performed on plutonium-contaminated Rocky Flats Plant soil and uranium-contaminated Fernald soil with a superconducting magnet. From predicted values[21] and surrogate work, increased decontamination levels at field strengths above 2 T are expected.

For chemical-processing applications, LANL has performed a series of HGMS experiments on contaminated graphite residues.[22] Copper oxide served as the surrogate for plutonium oxide. Particles were sized less than 90 μm. The series of tests have shown that HGMS performance is strongly influenced by the graphite solids content. The highest separation efficiency was observed under conditions of low solids fraction (<5% solids), with a fairly minor effect from external field strength. As discussed earlier, the effective particle size range for the roll separator on dry powders is greater than 90 μm. Thus, the coupling of HGMS with the roll separator would be advantageous for treatment of chemical-processing residues.

The removal of uranium from Magnox sludges with HGMS has been studied in small-scale tests.[23,24] Magnox sludge results from the mechanical breakdown of Magnox fuel. During the process, uranium-contaminated magnesium alloy cladding debris is generated. After storage under water, a sludge of magnesium hydroxide and uranium hydrous oxide is formed. The majority of the activity is located in the <100-μm fraction. The results indicated that with a well-dispersed system of uranium dioxide and magnesium hydroxide, separation efficiencies of more than 75% can be achieved for surrogate systems. For actual Magnox sludges, the efficiencies were lowered to 50%.

The application of HGMS for underground storage tank waste is being developed by LANL and WHC.[25,26] The potential exists for HGMS to take certain types of tank waste directly to nontransuranic (non-TRU) levels in one step. The initial tank waste target is neutralized cladding removal waste (NCRW). NCRW is formed from the chemical decladding of Zircaloy-clad metallic uranium fuel, which is made alkaline for storage. It contains zirconium hydrous oxide solids, TRU elements (~1000 nCi/g in sludge), and mixed fission products. Surrogate components for NCRW tank waste have been prepared. The material was generated from plutonium–uranium extraction (PUREX) plant reagents used in NCRW production. Uranyl nitrate or iron nitrate was added to the acidic solution serving as the TRU component prior to precipitation of the hydrous oxides. HGMS tests of aged and unaged slurries containing precipitated iron and zirconium have been conducted at a 1 T field. The results indicated that 84% of the aged ferric hydrous oxides can be captured from a slurry. With this favorable result using the iron solutions, HGMS tests are planned on zirconium/uranium hydrous oxide solutions. In addition, WHC has prepared a simulated NCRW sludge by the PUREX dissolution of an unirradiated fuel pin.[26] The slurry contains zirconium hydrous oxide solids with low level of uranium particles. HGMS tests for this material are under way.

There are more than ten variables that independently affect the magnetic separation process. LANL has developed an analytical model for the HGMS process that provides guidance in selecting the appropriate bench-scale experiments to perform and assist in analyzing the resulting data.[21] Much work[14,27-31] has been done to model the behavior of the paramagnetic particles as they interact with the magnetized matrix. Dynamic effects have been investigated, and the influence of particle buildup on the matrix elements have been studied. In general, if the particles are physically liberated from the host material and are not electrically charged, the principal forces governing their behavior are magnetic, viscous, and gravitational. The performance of the magnetic separator is modeled using a static force balance on an individual paramagnetic particle in the immediate vicinity of a matrix element, as shown in Figure 5. The model assumes that if the magnetic capture forces are greater than the competing viscous drag and gravitational forces, the particle is captured and removed from the flow stream.

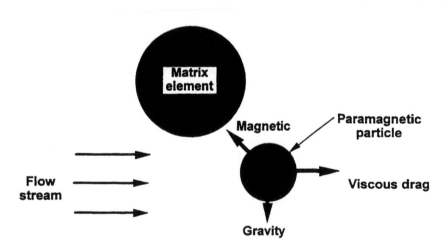

Figure 5 Force balance on a paramagnetic particle in the vicinity of a matrix element.

Table 2 Electrical Conductivity and Specific
Gravity of Selected Compounds and Elements

Compound/element	Electrical conductivity $\times 10^6$ (1/Ω•m)	Density (g/cm³)
Gold	42.55	19
U	3.33	19
Pu	0.71	19.8–16.5
AmO_2	—	11.7
PuO_2	—	11.5
UO_2	—	10.9
UO_3	—	7.3
Zr/ZrO_2	2.50	6.5–5.6
MgO	—	3.6
SiO_2	10^{-4}–10^{-8}	2.2
Mg	22.47	1.7
Graphite	0.07	1.6–2.7

III. ELECTROSTATIC SEPARATION

The principle of electrostatic separation is based on exploiting the differences in the electrical conductivities of the materials in a mixture.[1] The separation is based on the fact that if one or more of the materials in a heterogeneous mixture can receive a surface charge on or before entering an electrostatic field, the particles of that material will be repelled from one of the electrodes and attracted toward the other, depending upon the sign of the charge on the particle. The electrical conductivity of different elements and materials is included in Table 2. The mining industry has used the technique for ore processing such as concentrating hematite, separating quartz, feldspar, and mica, and concentrating ilmenite, rutile, and zircon from bulk materials.[1] The application of this method for radionuclides has been used for processing actinide process residues[12,32] and uranium-contaminated sand.[9]

Several forms of electrostatic separators exist that differ by the electrification mechanism. These include contact electrification, electrification by conductive induction, and electrification by ion bombardment. The separation technique based on particle charging by conductive induction is depicted in Figure 6. In general, a feed mixture is passed over a grounded rotating drum. The conductive particles will assume the potential of the rotor, which is opposite to that of the active electrode or banks of ionizing electrodes, and will therefore be attracted toward the active electrode. The nonconductive particles will become polarized and will be attracted to the rotor and repelled by the electrode. A splitter and collection trays complete the process, resulting in the final product separation.

Argonne National Laboratory researchers in Idaho are investigating waste minimization and metal recovery techniques for the Integral Fast Reactor program,[11,32] as discussed earlier. Quartz molds are used for casting fuel pins for the reactor. In the demolding process, uranium-contaminated residues are generated, such as mold scraps, metal alloy fines, and mold washes that contain ~6% of the finished uranium product. The averaged material content of the waste stream is 21% uranium, 4% zirconium, and 75% quartz by weight. An electrostatic separator was investigated for the recovery of uranium from the molding residues. In general, a multiple-stage process was used for the uranium recovery;

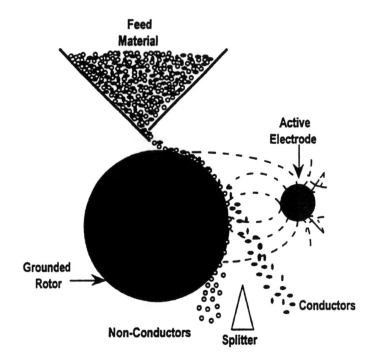

Figure 6 Simplified electrostatic separator.

the conducting phase (metal-containing phase) was passed through the separator two additional times. The additional passes were needed to remove sufficient quartz from the mold scraps. Several tests have been done on sized and unsized material. Uranium recovery values varied from 78 to 96%, depending on the particle size distribution of the feed stock. An increased performance was observed when feed material was screened greater than 4.75 μm. The electrostatic method was effectively demonstrated as a viable option for the recovery of uranium from scrap material.

Preliminary electrostatic separation tests were conducted by K. D. Engineering for ORNL on depleted uranium fragments from gun test catchments.[9] The uranium was dispersed in soil/sand hillsides. Electrostatic separations indicated the best results for sized fractions. For example, particles that were sieved less than 1.7 mm and greater than 850 μm showed that 96% of the uranium was captured in 43% of the initial bulk material. For material sized less than 500 μm, a minimal separation was observed.

IV. SEPARATION BASED ON SPECIFIC GRAVITY

The difference in specific gravity between materials has been used for centuries to effect material separations. Particle size and shape are other parameters that are factors in a separation. Simplistically, the process uses Stokes' law for differential settling of heavy fractions from the light ones. The density of selected materials and elements is included in Table 2. The density-based separation has been widely used for different minerals, including gold.[2] The gravity separation has limitations with respect to particle size; it is ineffective for fine particles.

Through the mining industry, several types of gravity separators have been developed. They are classified by the means in which a separation is achieved.[1] Jig-type separators depend on a vertical "jigging" motion; shaking concentrators and tables apply a horizontal motion, and gravity flow concentrators, such as sluices and troughs, move the slurry down an inclined surface. Classification is a third area where the separation of particles is based according to their settling rate in a fluid, depending on density and size. The three types of classifiers are hydraulic, mechanical, and nonmechanical. Hydraulic classifiers (i.e., cone, bowl, and cylindrical tank) introduce air or water so that the direction of flow opposes the settling particles; mechanical classifiers (i.e., rake, spiral, drag, and counter-current) have particles with a slow settling velocity that are carried away in a liquid overflow and particles with a high settling velocity that are deposited on the bottom; and nonmechanical classifiers (i.e., hydrocyclone, settling cone, and elutriator) rely on gravitational or centrifugal force to separate particles. These separating devices have been extensively used in the mining industry[1] and more recently in soil decontamination efforts, where they are one of several unit processes forming complete soil-washing processes.[4,9,33,34] Yet, for applications in the nuclear industry, equipment from the mining industry is currently not being used or developed.

Preliminary investigations have been performed at LANL to explore the separation of plutonium from graphite-casting residues[35] using a density-based technique. Graphite and plutonium have significantly different densities (Table 2), which would form a basis for separation if the plutonium could be made discrete from the graphite residue. One technique is the formation of a molten salt. Several feasibility experiments have been done in a calcium chloride molten-salt media. In general, the plutonium-contaminated graphite residues are crushed to reduce the large chunks of material and then heated to ~860°C to generate a calcium chloride melt. Sodium sulfate and oxygen are added to the system, which oxidizes graphite. As the material cools, three layers are generated: a top, dark, graphite-rich layer; a middle, white, salt-rich layer; and a bottom, plutonium-rich layer. The top two layers have been shown to contain ~4% of the total plutonium in ~76% of the total bulk material. The top layers of material can be discarded without further processing. The bottom layer contains ~95% of the plutonium in ~24% of the total bulk. The success of these preliminary feasibility tests have generated further investigations into the technique.

The scope of this text is not to give a detailed coverage of all soil decontamination efforts. Yet, for completeness, several techniques will be highlighted that have demonstrated the density-based segregation of radionuclides from soil matrices. Lockheed Environmental Systems and Technologies (LESAT) has developed a soil-washing system (TRUclean) that is based on gravimetric separation or a mineral jig device.[36] The system has been demonstrated on a large-scale basis at Johnston Atoll and at the Nevada Test Site, where plutonium- and americium-contaminated soils exist. Soil was decontaminated to less than 10 pCi/g with a volume activity reduction of 90 to 98%. An effort was conducted at Rocky Flats Plant that also tested a mineral jig on Rocky Flats soil.[37] The results showed a segregation of plutonium. The <0.42-mm fraction of soil (37% of the initial bulk material) was passed to the mineral jig where two output fractions resulted: <0.42 to 0.25 mm, which was 4% of the initial bulk, and <0.250 mm, 33% of the initial bulk. Less than 2% of the plutonium was in the <0.42- to 0.25-mm fraction while ~90% of the plutonium was in the <0.250-mm fraction after the jigging. The method was effective in segregation of plutonium, but a significant bulk reduction was not observed. Earlier, Rocky Flats Plant conducted tests on a soil-washing system that contained a hydrocyclone classifier.[38] These tests were performed on plutonium-contaminated Rocky Flats Soil. They observed a 60 to 80% activity reduction in the soil after treatment, depending on the origin of the soil. The soil remediation arena is a growing field, as evidenced by the variety of industrial soil-washing systems the are being developed.[4,33,34]

A new technology called the Cambell Centrifugal Jig, developed by TransMar, Inc., is being tested for soil remediation.[39] The process is a combination of jigging and centrifuging, both widely used methods for treatment of dense particle separation. The Cambell Centrifugal Jig combines the effectiveness of continuous flow and pulsating bed of the standard mineral jig with the high gravitational forces of the centrifuge. Preliminary results on surrogate materials indicate that 70 to 90% of the contaminant (bismuth as a surrogate for plutonium) was removed from a spiked host soil.

V. AQUEOUS BIPHASIC EXTRACTION

Aqueous biphasic extraction systems are being developed by Argonne National Laboratory to treat radioactive residues.[40] The separation technique is based on the selective partitioning of ultrafine or colloid-size particles between two immiscible aqueous phases. A separation relies on the difference in the surface properties of the contaminant and the host. The aqueous/aqueous phases are generated from mixtures of unlike aqueous solutions, typically (1) a straight-chain polymer (polyethylene glycol, PEG) and (2) an inorganic salt or a branched polymer (dextran) will form a two-phase system. For example, if equal volumes of a 15 wt% solution of sodium sulfate and a 30 wt% solution of PEG are mixed, a two phase system is generated with the salt concentrated in the bottom layer and the polymer in the top layer. When small particles of different surface characteristics are introduced into the biphase system, they partition selectively to the layer that generates attractive forces between the specific particle and the surrounding solvent. In an UO_2/clay mixture, UO_2 particles can be isolated in the lower salt layer while clay particles are isolated in the upper polymeric layer. The suspended solids can be removed by centrifugation or the polymer can later be isolated by ultrafiltration, allowing the aqueous phases to be recycled. Advantages of the technique are that submicrometer contaminant particles can be isolated. The broad particle size range that has been effectively tested is 0.020 to 40 μm.

Contaminated chemical-processing residues have been tested in a feasibility study for the aqueous biphasic extraction technique.[40a] The goal is to produce a plutonium concentrate that would integrate with existing processes and also to produce a non-TRU waste stream. Contaminated residues examined include ash heels, LECO crucibles, incinerator ash, SS&C, and graphite. Preliminary studies have centered on surrogate metal oxides (Fe_2O_3, Al_2O_3, and TiO_2) or plutonium-spiked graphite and sand. In general, the particle size of the materials was less than 1 μm; thus, actual plutonium-contaminated residues need to be pretreated by ultrafine grinding prior to the separation. The grinding step will also aid in liberation of contaminant particles from the host material. Results from partition coefficient studies are listed in Table 3.[40a] The results illustrate that plutonium oxide can be selectively isolated in a lower sulfate layer and materials such as graphite and silica can be isolated in an upper PEG layer. Additional studies have examined the use of surfactants or complexants for their ability to completely extract plutonium particles in a specified phase. This will allow selective control over the partitioning of plutonium from residues.

Chaiko et al. have also explored the application of this method for uranium-contaminated soils.[40b] The initial soil was collected on the Fernald site at two locations: near the

Table 3 Partitioning of Metal Oxides and
Selected Materials in a Sodium Sulfate/PEG
Aqueous Biphasic System[a,b]

| Compound/element | Partition coefficient[c] | |
	Metal oxide	Host material
Graphite	na[d]	>200
Silica	na	>200
Quartz	na	>200
Monomeric PuO_2	0.17	na
Polymeric Pu(IV)	0.001	na
PuO_2/SiO_2	0.009	>10^4
PuO_2/graphite	4.0	>10^3
PuO_2/graphite[e]	0.04	>10^3
UO_2/kaolinite	0.001	34

[a] Data is from Chaiko et al.[40]
[b] Biphasic solution is composed of 15% PEG/7.5% sodium sulfate at pH 3.
[c] The partition coefficient is determined by the quantity of a specified material in the top PEG phase divided by the quantity of the same material in the bottom sulfate phase; i.e., for values >200, the material is located in the top phase.
[d] None added to system.
[e] 0.01% of dispersant, aerosol OT (sodium *bis*(2-ethyl-hexyl)sulfosuccinate), was added.

waste incinerator area (A-14) and near the Plant 1 storage pad area (B-14). Prior to the treatment of real soil samples, surrogate work was done on UO_2 in kaolin clay mixtures. In these tests, the uranium concentration was reduced from 200,000 ppm to less than 50 ppm. The partition coefficients, included in Table 3, resulted in an overall single-stage separation factor of 34,000. Tests from the two types of Fernald soils showed mixed results. For soil that originated near the Plant 1 storage pad, no selectivity with regard to uranium partitioning was found. This suggests that the uranium is adsorbed into the soil particles. For the incinerator pad soil, partitioning results were more positive, showing a definite concentration of uranium. For example, in a 15% PEG/10% Na_2CO_3 biphasic system, a feed sample was introduced that contained 545 ppm uranium. After phase separation, 904 ppm uranium was concentrated in 1.1% of the bulk material in the sulfate phase, and in the PEG phase, 138 ppm uranium was in 65% of the bulk material. Complications existed due to the presence of uranyl (UO_2^{2+}) species in the soil. The uranyl species was found to dissolve during the separation tests. This reduced the amount of uranium recovered in the solid concentrate and increased the quantity of uranium for secondary liquid waste treatment. Scale-up tests are being performed with a Karr column. The system is capable of handling a broad distribution of pariticle sizes (from <1 to 53 μm) and a solids throughput of 175 g/h.

VI. APPLICABLE TECHNOLOGIES

The physical separation techniques described above have been demonstrated for the treatment of the nuclear industry contaminated residues and waste streams. Several methods have proven to be effective for separating actinides from contaminated soil matrices and have not yet been tested in the nuclear industry waste-processing schemes.

A. SEGMENTED GATE SYSTEM

One method that has proven effective for the removal of plutonium and americium from a contaminated coral matrix is the segmented gate system.[41] Johnson Atoll contains a small coral island that has been contaminated with plutonium and americium. TMA/Eberline has been contracted by the Defense Nuclear Agency to remediate 100,000 yd³ of contaminated coral. The segmented gate system uses arrays of sensitive radiation detectors that are computer interfaced to the segmented gate system. Contaminated soil is diverted from a moving feed supply when radioactive particles are detected. More specifically, 15 NaI detectors are arranged in two overlapping rows that are sensitive to the low-energy, 60-keV gamma radiation emitted by americium-241 (a decay product of plutonium-241). A conveyer belt carries feed material (30 ft/min) under the detectors. When radioactive particles are detected, one or more of the eight diversion segmented gates located at the end of the conveyor belt is electronically directed to divert the contaminated material. The contaminated soil is either diverted to a steel drum if the activity is greater than 135 nCi, or to a supplementary soil-washing process designed to remove dispersed low-level contamination from a soil fraction of small particles.

The results have been extremely promising. Over 4500 t have been processed and 10,800 μCi of contamination have been removed from the feed material. Volume reductions range from 93 to 99% per day. Soil-processing operations are expected to continue until 1998.

B. FLOTATION

Froth flotation is one of the most widely used techniques in the mineral and kaolin industries and is used for separating finely divided solids.[1] Flotation separation is possible when solids differ in surface wettability or surface chemistry, and is applicable to very small particles, 0.1- to 0.01-mm size range. The solids are suspended in a liquid (water) and are contacted by a chemical or a promoter that causes specified particles to become hydrophobic and floatable. The solids are also treated with reagents so that other specified particles become hydrophilic. The slurry is treated with flowing bubbles, and the hydrophobic particles (usually the contaminant) become attached to the bubbles where they collect in the froth and are skimmed off. The nonfloatable solids (usually the host material) are discharged from the cell bottom. The process can be made very selective by the appropriate selection of reagents.

Froth flotation has been integrated into several commercial soil-washing systems for treating actinide-contaminated soils.[4,33,34] Decontamination results for froth flotation equipment as part of soil-washing systems were not available. A lab-scale froth flotation study was done on surrogate spiked Nevada Test Site soil.[42] Results using CeO_2 or TiO_2 for plutonium have shown that more than 90% of the surrogates could be separated from the soil.

A fairly new technology that combines froth flotation principles with the flow characteristics of a hydrocyclone is being developed at the University of Utah and Advanced Processing Technologies, Inc.[39] The technology, or the Air-Sparged Hydrocyclone (ASH) flotation system can perform flotation separations in less than a second. This feature provides a high processing capacity, 100 to 600 times greater than the capacity of conventional flotation apparatus.

Table 4 Actinide Removal by Wet Screening with Rocky Flats Soil

Screening test	Particle size range (mm)	Percent of initial soil	Activity in screened soil (pCi/g Pu)
1A[a]	>4.2	65%	<6
1B	<4.2, >2.4	<15%	300
2A[b,c]	>4	66%	1.4
2B	<4.0, >2.4	0.8%	22.4
2C	<2.4, >0.42	4.9%	43.7
2D	< 0.42	27.7%	441.5

[a] Navratil, J.; Kochen, R. L., Reference 46.
[b] Kochen, R. L. et al., Reference 37.
[c] Initial feed was 500 g at 125 pCi/g of plutonium.

C. SCREENING AND SIEVING

Screening is the mechanical separation of particles on the basis of size. Screening can process an extremely wide particle size range from talc-sized particles (45 μm) to boulders (18 in.). Several classifications of screens exist, depending on the motion of the screen with respect to the particles.[1,43] Major types of screens developed for industry are the stationary grizzly; roll grizzly; sieve bends; trammel, centrifugal, and revolving screens; inclined, horizontal, and probability revolving screens; shaking screens; and reciprocating and gyrating screens. An advantage of screening is that the method is inexpensive. The problems are that the screens can become plugged, fine screens are fragile, and dry screening can produce dust.

Screening has been applied alone or as a unit in soil-washing systems for soil decontamination.[4,33,34] Several studies have indicated that the contamination in many soil sites is located in the fine particles.[34,42,44-46] Contamination that is located in large-sized fractions can be pretreated with ball milling or attrition scrubbing to reduce the particle size or liberate the contaminant from the large host material. A study was done on plutonium-contaminated soil at Rocky Flats that showed the merits of wet and dry screening for segregation of plutonium.[46] Dry screening showed no segregation of the plutonium. Yet, wet screening (see Table 4) was able to decontaminate 65% of the soil (>2.4-mm fraction) to <6 pCi/g plutonium. The 2.4- to 0.42-mm fraction was decontaminated to 300 pCi/g plutonium. Later studies again showed positive results with wet sieving.[37] The results are tabulated in Table 4. In summary, the majority of the plutonium was concentrated in a small percentage of the initial feed material. Characterization studies have also been done on Nevada Test Site soils that showed that the majority of the plutonium contamination is located in the 53- to 5-μm particle size range.[47] The bulk of the soil is in a particle size range less than 300 and greater than 150 μm. Wet sieving was proposed for concentrating the plutonium in a <75-μm sized fraction.

VII. CONCLUSIONS

Historically, processing of actinide residues has been carried out with little or no full-scale physical beneficiation prior to chemical processing. Many actinide residues generated during fuel-reprocessing and metal recovery operations are heterogeneous materials. The actinides are distributed as discrete grains or particles within these host matrices. Thus, many of the physical separation methods discussed above are ideal technologies for actinide concentration and segregation. The most efficient separations occur when the radionuclide particles are easily liberated from the host material. These technologies are applicable to specific physical properties and particle size ranges. Thus, many of the techniques can feasibly be coupled together for an efficient operation.

Application of physical separation methods for the treatment of nuclear waste streams offers reduced volumes of contaminated material for eventual chemical processing. The contaminant can be concentrated in a reduced volume and the clean fraction is reduced to the next lower waste category, i.e., TRU to low-level waste.

ACKNOWLEDGMENTS

The author wishes to thank Larry Avens for helpful discussions and the opportunity to write this chapter.

REFERENCES

1. **Perry, R. H.; Chilton, C. H.,** *Chemical Engineers' Handbook, 5th Ed.*, McGraw-Hill Book Company, New York, 1973, chap. 21.
2. **Taggart, A. F.,** *Handbook of Mineral Dressing,* John Wiley & Sons, New York, 1945.
3. **Roberts, E. J.; Stavenger, P.; Bowersox, A. K.; Walton, A. K.; Mehta, M.,** "Solid/Solid Separations," *Chem. Eng.,* Feb. 15, 89, 1971.
4. **Devgun, J. S.; Natsis, M. E.; Beskid, N. J.; Walker, J. S.,** "Soil Washing as a Potential Remediation Technology for Contaminated DOE Sites," U.S. DOE Report ANL/CMT/CP-78935, Waste Management '93, Tucson, March 1993.
5. **Oberteuffer, J. A.,** "Magnetic Separation: A Review of Principles, Devices, and Applications," *IEEE Trans. Magn.,* MAG-10, 223, 1974.
6. **Lyman, W. J.,** "High-Gradient Magnetic Separation" *Unit Operations for Treatment of Hazardous Industrial Wastes,* De Renzo, D. J. Ed., Noyes Data Corp., Park Ridge, 1978, 590.
7. **Avens, L. R.; Gallegos, U. F.; McFarlan, J. T.,** "Magnetic Separation as a Plutonium Residue Enrichment Process," *Sep. Sci. Technol.,* 25, 1967, 1990.
8. (a) **Hoegler, J. M.,** "Magnetic Separation of Uranium from Waste Materials," *Hazardous and Industrial Solid Waste Minimization Practices,* Conway, R. A.; Frick, J. H.; Warner, D. J.; Wiles, C. C.; Duckett, J. E., Eds., ASTM, Philadelphia, 1989, 172. (b) **Hoegler, J. M.; Bradshaw, W. M.,** U.S. DOE Report ORNL TM-11117, Oak Ridge National Laboratory, Oak Ridge, TN, 1989.
9. **Wichner, R. P.; Khan, A. A.; Hoegler, J. M.,** U.S. DOE Report ORNL TM-11141, Oak Ridge National Laboratory, Oak Ridge, TN, 1989.
10. **Avens, L. R.; Worl, L. A.; deAguero, K. J.; Prenger, F. G.; Stewart, W. F.; Hill, D. D.,** "Opportunities for Magnetic Separation in Complex 21," *Nuclear Materials Technology Annual Report 1992,* U.S. DOE Report LALP-92-41, Los Alamos National Laboratory, Los Alamos, NM, 1992, 69.
11. **Ramsey, K. B.,** Los Alamos National Laboratory, Los Alamos, NM, personal communication, 1993.
12. **Westphal, B. R.; Benedict, R. W.,** "Actinide Recovery Techniques Utilizing Electromechanical Processes," submitted to International Symposium on Actinides: Processing & Materials 1994 TMS Annual Meeting, San Francisco, CA, February 27–March 3, 1994.
13. **Winters, A. J.; Selvaggi, J. A.,** "Large-Scale Superconducting Separator for Kaolin Processing," *Chem. Eng. Prog.,* 37, January 1990.
14. **Watson, J. H. P.,** "Magnetic Filtration," *J. Appl. Phys.,* 44, 4209, 1973.
15. **Kolm, H. J.; Oberteuffer, J.; Kelland, D.,** "High Gradient Magnetic Separation," *Sci. Am.,* 223 (5), 47, 1975.

16. **Watson, J. H. P.; Boorman, C. H.,** "A Permanently Magnetized High Gradient Magnetic Filter for Glove-Box Cleaning and Increasing HEPA Filter Life," in *Proceedings of the 21st DOE/NRC Nuclear Air Cleaning Conference, Vol. 2,* First, M. W., Ed., San Diego, CA, 1990, 762.
17. (a) **Emory, B. B.,** "Radionuclide Removal from Reactor Wastes by HGMF," *IEEE Trans. Magn.,* MAG-17, 3296, 1981. (b) **Emory, B. B.,** "Pilot Plant Testing of Magnetic Filters for the N-Reactor Primary Cooling Circuit," *Proc. Int. Water Conf. Eng. Soc.,* 307, 1982.
18. (a) **Kochen, R. L.; Navratil, J. D.,** "Ferrite Treatment of Actinide Aqueous Waste," *Lanthanide Actinide Res.,* 2, 9–22, 1987. (b) **Boyd, T. E.; Cusick, M. J.; Navratil, J. D.,** "Ferrite Use in Separation Science and Technology," *Recent Developments in Separation Science, Vol. 8.,* Li, N. N.; Navratil, J. D., Eds., CRC Press, Boca Raton, FL, 1986, chap. 6.
19. (a) **Harding, K.; Baxter, W.,** "Application of High Gradient Magnetic Separation to Ferric Hydroxide Filtration," *IEEE Trans. Magn.,* MAG-17, 2795, 1981. (b) **Harding, K.,** "Precipitated Magnetite in Alpha Effluent Treatment," *Inorganic Ion Exchangers and Adsorbents for Chemical Processing in the Nuclear Fuel Cycle, Proceedings of a Technical Committee Meeting of Inorganic Ion Exchangers and Adsorbents for Chemical Processing in the Nuclear Fuel Cycle,* International Atomic Energy Agency, Vienna, 1985, 97.
20. **Avens, L. R.; Worl, L. A.; deAguero, K. J.; Padilla, D. D.; Prenger, F. C.; Stewart, W. F.; Hill, D. D.; Tolt, T. L.,** "Magnetic Separation for Soil Decontamination," U.S. DOE Report LANL LA-UR-93-229, Waste Management '93, Tucson, AZ, March 1993.
21. **Prenger, F. C.; Stewart, W. F.; Hill, D. D.; Avens, L. R.; Worl, L. A.; Schake, A.; deAguero, K. J.; Padilla, D. D.; Tolt, T. L.,** "High Gradient Magnetic Separation Applied to Environmental Remediation," U.S. DOE Report LANL LA-UR-93-2516, Cryogenic Engineering Conference, Albuquerque, NM, July 1993.
22. **Avens, L. R.; Worl, L. A.; Schake, A. R.; Padilla, D. D.; Prenger, F. C., Hill, D. D.; Stewart, W. F.,** Los Alamos National Laboratory, Los Alamos, NM, unpublished results, 1993.
23. **Williams, J. A.; Leslie, C. M.,** "High Gradient Magnetic Separation in the Nuclear Fuel Cycle," *IEEE Trans. Magn.,* MAG-17, 2790, 1981.
24. **Harding, K.,** AEA Technology, Decommissioning and Waste Management, Dorset, United Kingdom, personal communication, 1993.
25. **Worl, L. A.; Avens, L. R.; deAguero, K. J.; Padilla, D. D.; Prenger, F. C.; Stewart, W. F.; Hill, D. D.,** "Remediation of Hanford Tank Waste Using Magnetic Separation," U.S. DOE Report LANL LA-UR-92-3977, Waste Management '93, Tucson, March 1993.
26. **Emory, B. B.,** Westinghouse Hanford Company, Richland, WA, personal communication, 1993.
27. **Akoto, I. V.,** "Mathematical Modeling of HGMS Devices," *IEEE Trans. Magn.,* 13, 1486, 1977.
28. **Friedlander, F. J.; Takayasu, M.; Rettig, J. B.; Kentzer, C. P.,** "Particle Flow and Collection Processes in Single Wire HGMS Studies," *IEEE Trans. Magn.,* 14, 1158, 1978.
29. **Lawson, W. F.; Simons, W. H.; Treat, R. P.,** "The Dynamics of a Particle Attracted by a Magnetized Wire," *J. Appl. Phys.,* 48, 3213, 1977.
30. **Luborsky, F. E.; Drummond, B. J.,** "Buildup of Particles on Fibers in a High Field-High Gradient Separator," *IEEE Trans. Magn.,* 12, 463, 1976.
31. **Stekly, Z. J. J.; Minervini, J. V.,** "Shape Effect of the Matrix on the Capture Cross Section of Particles in High Gradient Magnetic Separation," *IEEE Trans. Magn.,* 12, 474, 1976.
32. **Westphal, B. R.; Benedict, R. W.,** "Process Evaluations for Uranium Recovery from Scrap Material," *Residues and Effluents — Processing and Environmental Considerations,* Reddy, R. G.; Imrie, W. P.; Queneau, P. B., Eds., The Minerals, Metals and Materials Society, Warrendale, PA, 1992, 681.
33. **Eagle, M. C.; Richardson, W. S.; Hay, S. S.; Cox, C.,** "Soil Washing for Volume Reduction of Radioactively Contaminated Soils," *Remediation,* Summer, 327, 1993.
34. **Gerber, M. A.; Freeman, H. D.; Baker, E. G.; Riemath, W. F.,** "Soil Washing: A Preliminary Assessment of Its Applicability to Hanford," U.S. DOE Report PNL-7787 UC-902, September 1993.
35. **West, M.; Bird, G.; Vargas, A.,** Los Alamos National Laboratory, Los Alamos, NM, personal communication, 1993.

36. (a) **AWC Lockheed,** "TRUclean A Process for Decontaminating Radioactive and Hazardous Soils," AWC, Inc., Las Vegas, NV, 1990. (b) **AWC Lockheed,** "Demonstration of the Use of the TRUclean Process to Remove Radium Contaminants from Soils at Superfund Sites," AWC, Inc., Las Vegas, NV, 1990.

37. **Kochen, R. L.; McGlochlin, S. C.,** "Actinide Removal from Soil at Rocky Flats," unpublished presentation, 1990.

38. **Garnett, J. E.; Mitchell, D. L.; Faccini, P. T.,** "Initial Testing of Pilot Scale Equipment for Soil Decontamination," U.S. DOE. Report RFP-3022, Rockwell International, Rocky Flats Plant, CO, 1980.

39. **Mathur, S. P.,** "Heavy Metals in Contaminated Soil Treatability Project," U.S. DOE, Office of Technology Development, Environmental Restoration and Waste Management, personal communication, 1993.

40. (a) **Chaiko, D. J.; Mensah-Biney, R.; Mertz, C. J.; Rollins, A. N.,** "Actinide Recovery Using Aqueous Biphasic Extraction: Initial Developmental Studies," U.S. DOE Report ANL–92/36, Argonne National Laboratory, Argonne IL, August, 1992. (b) **Chaiko, D. J.; Mensah-Biney, R.; Van Deventer, E.,** "Soil Decontamination by Aqueous Biphasic Extraction," U.S. DOE Report ORNL-6762, Oak Ridge National Laboratory, Oak Ridge, TN, September 1993.

41. **Moroney, K. S.; Moroney III, J. D.; Johnson, N. R.,** "Remediation of Transuranic-Contaminated Coral Soil at Johnston Atoll Using the Segmented Gate System," in *Waste Management '93*, Post, R. G., Ed., WM Symposia, Inc., 1993, 848.

42. **Misra, M.; Neve, C.; Raichur, A.,** "Characterization and Physical Separation of Radionuclides from Contaminated Soil," presented at TNO Fourth International Conference on Contaminated Soil, ICC, Berlin, Germany, May 3–7, 1993.

43. **Matthews, C. W.,** "Screening," *Chem. Eng.*, Feb. 15, 89, 1971.

44. **Miller, J. D.; Hupka, J.; Weidner, J. R.,** "Particle Characterization in Contaminated Soil," U.S. DOE Report INEL EGG-WTD-9736, May 1991.

45. **Tamura, T.,** "Plutonium Association in Soils," Transuranics in Natural Environments, Nevada Applied Ecology Group, NVO-178, 97, 1976.

46. **Navratil, J. D.; Kochen, R. L.,** "Decontamination of Soil Containing Plutonium and Americium," U.S. DOE Report RFP-3139, Rockwell International, Rocky Flats Plant, Golden, CO, 1982.

47. **Murarik, T. M.; Wenstrand, T. K.; Rogers, L. A.,** "Characterization Studies and Indicated Remediation Methods for Plutonium Contaminated Soils at the Nevada Test Site," *Spectrum '92*, International Topical Meeting, Nuclear and Hazardous Waste Management, Boise, ID, Aug. 23–27, 1992.

Section III

Treatment of Process Wastes and Gases

The concentrated radionuclides that result from separations processes must be immobilized for management, interim storage, or disposition. The ultimate disposition of high-level and high-activity wastes (i.e., concentrated fission products) will be in a mined geologic repository. Vitrification to a glass is a technology for HLW immobilization that has been developed and put into use in Europe; it is also being developed and facilities being readied for operation in the United States. In the United States, at the Idaho Chemical Processing Plant, the high-level and high-activity wastes are currently being calcined to a granular solid, which includes some powder, in a fluidized-bed calciner. This calcine, which is being stored in stainless steel bins inside concrete vaults, may eventually be converted to a glass or glass-ceramic form for disposal. In the future, the high-activity sodium-bearing wastes may be directly vitrified.

Separated and concentrated transuranic wastes may also be immobilized by vitrification for engineered storage or repository disposal.

These solidification processes generate off gases that must be treated to prevent release of airborne radionuclides in the form of vapors or particles and, in some instances, to contain hazardous chemical species.

Discussions of off-gas characteristics, treatment technologies for off-gases from calcination and vitrification processes, and operating experiences are provided in the chapter following this introduction. Brief descriptions of the processes are given in the lead-in to the descriptions of the off-gas treatment systems.

Jerry D. Christian

Chapter 13

Treatment of Off-Gases from Process Operations

Jerry D. Christian and Thomas R. Thomas

I. INTRODUCTION

The major processes considered that require extensive off-gas treatment are those used for the solidification of the concentrated radioactive process wastes. The separations processes themselves generally require only filtration, along with any concomitant cooling, heating for dehumidification, etc. All off-gas treatment systems (OGTSs) are operated under a negative pressure relative to the process equipment and cell pressures. Draft fans for this purpose are generally placed at the end of the off-gas treatment system; sometimes additional high-efficiency particulate air (HEPA) filters follow the fans prior to stack release of the process off-gas. Treatises on draft fan technologies are available,[1,2] and they are not treated here.

Because of the general requirements for filtration in nuclear process systems and the generic features of filtration systems, a separate discussion is given of filters. Many thorough reviews are available on filtration for nuclear process applications; therefore, only an overview is provided. The status of off-gas treatment systems is described in detail for calcination and vitrification of radioactive wastes.

The waste feed streams to the waste solidification processes considered generally contain aqueous nitrate solutions; some contain fluorides. The solidification processes generate water vapor and NO_x as the waste is calcined or vitrified; these gases contribute to the burden of the off-gas system along with process air or other carrier gases. Other significant gas stream constituents include entrained small particles and aerosols, which may contain dissolved radioactive constituents, and volatilized radioactive ruthenium and cesium. Technetium is partially volatilized during high-temperature vitrification.

Removal of these constituents is required to protect the environment and, depending on the quantities and nature, to avoid equipment operating problems. The overall removal efficiencies required depend on the type of constituent (radionuclide, chemical form, physical form, etc.), its concentration in the off-gas stream, and the release concentration limits specified by regulatory bodies (NRC, EPA, states).

Some wastes being solidified contain salts of mercury. Vitrification of these may cause volatilization. A brief review of available information is given. Because mercury is a hazardous material, consideration of its removal from the off gas may be important.

The primary objective is to describe the general off-gas cleanup requirements for radioactive processes, the considerations for approaches to achieving the cleanup, and the technologies of specific unit operations. As such, regulatory limits for releases are not addressed in detail; only a brief summary of target decontamination factor goals for processing light-water reactor (LWR) high-level waste (HLW) is given.

In the 1960s and 1970s, a number of high-level waste solidification processes were investigated. High-level liquid wastes (HLLWs) are the first-cycle extraction raffinates that are generated from the processing of spent nuclear fuels for the recovery of uranium and plutonium. Their characteristics influence the type and chemistry of the solidification process and affect the nature of the off-gas that must be treated. There are, in general, four types of HLLW: acidic or basic and concentrated in fission products or diluted with salts of nonradioactive "inert" constituents. Those generated from the processing of commercial LWR fuels from the chop-leach PUREX process are highly concentrated in fission product radionuclides. The raffinate that is produced is acidic in nitric acid and that is the primary liquid waste form. In the case of the West Valley (U.S.A.) wastes, the raffinates have been neutralized and made basic for interim storage and contain sludge. Defense HLW from the processing of naval Zircaloy fuels at the Idaho Chemical Processing Plant (ICPP) in the U.S. are acidic nitrate solutions with complexed fluoride. Dissolved nonradioactive fuel constituents and associated process chemicals make up the bulk of the liquid waste solutes; the activity levels of these wastes are orders of magnitude less than those of commercial LWR wastes. Other ICPP wastes, primarily from processing aluminum and stainless steel fuels, also contain substantial nonradioactive diluents, although not as much as the zirconium wastes. Defense waste solutions at Hanford and Savannah River have been neutralized and made basic.

Of the waste solidification processes investigated, three have emerged as practical systems: fluidized-bed calcination in use at the ICPP in the United States, rotary-kiln calcination with vitrification developed in France (AVM process) and implemented in France and the United Kingdom, and liquid-fed ceramic melter developed in Germany and the United States and built at the Savannah River Site and at West Valley in the United States. Preconcentration of wastes by evaporation prior to solidification is sometimes used. Off-gas treatment systems for these solidification processes are described.

II. GOALS FOR RADIONUCLIDE DECONTAMINATION FACTORS

Only commercial LWR wastes are considered here; they generally bracket defense wastes in terms of radioactive source terms and control requirements for off-gas treatment systems of HLW treatment processes. Regulations and standards for the United States are used as a basis for estimating off-gas treatment performance requirements.

Two regulatory controls affect design performance requirements for off-gas treatment systems of commercial facilities. The U.S. Nuclear Regulatory Commission has published (10CFR20) effluent concentration limits in uncontrolled areas for individual radionuclides that may be emitted from NRC licensee facilities,[3] and the U.S. Environmental Protection Agency has promulgated standards (40CFR190) for dose limits and specific radionuclide controls for uranium fuel cycle facilities based on a curie per gigawatt-electrical year release limit.[4] In some instances, it is more restrictive than the NRC limits, especially for LWR HLW that contains ^{244}Cm (see discussion below).

The maximum permissible concentrations in the gas released from the stack may be related to those specified for the site boundary in 10CFR20 by the dilution factor that results between the release point of the stack and the site boundary. The dilution factor, D, is related to the atmospheric dispersion, χ/Q, and the stack gas flow rate, F, by the equation

$$D = 1/[(\chi/Q)F] \qquad (1)$$

where χ is the concentration at a point on the ground resulting from the release rate Q at the plant stack. The maximum permissible release rates may thus be derived relative to the stack gas flow rate, and the required decontamination factors calculated based on the processing rate and associated source terms.

Table 1 Feed-to-Stack Radionuclide DFs
Needed for LWR HLLW Solidifier[a]

	Ten times minimum required DF	
Radionuclide	10CFR20 basis	40CFR190 basis
Potentially volatile		
^{99}Tc	1.1	
^{106}Ru	7×10^5	
^{127m}Te	8×10^1	
^{125}Sb	6×10^2	
^{134}Cs	4×10^4	
Limiting of potential solids		
^{244}Cm	2×10^6	6×10^9

a   See text for assumptions and conditions. 10CFR20 limits are based on the
worst class of compound, i.e., oxide. The limiting radionuclide of an element
is listed.

Maximum permissible release quantities of the major radionuclides in LWR HLLW,
and the associated minimum required feed-to-stack decontamination factors (DFs) for a
solidification process, based on meeting the 10CFR20 concentration limits in effect in
1977, have been derived and tabulated by Christian.[5] The assumed conditions are HLLW
feed rate from $6^2/_3$ MTIHM/d (MTIHM = metric tons initial heavy metal) of fuel
irradiated to 28,700 MWd/MTIHM and cooled $1^1/_2$ years, 8% of the tritium and 0.1% of
the ^{129}I from the spent fuel are in the HLLW, stack gas flow rate of 52 m³/s (110,000 cfm),
and annual average χ/Q of 1×10^{-7} s/m³. These conditions correspond to an atmospheric
dilution factor of 2×10^5. In addition, a factor of 10 less than the release limits was
imposed for conservativeness; in practice, even more conservative factors may be desirable. For a combination of radionuclides, the sum-of-fractions rule must be adhered to.
This requires that the sum of the ratios of concentration to the effluent concentration limit
for all the released radionuclides be less than one.

The assessment resulted in the determination that all of the volatile tritium and ^{129}I in
the HLLW could be released and that the limiting radionuclide for particulate control
requirements is ^{244}Cm. Of potentially volatile (semivolatile) radionuclides, some containment is required for Ru, Te, Sb, and Cs, the greatest being for ^{106}Ru.

Effective January 1, 1994, the 10CFR20 effluent concentration limits were updated
and modified to restrict doses from annual exposure via inhalation or ingestion to 50
mrem total effective dose equivalent. The control requirements derived by Christian[5] are
updated here to reflect these changes, retaining the same assumptions. The results, again
providing a factor of 10 increase to the minimum required DFs, are summarized in
Table 1. The potentially volatile radionuclide ^{179}Se may be completely released within
these guidelines.

The standards of 40CFR190, which applies to HLLW processing, limit combined
releases from fuel cycle facilities of all alpha-emitting transuranic elements with half-
lives greater than 1 year to 0.5 mCi/GW$_e$ year. This is much more restrictive than
10CFR20 for ^{244}Cm. The corresponding required feed-to-stack DF, with a safety factor
of 10 and given in Table 1, is 6×10^9.[5]

As indicated above, the assumed processing rate is HLLW from $6^2/_3$ MTIHM/d. To
provide information for deriving capacity needs of individual off-gas cleanup devices
(that depend on fractional releases from the solidifier and performance of devices ahead

of the unit), the elemental masses of the radionuclide constituents, in g/MTIHM, are Tc, 7.4×10^2; Ru, 2.0×10^3; Te, 5.2×10^2; Sb, 15; Cs, 2.4×10^3; total solids (in calcined form), 6.5×10^4.[5]

Note that 40CFR190 does not apply to defense wastes. For defense HLLW generated from high-burnup highly enriched uranium (HEU) fuels, ^{144}Cm will not be limiting. For example, for the liquid-fed ceramic melter at Savannah River (the Defense Waste Vitrification Facility), ^{137}Cs is the radionuclide of greatest concern; based on 10% of the cesium volatilizing, an off-gas treatment DF of 8×10^8 was used as the design basis to meet allowable release limits (see Section V.C.2).

III. FILTRATION OF AIRBORNE PARTICLES

A. INTRODUCTION

Solid and liquid airborne aerosols generated by process, vessel, and ventilation off-gas streams fall into two broad categories. Particles larger than 3 µm diameter generally account for more than 96% of the mass of airborne aerosols and are relatively easy to remove by pretreatment with scrubbers, demisters, shakeable bag filters, cyclones, and electrostatic precipitators. These types of gas filters are treated in detail in the *Chemical Engineers' Handbook*.[6] Particles smaller than 3 µm account for more than 99.999% of the number of particles or aerosols present and are normally removed after pretreatment by prefilters and HEPA filters. An overview of prefilters and HEPA filter technology is given here. Comprehensive technology reviews have been published elsewhere.[1,7-9]

B. HEPA FILTERS

The HEPA filter is defined[10] as a throwaway, extended medium, dry-type fiber filter in a rigid frame, having a minimum aerosol collection efficiency of 99.97% for monodispersed 0.3-µm thermally generated dioctylphthalate (DOP) aerosols and a maximum clean-filter pressure drop of 2.54 cm of water at rated airflow capacity. Aerosols down to 1 µm in diameter are collected on filter fibers by impaction. Smaller aerosols, which tend to behave like gases, diffuse to the fiber and are held by electrostatic forces. The aerosols between 0.1 and 1.0 µm in diameter have both particulate and gaseous behavior and are considered the most difficult size to collect.

HEPA filters are available in plywood, particle board, cadmium-plated carbon steel, and stainless steel cases. The most common filter medium is a continuous web of glass fiber paper pleated back and forth across corrugated aluminum separators. Variations in the filter media include adding up to 5 wt% acid-resistant fibers (e.g., DuPont Nomex or Kevlar) to the glass fibers, and using acid-resistant separators made of Teflon, epoxy resin, or thermoset vinyl, or no separator at all.[11] Nonflammable or self-extinguishing adhesives are used to seal the filter pack into the case, and synthetic rubber, neoprene, mineral fiber, or nonflammable high-viscosity fluids are used to seal the filter case to its mounting frame.

Operational parameters vary depending on the composition and size of the HEPA filter. Table 2 lists some of the typical operating parameters of HEPA filters. Special considerations for use of HEPA filters include choosing filters with the proper materials of construction for the temperature and gas composition they will be exposed to and avoiding condensation and pressure transients during operation. HEPA filters should not be operated beyond their prescribed pressure drop limit and potentially explosive mixtures of organic compounds and condensed nitric acid should not be allowed to build up on the filter.

Table 2 Operating Parameters for HEPA Filters[12,13]

Parameter	Range	Comment
Temperature	120°C maximum	Particle board frame and rubber-based adhesive
	260°C maximum	Steel frame and silicone adhesive
	540°C maximum	Steel frame and glass packing seal
Flow rate	42–3400 m³/h (50–2000 ft³/min)	Rated flow for filter sizes from 20 × 20 cm to 61 × 61 cm face dimension
Pressure drop	2.5 cm water	Clean pressure drop at rated flow
	25 cm water	Must withstand this pressure difference for 60 min without visible evidence of damage
Humidity	0–99%	Condensation should be avoided
Particulate load	Up to 2 kg on 1700 m³/h rated filter	Depends on particle size and relative humidity
Removal efficiency	≥99.97%	In acceptance test with DOP or similar aerosol
Corrosive service	Up to several % vol. of NO$_x$, HNO$_3$ and HF in gas stream	Stainless steel cases, acid resistant fibers, separators and sealants are used

The performance of HEPA filters or filter banks *in situ* can be measured by testing with DOP, sodium chloride, paraffin vapor, tagged aerosols, and uranine aerosols.[14] In the U.S., HEPA filter banks used in a nuclear application must be tested *in situ* with DOP after installation and at 18-month intervals and show a removal efficiency of 99.95% or greater.[15] A filter bank that passes the test is assumed to have an efficiency of only 99%.

The service life of a HEPA filter varies with the type of gas stream it is exposed to and the particulate loading of the gas stream. High humidity, corrosive gases, and high particulate loading in the gas stream can reduce service life to a few weeks. In dry air streams with small particulate loadings, HEPA filters will maintain their efficiency beyond 10 years. A survey of over 9000 filter applications in U.S. Department of Energy facilities indicated that the average service life of HEPA filters was about 3 years.[16] The minimum service life was found to be about 2 months and the maximum about 10 years. Environments containing high dust loads, grease, oil, and acid vapors caused the shortest service life. Of the approximately 960 filter applications indicated in a proposed Exxon Fuel Reprocessing Plant,[17] about 250 would be changed out in 1 year, about 570 would be changed out between 1 and 5 years, and 140 would be changed out every 10 years.

C. PREFILTERS

The principle of recovery, materials of construction, and operating parameters for prefilters are similar to those for HEPA filters. The main difference is the aerosol removal efficiency vs. aerosol size as indicated in Table 3.

Group I prefilters are a viscous impingement, flat panel type filter and group II and III are extended-media, dry-type filters. Group III type prefilters with the same geometry and materials of construction (e.g., fiberglass medium, aluminum separators, and wooden case) as HEPA filters are routinely placed upstream of the HEPA filters used for radioactive cleaning applications.

Table 3 Prefilter Removal Efficiency vs. Particle Sizes[1]

| Group | Efficiency | Removal efficiency (%) vs. particle size | | |
		0.3 μm	1.0 μm	5 μm
I	Low	0–2	10–30	40–70
II	Moderate	10–40	40–70	85–95
III	High	45–85	75–99	99–99.9
HEPA	Extreme	>99.9	99.99	100

The prefilter may have a clean-air resistance as low as 1 cm of water and a dust loading capacity up to 450 g. Prefilters are not subjected to acceptance tests. Filter efficiency tests for Groups I to III filters are made on prototypes only and the results extrapolated to the variously sized filters of similar design. A common filter efficiency quoted by American vendors is 80 to 90% for an all-glass medium filter in a particle board frame with polyurethane foam or rubber sealant.

IV. BEHAVIOR, CONTROL, AND REMOVAL OF VOLATILE AND SEMIVOLATILE RADIONUCLIDES AND HAZARDOUS SPECIES

A. INTRODUCTION

When uranium fissions, nuclides of masses near 95 and 140 amu are most predominantly produced. Two of these, ruthenium and cesium, can volatilize from the HLLW solution as it is dried and solidified to a powder or vitrified into a glass. The sheer mass of total ruthenium can cause plugging of off-gas pipes if it is allowed to volatilize from the waste-processing system and is then deposited in the off-gas system. Although most of the fission product ruthenium is stable (nonradioactive), ^{106}Ru, with a half-life of 1.02 years, is present in sufficient quantities to impart high radiation fields wherever ruthenium may be deposited. This can result in difficulties in decontaminating equipment when contact maintenance is required. Ruthenium can adhere quite tenaciously to metal surfaces.

Cesium volatilization is not a problem at calcination temperatures of 400°C and may very slightly volatilize at 500°C. At glass-forming temperatures, however, volatilization can be a problem.

Technetium-99, again formed in significant quantities during the fission process, is of particular concern because of its long half-life (2.13×10^5 years) and its relative mobility in the environment. As indicated in Table 1, for LWR HLLW, the NRC effluent concentration limits could be met with virtually complete release of technetium. Its behavior in the solidification process is of interest because of the as-low-as-reasonably achievable (ALARA) aspects and operational considerations, given the fairly large quantities that are processed with the HLLW and the ability of technetium to volatilize at high temperatures.

The behavior and control of these radionuclides during high-temperature processes are discussed here. Considerable development has been made to techniques to remove volatilized ruthenium from off-gas streams; this is briefly reviewed.

Mercuric nitrate has been used as a catalyst for the dissolution of aluminum-clad and alloy fuels at the U.S. Department of Energy ICPP and Savannah River Site. The catalyst is also for dissolution of aluminum fuels for recovery of medical isotopes at Canada's AECL facility in Chalk River. Mercury salts are both volatile and toxic at elevated temperatures. Therefore, attention has to be given to mercury in HLW waste solidification processes where appropriate.

Specific discussions of characteristics of ruthenium, technetium, and mercury are provided here. Cesium is considered only in the descriptions of the integrated off-gas treatment systems for vitrification processes. Because of its significance in terms of operational impacts and the extensive characterization of ruthenium relative to high-temperature processes, more emphasis is given to ruthenium. The discussions are meant to indicate considerations that may be important in assessing or designing a waste concentration or solidification system, rather than be exhaustive. More thorough treatments of specific issues can be found in the referenced and recent literature, and need not be repeated here.

B. RUTHENIUM

Thorough reviews of the behavior and control of ruthenium and of other so-called semivolatile radionuclides during high-temperature processes have been given by Christian,[18] Christian and Rhodes,[19] Christian and Pence,[5] Klein and Deuber,[20] and Klein.[21] A summary of salient features is given here. The interested reader should consult the above references for specific and detailed information.

Concentrations of ^{106}Ru and total ruthenium in reprocessing wastes from DOE fuels (e.g., at the ICPP) are sufficiently small that simple technologies enable removal of any volatilized material from the off-gas stream of a waste treatment process. The concentrations of ^{106}Ru and total ruthenium in ICPP wastes are approximately 0.6 Ci/gal (0.16 Ci/l) and 2×10^{-4} M, respectively.[18] In 1-year-cooled first-cycle raffinate HLLW from the processing of commercial LWR fuels that have been irradiated to 35,000 MWd/MTIHM, ^{106}Ru is present at a concentration 1.7×10^3 Ci/gal (2.5×10^5 Ci/MTHIM) and constitutes 3.7% of total ruthenium present. The total ruthenium concentration is 0.04 M.

The much higher level of ^{106}Ru in commercial fuel HLLW places severely greater restrictions on the fraction of the ruthenium that is allowed to volatilize during a waste solidification process, or else extensive use of expensive ruthenium removal systems in the off-gas stream will be required. Furthermore, the large increase in total ruthenium would place substantial burdens on the capacity of any gas cleanup device if a significant fraction of the ruthenium volatilizes.

Ruthenium exists in nitric acid solutions principally as complexes of nitrosyl ruthenium cation, RuNO(III). The entity is quite stable and has, in addition to the one position occupied by NO, five coordination positions available for complexing H_2O, NO_2, NO_3^-, or, at high pH, OH^-. Ruthenium volatilizes from evaporating acidic solutions when the nitrate concentration increases.[18,22,23] This causes the complex to become completely nitrated and eventually results in the RuNO(III) becoming oxidized to the volatile Ru(VIII) form. Volatilization does not begin until the nitric acid concentration exceeds 2.5 M. The oxidation rate becomes very rapid above 8 M HNO$_3$.

The Ru(VIII) vapor species resulting from evaporation of nitric acid solutions in air and at temperatures above about 130°C is a nitrosyl ruthenium complex.[18] Under certain conditions, in the absence of air, RuO$_4$ vapor may form at lower temperatures, but it is unstable at higher temperatures. Observed partial pressures of ruthenium in calcination processes at temperatures below 600°C exceed those that can be attributed to RuO$_4$ formation from oxidation of RuO$_2$.[18] UV-visible and IR spectral observations of RuO$_4$ and NO$_2$ gases isolated together show the formation of a complex species.[18] When measured concentrations of the gases were placed in an evacuated cell and allowed to interact at room temperature, the UV-visible spectra of both RuO$_4$ and the NO$_2$–N$_2$O$_4$ mixture decreased in intensity from the calculated values and a new charge transfer band appeared in the UV region. The new band did not correspond to that for any NO_x species. An IR spectrum of the system showed only two intense bands centered at 1634 and 1605 cm^{-1} that cannot be attributed to RuO_4 or any NO_x species.

Igarashi et al. have observed the IR spectrum of nitrato nitrosyl ruthenium, $Ru(NO)(NO_2)_2(OH) \cdot 2H_2O$, in the off-gas condensate from calcination of simulated HLLW, suggesting the formation of a gas-phase complex[24] (the condensate sample for the IR absorption spectrum was dried with potassium bromide at 100°C before the analysis). Of course, the complex could have formed in the condensed solution, but the results are not inconsistent with the hypothesis of a gaseous nitrosyl ruthenium species. The same authors did, in a separate study, observe the same IR spectrum on black deposits formed on a room-temperature glass wall when NO or NO_2 was added to $RuO_4(g)$ in a 40°C water-saturated air stream.[25]

When fluoride is present in the acidic waste solution, both the percent of ruthenium volatilized and the partial pressure of ruthenium in the off gas during calcination in the 450 to 600°C range are greater than from corresponding solutions without fluoride, indicating the formation of a stable fluoride or oxyfluoride gas species of ruthenium.[18] The presence of sulfate in wastes may enhance the ruthenium volatility, also.

The percent volatility of ruthenium from evaporating nitric acid solutions increases as temperature is increased to a maximum at about 300°C. Above this temperature, it decreases.[24] Similarly, the percent volatility during calcination can be near 100 up to 350–380°C, above which temperature the volatility decreases. When the temperature is increased above 550–600°C, the volatility begins to increase from the minimum as a result of the increasing formation of $RuO_4(g)$ above RuO_2 in the presence of O_2.[18] Between 800 and 1200°C, a mixture of $RuO_4(g)$ and $RuO_3(g)$ exists; at 1200°C, RuO_3 is the predominant oxide gas species.

The mechanism of the decrease of ruthenium volatility with increasing calcination temperature is not clearly understood. It may be attributed to dissociation of the nitrosyl complex vapor species as temperature is increased. Spectral observations on the complex species formed between $RuO_4(g)$ and $NO_2(g)$ from 25 to 55°C show that the complex formation increases reversibly as temperature is increased.[18] But, at 400°C, the ruthenium species dissociated to RuO_2. Another possible mechanism may involve the formation of nitronium ions, NO_2^+, as the nitric acid concentration is increased during the flash evaporation process. Ortins de Bettencourt and Jouan[26] observed a decrease in ruthenium volatility from distilling nitric acid solutions, from the maximum in the range of 8 to 12 M HNO$_3$, at the nitric acid azeotrope concentration of 14.8 M. This may be related to the formation of NO_2^+, which Fletcher[27] believed caused reduction of RuO_4 to Ru(VI). A similar process may be occurring during calcination, in which the nitronium ion formation rate, and subsequent reduction of Ru(VIII), may increase relative to the evaporation rate of ruthenium with increasing temperature.

When a fluidized-bed calciner is heated internally by the burning of kerosene, the ruthenium volatility is greatly decreased.[18] Partially unburned hydrocarbons or CO probably reduce the oxidation potential of the system and reduce the ruthenium volatility.

In a liquid-fed ceramic melter, ruthenium may enter the off-gas system as oxide gas species or entrained as the dioxide with other aerosols. As the off gas is cooled, any ruthenium oxide vapor will dissociate to the solid dioxide and form a secondary aerosol. The presence of a cold cap of liquid feed on the molten glass provides a barrier for ruthenium losses. The cold cap can reduce losses to the off gas from 18–22% down to 2–4%.[28]

Control of ruthenium volatility during evaporation and solidification of waste solutions involves controlling nitrate concentration and operating temperature, as discussed above, and oxidizing potential of the solution by means of adding reducing agents. Removal of ruthenium from the feed prior to thermal treatment, of course, can eliminate difficulties associated with its presence. However, such approaches are not generally applied. Volatilization and precipitation methods have been investigated. The interested reader is referred to the review provided by Christian.[18]

Ruthenium volatilization from distilling solutions can be suppressed by addition of $NaNO_2$ or NO_2 gas (which forms nitrous acid by reaction with nitric acid).[23] Phosphite or hypophosphite additives can suppress ruthenium volatility.[29,30] Sugar addition is another possibility; it was planned for the high-activity waste evaporator tank at the Barnwell, South Carolina plant in its design stages. The conditions for sugar addition were 0.033% of the flow of HNO_3 and 15% of the flow of ruthenium into the concentrator.

Sugar and formic acid addition to HLLW before or during calcination or vitrification can be effective in suppressing ruthenium volatility. Summaries of the results of a number of studies involving reducing agents and other additives in waste solidification processes may be found in reviews by Christian[18] and the IAEA.[20,31]

Ruthenium that is volatilized into the off gas of a waste concentration or solidification process is partially removed by particulate wet scrubbing devices, condensers, and NO_x removal operations. If the quantity volatilized is substantial, it may be necessary to incorporate specific polishing removal devices for the ruthenium. Specific unit operations for volatile ruthenium removal are silica gel beds and heated Fe_2O_3, sometimes on glass beads. Typical decontamination factors for particles and for volatilized ruthenium of individual off-gas treatment devices are summarized in Table 4. The efficiency of a device may depend on its location in the system or on the solidification process and, therefore, off-gas composition. Thus, the estimates are only approximate. Reviews of operating experiences with specific ruthenium removal technologies, and the conditions of operation in applied systems, are available.[5,18,20,21,31] These and the primary references cited in them should be consulted for design and operating details.

The literature delineates process conditions necessary for successful use of sorbent materials for volatilized ruthenium. Silica gel has been shown to provide decontamination factors of 1000; total loadings as high as 58 g Ru per cubic foot (2.05 mg/cm^3; approximately 2.8 mg Ru per gram adsorbent) have been obtained without breakthrough at 80 to 85°C (10° above the dew point).[32-34] Klein and co-workers[21] measured an absorption capacity at 22°C for RuO_4 gas of 4.3 mg Ru per gram adsorbent when the vapor concentration of RuO_4 was 500 mg/m^3; they observed a dependence on vapor concentration. When NO_x vapors were present, which results in formation of a ruthenium nitrosyl complex gas species, a capacity of greater than 0.3 mg Ru per gram adsorbent was measured, with a DF of greater than 100 at 100°C. That adsorption capacity is stated to be two orders of magnitude greater than for RuO_4 vapor at the same conditions. A silica gel bed should be operated at as low a temperature as permitted that keeps it above the dew point; the DF is greatly increased as temperature is decreased.

Of a number of materials investigated, Newby and Barnes[35] found that silica gel, hydrous zirconium oxide, and ferric oxide provides DFs of 1000 or greater. They recommend silica gel for use in a fluidized-bed facility because of its regenerability (by washing with hot water) and low temperature of operation.

Table 4 Typical Decontamination Factors across Off-Gas Cleanup Devices[5]

Component	DF	
	Particles	Volatilized ruthenium
Cyclone	10	1
Venturi scrubber	100–600	10
Tube and shell condenser	10^2–10^3	2×10^2
NO_x absorber	10	10
Brink fiber mist eliminator	10^2	1–2
Packed spray tower	10^3	10^2
NO_x selective catalytic reduction	2	3.8×10^2
Ruthenium sorber		
Silica gel	8	10^3
Fe_2O_3 on glass	2	$(1–5) \times 10^2$
Sintered metal filter	10^3	1
HEPA filter	1	1

If silica gel containing ruthenium is deposited in a storage canister with calcine, it will be heated by the radionuclides. The potential for release of the ruthenium was investigated in the laboratory;[18] none was volatilized at temperatures up to 600°C.

Rastogi et al.[36] obtained removal efficiencies of 99.9% (DF = 1000) for iron oxide-coated silicious bed materials at 350°C and a flow velocity of 0.15 ft/s. Ortins de Bettencourt and Jouan[26,36] recommend use at 450 to 500°C of glass granules coated with ferric oxide for volatile ruthenium trapping. Ferric oxide functions as a catalyst for the reduction of ruthenium-containing vapor species to RuO_2. Its effectiveness tends to increase as RuO_2 is deposited, because RuO_2 is a catalyst for reducing ruthenium, i.e., the vapor-phase decomposition is autocatalytic. The capacity of ferric oxide, and other catalyst-based beds, is very large, greater than 50 g Ru per gram adsorbent.[21,38]

C. TECHNETIUM

Under oxidizing conditions, technetium forms a stable +7 valence. The stable oxide, Tc_2O_7, boils at 310.6°C. Stable alkali and alkaline-earth pertechnetates can form. Thus, the volatility of technetium is low at calcination temperatures and becomes significant as the temperature increases above about 550 to 600°C. Rimshaw and co-workers[39,40] studied the volatility of technetium during the flash evaporation of HLLW solutions in a laboratory pot calciner as a function of temperature. For wastes from the reprocessing of aluminum fuels in nitric acid,[39] the technetium volatility was less than 1.1 % from 250 to 600°C. The authors attribute the low volatilities to formation of alkaline-earth pertechnetates. For zirconium fuel wastes containing fluoride and nitric acid with aluminum ions, but with no $Ca(NO_3)_2$ added to hold the fluoride from evaporating as is done in the ICPP fluidized-bed calciner, the technetium volatility at temperatures up to 350°C was similarly small, but ranged from 9.2% at 450°C to 21% at 600°C.[39] It is evident that the presence of fluoride may result in formation of volatile technetium species. It is unknown what the effect of added $Ca(NO_3)_2$ to the calciner feed would be. For nitric acid PUREX wastes, the volatility was 0.2 to 1.4% at temperatures from 250 to 600°C,[40] similar to that for the nitric acid aluminum fuel wastes.

Halaszovich, Dix, and Merz[41] report volatilization of technetium as a drum-dried waste is heated up to 1150°C. Volatilization begins at 550°C and increases rapidly above 900°C. As the temperature is held at 1150°C, additional volatilization, beyond the total

60% observed as the waste is heated to that temperature, does not occur. This is because the technetium is, at that point, incorporated into the glass phase in a +4 valence state. Denitration of the liquid does not affect the observed volatility. Most of the volatilization occurs after the nitrates are thermally dissociated.

Cains and co-workers[42,43] found that, at 600°C, evaporative losses of cesium were proportional to the concentration of technetium in the liquid waste, suggesting volatilization of $CsTcO_4$. During vitrification, cesium evaporative losses without technetium present were $1.3 \pm 1.1\%$, but an order of magnitude higher when technetium was present. This, again, is evidence for the formation of mixed volatile species.

Technetium desublimation from the gas phase occurs as a black shiny deposit immediately at the cooler parts of the melter outlet.[41] Volatilized technetium can be scrubbed from the off gas.[41]

D. MERCURY

Some defense wastes contain significant quantities of mercury. Hanford wastes for feeding to the planned Hanford Waste Vitrification Plant (HWVP) will contain 5.3×10^{-3} M Hg. First-cycle extraction raffinates from processing aluminum fuels at the Idaho Chemical Processing Plant contain 8×10^{-3} M $Hg(NO_3)_2$. The AGNS Barnwell wastes projected from processing commercial fuels would have contained 7.6×10^{-3} M Hg^{+2}. The volatile nature of mercury and its compounds at high temperatures results in its release into the process off-gas systems. The characteristics are summarized here.

Goles, Sevigny, and Anderson[44] characterized the nature and behavior of elemental mercury in simulated Hanford sludge fed to a liquid-fed ceramic melter. The feed contained 0.0041 M Hg; chloride was present as NaCl at a concentration of 0.050 M. Essentially all of the mercury fed to the melter was lost to the process off-gas system in the gaseous state. The off-gas system consisted, in series, of an ejector venturi (quench) scrubber (EVS), a water-cooled condenser, and HEPA filter. The EVS removed 97% of the mercury (DF of 40), mostly in the form of insoluble Hg_2Cl_2. The final HEPA filters collected 1.9% of the total mercury processed. Accumulation of the mercury chloride deposits in the piping entrance to the scrubber could create operating problems. Air or steam injection to boost melter exhaust velocities and provide dilution is suggested as a possible means to overcome plugging problems.

In the WCF fluidized-bed calciner at the ICPP, mercury volatilized into the off-gas system.[45] The venturi acidic scrub system removed 98% of the mercury; this is similar to the results of Goles et al. above. The scrub solution was recycled to the calciner feed inlet. At the end of a campaign, the loaded scrub solution was returned to the HLW tanks. Silica gel adsorber beds downstream of the scrubber collect 1.4% of the mercury. These beds were periodically regenerated by washing; the washings were recycled to evaporators or routed to the HLW tank farm and then calcined with the HLLW. Provisions were in place for the silica gel to be changed out and the spent material placed in the calcine storage bins; however, this was never necessary. HEPA filters beyond the adsorbers collected 0.1% of feed mercury and 0.02% of the mercury, about 6 g/d, was vented to the main stack.

Osteen and Bibler have investigated the removal of mercury from radioactive waste solutions.[46] The resin Duolite™ GT-73 was shown to be effective for the removal of mercury in its three common forms (Hg_2^{2+}, Hg^{2+}, and elemental) from a variety of aqueous acidic solutions. Once sorbed, mercury does not leach from the resin except under extraordinary chemical conditions, so the resin could be disposed of as a nonhazardous waste. The resin has a high specificity for mercury. Therefore, elution of radionuclide ions from the resin would be possible.

V. INTEGRATED OFF-GAS TREATMENT SYSTEMS

A. INTRODUCTION

Design and performance characteristics of OGTSs for the three major types of operating and developed HLW solidification processes and plants are summarized here. These are fluidized-bed calcination and vitrification, with subsets of the latter being a liquid-fed ceramic melter and a rotary-kiln calciner with glass melter.

B. FLUIDIZED-BED CALCINATION

1. Introduction

Fluidized-bed calcination of U.S. defense wastes has been done at the Idaho Chemical Processing Plant at the Idaho National Engineering Laboratory since 1963. In this process, the acidic high-activity aqueous waste is sprayed through an atomizing nozzle into a fluidized bed of particles that are heated to 400 to 500°C. The particles are fluidized by a flow of air through a distributor plate with holes at the bottom of the bed. The superficial velocity is 0.18 to 0.36 m/s. The water and nitric acid are flash evaporated and the residual salts and oxides are deposited on the surface of the bed particles. Particle size is maintained by a balancing grinding action that is controlled by adjusting the nozzle air-to-feed ratio. The fluidization maintains a constant and uniform temperature throughout the bed. As the bed grows, it is drawn off and pneumatically transported to stainless steel storage bins in concrete vaults.

The composition of the solution must be adjusted to produce calcine particles of appropriate hardness that will not agglomerate or "clinker", but that will not be so soft as to grind to a powder. When fluoride wastes are calcined, $Ca(NO_3)_2$ is added to hold the fluorine in the bed as CaF_2.

A portion of the atomized liquid evaporates to a dry powder before striking the surface of a bed particle, and the particles undergo a grinding process. Thus, a fraction of the dried waste forms "fines", small particles that are entrained in the off gas. The calcine and powdery solids range in size from 0.05 to 5 mm; the mass mean particle size is nominally 3 mm. Also, ruthenium, under oxidizing conditions, can volatilize. An extensive off-gas treatment system is needed to remove these particulate and volatilized ruthenium species.

During the first three campaigns of the original Waste Calcining Facility (WCF), the fluidized bed was indirectly heated to 400°C by circulating hot NaK alloy through a tube bundle in the bed. These conditions resulted in virtually 100% volatilization of the ruthenium in the feed and required specific unit operations in the off-gas system for its removal. Starting with the fourth campaign, the heating system was converted to one of internal combustion of kerosene; this required an operating temperature of 500°C to maintain smooth combustion. This mode of operation was also designed into the New Waste Calcining Facility (NWCF) that began operating in 1982, replacing the WCF. The combination of higher temperature and partially reducing conditions in the off-gas resulted in some reduction in volatilization of the ruthenium from the bed; more significantly, that which did volatilize was quickly reduced by unburned hydrocarbons in the off-gas to particulate RuO_2 and was effectively removed by the scrubber.

The calciner is used to calcine HLLWs that are either nitric acid solutions or acidic fluoride–nitrate solutions. When fluoride–nitrate wastes are calcined, $Ca(NO_3)_2$ is added to the liquid feed and reacts with the fluorides to form stable, nonvolatile CaF_2. Thus, the primary off-gas constituents are fluidizing air, combustion products of kerosene, water vapor, oxides of nitrogen, and calcine fines, with small amounts of hydrocarbons.

The normal concentration of ruthenium in the high-level liquid waste is approximately 10^{-3} M or less. ^{106}Ru comprises 0.1 to 0.2% of the total ruthenium. A steady-state calcine bed in the NWCF weighs about 2300 kg and contains on the order of a kilogram of ruthenium and 5000 Ci ^{106}Ru (2.2 Ci/kg);[47] the total radionuclide activity concentration ranges from 8 to 80 Ci/kg.[48]

Laboratory studies[18] have shown that during calcination, 20 to 62% of the ruthenium is volatilized, the higher value occurring during smooth burning of the kerosene, as is characteristic of the full-scale calciner. The volatile form is most likely a nitrato or oxyfluoride species. Generally, gaseous RuO_4 cannot account for the partial pressure of ruthenium observed in calcination processes.[18] The partial pressure and, thus, volatility, decreases with increasing temperature, in contrast with the formation of RuO_4 over RuO_2.

With in-bed combustion heating at 500°C and a ruthenium volatility from the bed of approximately 60% during calcination of zirconium wastes with fluoride and nitrate, the partial pressure of ruthenium vapors entering the OGTS is approximately 5×10^{-6} atm.[18] Calcination of AGNS type commercial wastes would result in 0.002% volatilization of the ruthenium and a partial pressure of approximately 4×10^{-8} atm.

Approximately 8 to 20% of the solids content of the calcine solids formed were entrained as fines from the bed and through the cyclone into the OGTS of the WCF.[19,48] For the NWCF, improvements in design result in approximately 2% carryover of particulate materials through the cyclone.[31]

Operating data for the WCF are reported for campaigns 1 through 5[50-54] and 9.[55] No campaign reports are available for the NWCF.

2. System Design and Performance

The OGTS for the NWCF is state-of-the-art and is described here. It is essentially similar to that of the WCF, with some improvements, and is a combination of a wet–dry system consisting of a series of unit operations designed for progressively smaller particles. Operational data are presented for the similar WCF off-gas system because more information has been compiled and because effects of indirect heating vs. in-bed combustion of kerosene can be shown. Improvements in the NWCF individual component designs generally result in a better performance. Data during cold testing of the NWCF and early operational data are also provided. Schindler[49] discusses the specific design details of the individual off-gas components of the WCF and improvements for the NWCF that resulted from the WCF experience.

The NWCF system is depicted in Figure 1. The nominal gross feed rate is 570 l/h, the limit being established by particle entrainment and, in recent years, constraints on NO_x release rates imposed by agreement with the State of Idaho. The NWCF has operated, when conditions permit, at up to 900 l/h, the limit being the ability of the OGTS, primarily the blowers, to handle the flow. To minimize carryover of fine particles from the fluidized bed, the 1.5-m (5 ft) diameter bed section has a 1.7-m (7 ft) diameter expanded deentrainnment section above it as high as permitted by cell dimensions (about 2 to 3 m). The top of the deentrainment section has a double venetian-blind baffle for deflecting particles. A majority of entrained fine particulate material is removed from the 500°C off-gas by a high-efficiency cyclone fines separator at the exit of the calciner and transported pneumatically to the product storage bins. The off-gas is then cooled to 70 to 75°C and saturated in a spray quench tower prior to passing through a high-energy venturi scrubber where most of the remaining particles are removed along with some of the volatilized ruthenium. The scrub solution is typically 2 to 4 M HNO_3; this dissolves the scrubbed solids. Some condensation of water occurs in this step, generating a liquid recycle to the

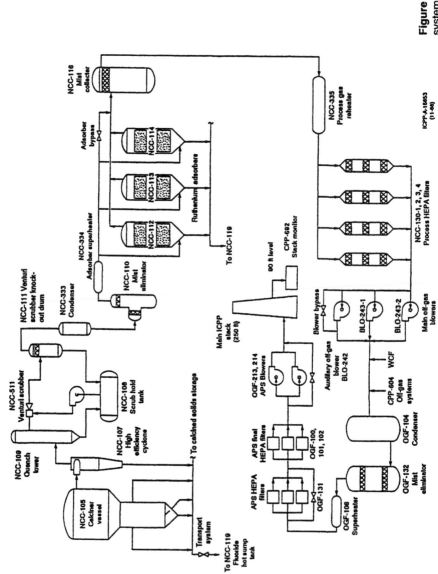

Figure 1 NWCF process off-gas system.

feed system of some 10%. The off gas then passes through a venturi scrubber knockout drum, where removed water aerosols are returned to the scrub hold tank for the quench tower, a condenser, and a mist eliminator.

The scrubber off gas exiting the mist eliminator passes through the adsorber superheater and three parallel silica gel ruthenium adsorber beds. The adsorbers each contain a lower and upper bed of silica gel. The lower bed acts primarily as a prefilter to remove residual entrained particles and hydrocarbons. The dry upper bed will remove any remaining ruthenium vapor species. The final off-gas cleanup step consists of four parallel banks of three series HEPA filters. Two banks are on-line during normal processing; one bank is on-line during shutdown periods. The whole off-gas system is maintained under a slight negative pressure by two off-gas blowers in parallel; a third parallel, auxiliary, blower is on standby.

The silica gel beds are periodically washed to reduce the pressure drop from accumulation of particles and grease. This will also regenerate the silica gel for capacity of adsorbing ruthenium, although it is never challenged with in-bed combustion heating of the calciner.

The NWCF process off gas is combined with other plant gases, and is finally treated with the atmospheric protection system (APS) HEPA filters. Only one of the two stages of HEPA filters shown in Figure 1 has been used.

The stack gases are sampled at the 90-ft level. The sampling system consists of a particulate filter and three gas-monitoring beds in series: charcoal, silver zeolite, and silica gel. The particulate filter is pulled daily for counting. It has a detection limit for ^{106}Ru of 0.63 µCi for the 24-h sample. The gas sample adsorber beds are pulled biweekly and have a detection limit of 0.41 µCi ^{106}Ru, equivalent to 0.029 µCi/24 h average. No ^{106}Ru gas species has ever been detected since operation of the NWCF began in 1982.

The NO_x, present in the approximately 2000 scfm (3400 Nm^3/h[*]) process off-gas at 1.5 to 3 vol%, is diluted by a factor of 50 in the stack by plant ventilation air and released. The concentration at the site boundary is generally two orders of magnitude below the clean air standard of 0.05 ppm. (Nevertheless, as mentioned earlier, the State of Idaho has recently imposed limits on NO_x emission rates, agreed to by the Department of Energy, that sometimes restrict the processing rate.) Iodine-129 quantities in the processed fuels have been sufficiently small that the site boundary doses resulting from releases during fuel dissolution and calcination of HLLW have been on the order of 0.1 mrem/year EDE, well within the 10 mrem/year DOE INEL limit. Thus, iodine removal technology has not been necessary.

Schindler[49] summarized the performance of the WCF OGTS for removal of ^{137}Cs and ^{90}Sr for eight campaigns; the report discusses the basis for the design of the NWCF OGTS in light of the WCF design and performance. Christian and Rhodes[19] tabulated the performance of the WCF OGTS components for ruthenium and particle removal during the first five campaigns. The results of these two compilations are summarized in Table 5. The waste feed types are aluminum fuel dissolution first-cycle extraction raffinate, which is a nitric acid solution, and Zircaloy fuel dissolution first-cycle extraction raffinate, which is an acidic fluoride–nitrate solution. In the first three runs, the WCF was operated at 400°C with indirect heating (by circulating hot NaK through a tube bundle in the fluidized bed). In subsequent runs, the bed was heated by in-bed combustion of kerosene at 500°C, which is the mode of operation for the NWCF.

---

[*] Refers to cubic of gas per hour at normal (N) conditions, i.e., standard temperature and pressure.

Table 5 Component Decontamination Factors for the WCF[19,49]

Campaign no.	Waste feed type	Nuclide removed	Calciner & cyclone	Scrubber system	Feed thru scrubber	Silica gel adsorbers	HEPA filters	Overall (feed-to-stack)[a]
						DF		
1	Al	^{90}Sr	6.6	600	3900	12	160	1×10^7
		^{137}Cs	5.9	690	4100	9	280	1.3×10^7
		^{106}Ru	2.1	10.8	23	240	1	$(2\text{–}5) \times 10^3$
2	Al	^{90}Sr			1000	10	1000–2000	$1 \times 10^7\text{–}5 \times 10^8$
		^{106}Ru			13	400–2000	1	$10^1\text{–}10^5$
	Zr	^{90}Sr			2750	10	3000	1.7×10^8
		^{106}Ru			100	1000	1	10^5
4	Zr	^{90}Sr			2800	8.0	630	1.4×10^7
		^{106}Ru			80	12.4	850	8×10^5
5	Zr	^{137}Cs	12	25	300	3 (wet)	1300	1.2×10^6
		^{106}Ru	1.4	30	42	3	1030	1×10^5
6	Zr	^{90}Sr			2300	3 (est)	75	5×10^5
		^{137}Cs			400	3 (est)	75	5×10^5
8	Zr	^{137}Cs				2–6 (wet)	10–30	9×10^4

[a] The overall DF is usually an independent measurement from the individual DFs and may be larger than the combined component DFs.

During the first three campaigns, the WCF used a "multiclone" with six elements operating and ten elements plugged. Then it was replaced with a cyclone and wire mesh demister.

In campaign 2, the efficiency of the scrubber system for removing ruthenium and for removing particles was greater for the Zr type waste than for the Al waste. The latter is representative of a commercial fuel dissolution HLW. In the later runs, the pressure drop across the scrubber was decreased from 13–17.5 kPa to 5–7.5 kPa, resulting in decreased efficiency for particle removal.

The behavior of cesium changed when the calciner was converted from indirect heating at 400°C to in-bed combustion heating at 500°C. In run 1, the DFs for ^{137}Cs and ^{90}Sr were essentially the same, indicating that cesium and strontium were both particulate solids. In run 6 at 500°C, the DF for cesium was a factor of six smaller than that for strontium and other particulate species. The cesium may be partially volatilized at the higher temperature and then condensed in the quench tower to form fine particles. In-bed combustion of kerosene may have an effect.

As discussed in the section on ruthenium, the conversion to in-bed combustion heating resulted in reducing the volatile form of ruthenium to particulate RuO_2 by the time it reached the scrubbers. Thus, after the third campaign, its removal efficiencies by the silica gel adsorbers and HEPA filters were the same as for other particulate species. In earlier campaigns, ruthenium was in the form of a vapor species and the silica gel provided a high DF for it while the HEPA filters provided no removal.

In the first four campaigns, separator type HEPA filters were used and provided an average DF of greater than 1000. During campaign 5, the filter type was changed to a separatorless filter that provided greater flow and solids capacities. However, they deteriorated rapidly in service from moisture, acid fumes, and pressure stresses. When freshly installed, DFs of 42,000 were measured. However, failed filter elements in a bank during the course of a run resulted in DFs decreasing to 75.

Specific conditions and details of the WCF off-gas equipment operation and performance may be found in Schindler's report.[49]

The WCF was replaced by the NWCF operations in 1982. Improvements in the design of specific off-gas treatment components, such as the cyclone and venturi scrubber, resulted in more efficient particulate removal. Provisions were incorporated to keep the HEPA filters dry.

Unpublished results of measurements of particle decontamination factors show a feed-to-filter DF of 1.0×10^5, based on filter leach analyses during hot operation,[56] and 0.83×10^5, based on extensive off-gas sampling during initial cold (nonradioactive) operation of the NWCF.[57] Also, cold tests of particle removal by the silica gel adsorbers showed an average DF of 13 (range 9.1 to 19; because of measurement uncertainties downstream of the adsorbers, an average DF of 10 is claimed).[57] Measurements of the off-gas particle loading just ahead of the HEPA filters showed an average and maximum value of 0.29 and 0.43 mg/Nm^3, respectively (a normal cubic meter of off gas corresponds to feeding of 0.33 l aqueous waste to the calciner).

Additional limited performance data for the NWCF OGTS for particles during cold testing and early hot operations were tabulated by the IAEA.[31] DFs are calciner and cyclone, approximately 50; quench tower and venturi scrubber, approximately 200; calciner through venturi scrubber, 10^4; silica gel adsorber bed, 10. The DF from feed through the adsorber bed is 1×10^5, consistent with the unpublished results reported above for feed-to-filter DFs. If one assumes 1000 for the HEPA filter DF, the calculated feed-to-stack DF is 10^8.

C. VITRIFICATION

1. Introduction

Vitrification of radioactive wastes is a process in which the waste liquids, slurries, and solids are added to molten glass in the temperature range of 1100 to 1200°C and fused into a glass waste product. Two types of vitrification processes, the liquid-fed ceramic melter (LFCM) and the French AVM (Atelier de Vitrification de Marcoule) melter, have been developed for industrial-scale application to liquid radioactive wastes.

Based on its early success, simplicity, and ease of operation, the LFCM process was adopted as the reference high-level liquid waste (HLLW) vitrification process in the U.S., Japan, and the Federal Republic of Germany.[58] In the U.S., the Defense Waste Processing Facility (DWPF) at the Savannah River Plant and the West Valley Demonstration Project (WVDP) at West Valley, New York have design liquid feed rates[59,60] of 200 and 130 l/h and design glass production rates[59,60] of 100 and 45 kg/h, respectively. The PAMELA plant located in Mol, Belgium (constructed by Germany) and the vitrification plant located in Tokai, Japan have design glass production rates of 30 and 9 kg/h, respectively. The PAMELA plant has vitrified 910 m³ of waste and is in standby condition.[61] The other three plants are in various stages of testing before full-scale hot startup.

The first industrial-scale radioactive vitrification facility was the Marcoule, France AVM plant, which began hot operation in 1978. The AVM plant, has a glass production capacity of 15 kg/h and has vitrified over 1500 m³ of waste as of 1992.[62] The AVM process uses two stages of calcination before adding the calcined oxide waste to the glass melter. The liquid waste is added to a slightly inclined rotating tube, which varies from 225°C at the inlet to 600°C at the outlet. The AVM process design has been used to build additional facilities, the R7 and T7 vitrification plants (which have 60 l/h per line liquid feed rates, 25 kg/h per line glass production capacities, and three parallel lines) at La Hague, France and the Windscale Vitrification Plant (WVP) in Windscale, England.[63]

2. Liquid-Fed Ceramic Melters

The off-gas effluent characteristics and off-gas cleaning equipment of the DWPF melter are assumed to be representative of LFCMs and a general description of the off-gas treatment equipment is provided below. Detailed information on off-gas equipment design can be found in papers by Kessler and Randall[64] and Randall and Sabatino.[65] The melter is operated with a cold-cap crust over the melt and the feed solution or slurry is introduced onto the cold cap, where water is evaporated and drawn into the off-gas system. The vapor space above the cold cap is held between 650 and 800°C and at a vacuum of about −11 mmHg. The off-gas equipment is designed for a flow rate of 30 m³/min at the exit of the melter and an overall DF of 4×10^6. The sand filter is designed to provide an additional DF of 200 for an overall DF of 8×10^8. The off-gas equipment is illustrated in Figure 2 and the treatment steps are as follows:

- Film coolers — The off-gas, which consists mostly of air and water vapor, exits the melter via an off-gas film cooler. The cooler consists of a slotted sleeve through which air or steam is passed to form a gas film on the walls of the exit pipe. The cooler lowers the off-gas temperature to about 400°C and the gas film reduces the tendency for solids to stick on the surfaces. A rotating Hastelloy wire brush is used periodically to clear the sleeve of deposits.

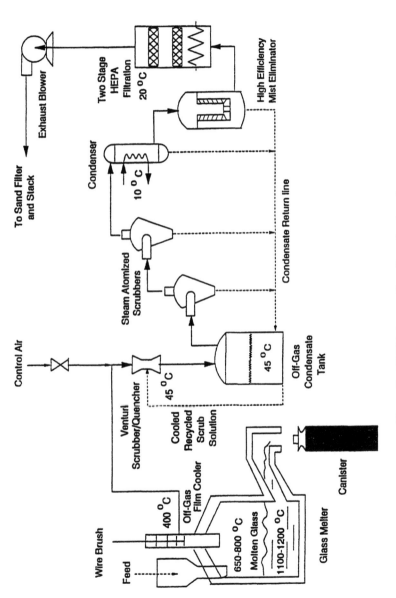

Figure 2 DWPF glass melter off-gas treatment.

Table 6 Off-Gas Decontamination
Factors for Cesium in the LSFM[64]

Component	DG	
	Goal	Achieved
Melter	10	40
Quencher	1	2
Steam atomized scrubber	50	60
High-efficiency mist eliminator	40	100
Tandem HEPA	2000	—
Sand filter	200	—

- Quencher/condensate tank — The quencher consists of a conventional jet venturi fume scrubber that cools the off-gas to about 45°C by direct mixing with cooled recirculated scrub solution. The bulk of the steam and larger aerosols are removed and collected in the off-gas condensate tank (OGCT).
- SAS scrubbers — Two steam-driven free-jet Hydro-Sonic® scrubbers (generically categorized as steam-atomized scrubbers [SAS]) in series follow the OGCT. The SAS scrubbers remove about 98% (design DF of 50) of the submicrometer aerosols. Atomized recirculated condensate serves as the scrub solution. Gas/liquid separation occurs in the attached reversing cyclone vessels.
- Condenser — The gas stream is cooled to about 10°C in a condenser downstream of the SAS scrubbers to further dry the gas stream.
- High-efficiency mist eliminator (HEME) — Further aerosol removal occurs in a glass-fiber HEME that has a removal efficiency of about 97.5% (design DF of 40).
- High-efficiency particulate air (HEPA) filter — The temperature of the off gas is raised to about 20°C (to avoid wetting of the filters) and passed through two stages of HEPA filtration that provide 99.95% removal efficiency (design DF of 2000) for particles and aerosols.

Because of the high-temperature overhead space of the melter, part of Cs, Sr, Ru, Ce, Pm, and Pu radionuclide inventory is volatilized and transported as aerosols in the off-gas. In tests using nonradioactive isotopes, cesium has been found to be the contaminant of greatest concern. About 10% of the feed inventory is assumed to volatilize (i.e., a design DF of 10 in the glass melter). On a curie basis, ^{137}Cs is predicted to be the most abundant, present mostly as submicrometer particles, and the most difficult to remove.[64] About 2.75, 0.267, and 0.202 Ci of ^{137}Cs, ^{90}Sr, and ^{106}Ru, respectively, per kilogram of glass produced would be generated in the off-gas. Based on the volatility of ^{137}Cs and meeting allowed release limits, the off-gas cleaning system was designed to achieve a DF of 8×10^8 for volatilized ^{137}Cs. An overall DF of 8×10^9 would be achieved based on ^{137}Cs retained by the melter.

A pilot plant called the Large Slurry Fed Melter (LSFM), which is designed to 43% capacity of the full-scale DWVP, has been used to obtain data on ^{137}Cs volatility and removal. The DFs for ^{137}Cs in various stages of the process are given in Table 6.

In all cases, the achieved DF exceeded the goal DF. Although the HEPA and sand filter DFs were not measured, the DFs for these filters were considered well established. Based on these tests, it was concluded that the overall DF for the ^{137}Cs would be a factor of 24 greater than the design DF.[64]

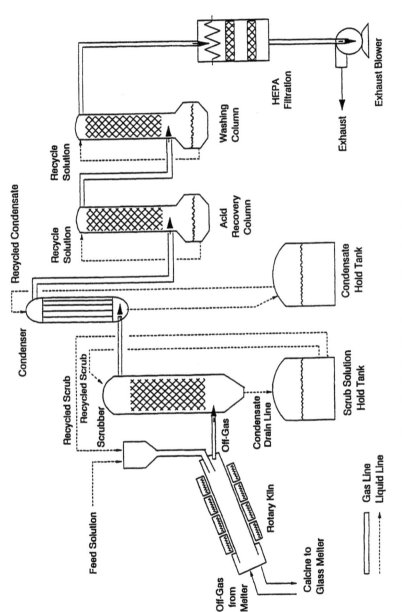

Figure 3 AVM calciner/melter off-gas treatment.

Table 7 DFs of Radionuclides in the AVM Off-Gas Equipment[66]

Nuclide	Calciner & melter	First scrubber	Condenser	Acid recovery column
^{106}Ru	5	3	280	3600
^{137}Cs	13	7	410	1600
^{144}Ce	22	3	470	190
^{90}Sr	27	26	4600	480

3. French AVM Melter

Both the off-gases from the calcination and vitrification processes flow in a countercurrent mode through the rotary tube and exit at the upper end of the calciner used in the French AVM vitrification process. The off-gas system illustrated in Figure 3 consists of a scrubber to retain entrained particulates, a condenser to remove steam and some nitric acid vapors, an acid recovery and washing column to remove the remaining nitric acid vapors, and finally HEPA filters. The scrub solution from the first scrubber is recycled to the rotary kiln. The average distribution of radionuclides in the off-gas treatment system based on the first two campaigns[66] is given in Table 7. Overall Dfs greater than 10^9 have been achieved for ruthenium and cesium.

In the first two campaigns, the loss of ^{106}Ru from the melter was 39 and 20%, respectively. The ruthenium loss can reduced by the addition of sugar to the feed solution. In the Atlas plant, which is half-scale AVM equipment, less than 1% of the ^{106}Ru volatilized with the addition of sugar.[66] Cold pilot-scale studies on the AVM process indicate that soluble and insoluble aerosols containing Fe, Al, Na, Mg, F, B, and Mo are carried by entrainment in the off-gas.[66] The insoluble aerosols are removed in the first scrubber and the soluble aerosols are found in both the scrubber and condensate, which are recycled as feed solution back to the rotary calciner. In addition to the aerosols, the off-gas treatment system removes oxides of nitrogen with an efficiency of about 90% in the condenser and about 82% in the acid recovery column.[66]

REFERENCES

1. **Burchsted, C. A., Kahn, J. E., and Fuller, A. B.,** Nuclear Air Cleaning Handbook — Design, Construction, and Testing of High-Efficiency Air Cleaning Systems for Nuclear Application, Second Edition, Report ERDA 76-21, U.S. Energy Research and Development Administration, 1976.
2. International Atomic Energy Agency, *Design and Operation of Off-Gas Cleaning and Ventilation Systems in Facilities Handling Low and Intermediate Level Radioactive Material,* Technical Reports Series No. 292, IAEA, Vienna, 1988.
3. U.S. Nuclear Regulatory Commission, *Code of Federal Regulations, Title 10, Energy, Part 20: Standards for Protection Against Radiation (10CFR20),* January 1, 1993 Edition, Appendix B to §§20.1001–20.2402.
4. U.S. Environmental Protection Agency, *Code of Federal Regulations, Title 40: Protection of the Environment, Part 190: Environmental Radiation Protection Standards for Nuclear Power Operations (40CFR190),* July 1, 1993 Edition (see Federal Register 2860, January 13, 1977).
5. **Christian, J. D. and Pence, D. T.,** Critical Assessment of Methods for Treating Airborne Effluents from High-Level Waste Solidification Processes, Report PNL-2486, June 1977.
6. **Perry, H. R. and Chilton, C. H.,** Gas–solids separations, in *Chemical Engineers' Handbook,* 5th Edition, McGraw-Hill Book Company, New York, 1973, 20-78 through 20-120.

7. U.S. Energy Research and Development Administration, Alternatives for Managing Wastes from Reactors and Post-Fission Operations of the LWR Fuel Cycle, Report ERDA 76-43, Vol. 2, 1976, chaps. 13 and 14.
8. U.S. Department of Energy, Technology for Commercial Radioactive Waste Management, Report DOE/ET-0028, Vols. 2 and 3, 1979.
9. First, M. V., Removal of airborne particles from radioactive aerosols, in *Treatment of Gaseous Effluents at Nuclear Facilities*, Goossens, W. R. A., Eichholz, G. G., and Tedder, D., Eds., Harwood Academic Publishers, Chur, Switzerland, 1991, chap. 2.
10. American Association for Contamination Control, Standard for HEPA Filters, Report AACC Standard CS-1, 1968, paragraph 4.a.
11. U.S. Department of Energy, Nuclear Standard Specifications for HEPA Filters used by DOE Contractors, Report NE F 3-45T, March 1985.
12. International Atomic Energy Agency, *Air Filters for Use at Nuclear Facilities*, Technical Reports Series No. 122, IAEA, Vienna, 1970.
13. Flanders Filters Inc., Air Filtration Products Catalog, P.O. Box 1219, Washington, North Carolina 27889.
14. International Atomic Energy Agency, *Testing and Monitoring of Off-Gas Cleaning Systems at Nuclear Facilities*, Technical Reports Series No. 243, IAEA, Vienna, 1984.
15. U.S. Nuclear Regulatory Commission, Design, Testing, and Maintenance Criteria for Normal Ventilation Exhaust System Air Filtration and Adsorption Units of Light-Water-Cooled Nuclear Power Plants, Report Regulatory Guide 1.140, October 1979.
16. **Carbaugh, E. H.,** A survey of HEPA filter experience, in *Proc. 17th DOE Nuclear Air Cleaning Conference*, U.S. Department of Energy, CONF-820851, 1982, 790–800.
17. Exxon Nuclear Company, Inc., Preliminary Safety Analysis Report, Nuclear Fuel Recovery and Recycle Center, Report XN-FR-32, NRC Docket Number 50–564, 1978.
18. **Christian, J. D.,** Process behavior and control of ruthenium and cerium, in *Proc. ANS-AIChE Topical Meeting on Controlling Airborne Effluents from Fuel Cycle Plants*, Sun Valley, Idaho, August 5 to 6, 1976, 2-1 to 2-34, American Nuclear Society, Hinsdale, Illinois.
19. **Christian, J. D. and Rhodes, D. W.,** Ruthenium Containment During Fluid-Bed Calcination of High-Level Waste from Commercial Nuclear Fuel Reprocessing Plants, Report ICP-1091, January 1977.
20. International Atomic Energy Agency, *Control of Semivolatile Radionuclides in Gaseous Effluents at Nuclear Facilities*, Technical Report Series No. 220, IAEA, Vienna, Austria, 1982.
21. **Klein, M.,** Retention of semivolatiles in high-temperature processes, in *Treatment of Gaseous Effluents at Nuclear Facilities*, Goossens, W. R. A., Eichholz, G. G., and Tedder, D., Eds., Harwood Academic Publishers, Chur, Switzerland, 1991, chap. 8.
22. **Ondrejcin, R. S.,** Volatile Ruthenium Chemistry, An Answer to RTA 925-3, Du Pont Savannah River Laboratory Internal Report DPST-80-254, January 31, 1980. See also, Christian, J. D., Behavior and control of ruthenium during operation of the New Waste Calcining Facility at the Idaho Chemical Processing Plant, in *Proc. 21st DOE/NRC Nuclear Air Cleaning Conference*, San Diego, California, August 13–16, 1990, Report NUREG/CP-0116, U.S. Nuclear Regulatory Commission, CONF-00813, U.S. Department of Energy, February 1991.
23. **Wilson, A. S.,** Ruthenium volatilization in the distillation of nitric acid, *J. Chem. Eng. Data*, 5, 521, 1960.
24. **Igarashi, H., Kato, K., and Takahashi, T.,** Effect of calcining temperature on volatilization of ruthenium in batch calcination of simulated high-level liquid waste, *Radiochim. Acta*, 60, 143, 1993.
25. **Igarashi, H., Kato, K., and Takahashi, T.,** Absorption behavior of gaseous ruthenium into water, *Radiochim. Acta*, 57, 51, 1992.
26. **Ortins de Bettencourt, A. and Jouan, A.,** Volatility of Ruthenium During Vitrification of Fission Products I. Distillation of Nitric Acid Solutions and Calcination of the Concentrates, Report CEA-R-3663(I), January 1969 (In French); *Chem. Abstr.* 70: 110946r; *INEL-tr-4*, Idaho National Engineering Laboratory, Idaho Falls, Idaho (English translation), May 1976.
27. **Fletcher, J. M.,** Complexes derived from (RuNO)III and RuIV, *J. Inorg. Chem.*, 1, 378, 1955.

28. **Weisenburger, S. and Weiss, K.,** Ruthenium volatility behavior during HLLW vitrification in a liquid-fed ceramic waste melter, in *Proceedings of Scientific Basis for Nuclear Waste Management,* Boston, 1979, Materials Research Society, Plenum Press, New York, 1980.

29. **Clark, W. E. and Godbee, H. W.,** Suppression of Ruthenium Volatilization in Evaporation and Calcination of Radioactive Waste Solutions, U.S. Patent No. 3,120,493, February 1964.

30. **Godbee, H. W. and Clark, W. E.,** The Use of Phosphite and Hypophosite to Fix Ruthenium from High Activity Wastes in Solid Media, Report ORNL-TM-125, January 1961.

31. International Atomic Energy Agency, *Design and Operation of Off-Gas Cleaning Systems at High Level Liquid Waste Conditioning Facilities,* Technical Reports Series No. 291, IAEA, Vienna, Austria, 1988.

32. **Hanson, D. A., Newby, B. J., and Rohde, K. L.,** The Adsorption of Ruthenium from Nitric Acid-Air Mixtures, Report IDO-14458, 1958.

33. **Rhodes, D. W.,** Adsorption of Volatile Ruthenium on Silica Gel, Report TID-7593, U.S. Atomic Energy Commission, 1959, 68–75.

34. **Rhodes, D. W. and Anderson, D. R.,** Capacity Test Data for the Adsorption of Volatile Ruthenium on Silica Gel, Report IDO-14510, June 1960.

35. **Newby, B. J. and Barnes, V. H.,** Volatile Ruthenium Removal from Calciner Off-Gas Using Solid Sorbents, Report ICP-1078, July 1975.

36. **Rastogi, R. C., Sehgal, J. D., and Thomas, K. T.,** Investigation of Materials and Methods for Fixation of Low and Medium Level Radioactive Waste in Stable Solid Media, Progress Report, Report A.E.E.T.-266, Atomic Energy Establishment Trombay, Bombay, India, 1966.

37. **Ortins de Bettencourt, A. and Jouan, A.,** Ruthenium Volatility During Vitrification of Fission Products. II. Fixation on a Steel Tube. Peroxide Decomposition, Report CEA-R-3663(II), January 1969 (In French); ABC-tr-7575 (English translation).

38. **Klein, M., de Smet, M., Goossens, W. R. A., and Baetsle, L.,** Filtration and Capture of Semivolatile Nuclides, Report IAEA-SM-245/51, International Atomic Energy Agency, Vienna, Austria, 1980.

39. **Rimshaw, S. J. and Case, F. N.,** Volatilities of ruthenium, iodine, and technetium on calcining fission product nitrate wastes, in *Proc. 16th DOE Nuclear Air Cleaning Conference,* San Diego, California, October 20–23, 1980, Report CONF-801038, U.S. Department of Energy, February 1991.

40. **Rimshaw, S. J., Case, F. N., and Tompkins, J. A.,** Volatility of Rhenium-106, Technetium-99, and Iodine-129, and the Evolution of Nitrogen Oxide Compounds During the Calcination of High-Level, Radioactive Nitric Acid Waste, Report ORNL-5562, February 1980.

41. **Halaszovich, St., Dix, S., and Merz, E. R.,** Studies of radioelement volatilization in the course of HLLW vitrification, in *Spectrum '86, Proc. Am. Nucl. Soc. Int. Topical Meeting on Waste Management and Decontamination and Decommissioning,* Niagara Falls, New York, September 14–18, 1986, Report CONF-860905, 1986.

42. **Cains, P. W.,** Semi-volatile radionuclides in high-level waste calcination and vitrification off-gasses, in *Proc. U.K. Filtration Conference on Gas Cleaning in the Nuclear Industry,* Manchester, England, March 12–13, 1987.

43. **Cains, P. W., Yewer, K. C., and Waring, S.,** Volatilization of ruthenium, caesium and technetium from nitrate systems in nuclear fuel processing and waste solidification, *Radiochim. Acta,* 56, 99, 1992.

44. **Goles, R. W., Sevigny, G. J., and Andersen, C. M.,** LFCM processing characteristics of mercury, in *Proc. 21st DOE/NRC Nuclear Air Cleaning Conference,* San Diego, California, August 13–16, 1990, Report NUREG/CP-0116, U.S. Nuclear Regulatory Commission, CONF-00813, U.S. Department of Energy, February 1991.

45. **Herbst, A. K.,** Mercury Concentrations in the Effluent Streams of the ICPP, Report ACI-378, May 1979.

46. **Osteen, A. B. and Bibler, J. P.,** Treatment of radioactive waste for mercury removal, *Water Air Soil Pollut.,* 56, 63, 1991.

47. **Christian, J. D.,** Behavior and control of ruthenium during operation of the new waste calcining facility at the Idaho Chemical Processing Plant, in *Proc. 21st DOE/NRC Nuclear Air Cleaning Conference*, San Diego, California, August 13–16, 1990, Report NUREG/CP-0116, U.S. Nuclear Regulatory Commission, CONF-00813, U.S. Department of Energy, February 1991.

48. **Berreth, J. R. and Dickey, B. R.,** High level waste management at the Idaho Chemical Processing Plant, in *Proc. Am. Nucl. Soc. Topical Meeting on Treatment and Handling of Radioactive Wastes*, April 1982, 449.

49. **Schindler, R. E.,** Removal of Particulate Solids from the Off-Gas of the WCF and NWCF, Report ICP-1157, June 1978.

50. **Commander, R. E., Lohse, G. E., Black, D. E., and Cooper, E. D.,** Operation of the Waste Calcining Facility with Highly Radioactive Aqueous Waste, Report IDO-14662, June 1966.

51. **Lohse, G. E. and Hales, M. P.,** Second Processing Campaign in the Waste Calcining Facility, Report IN-1344, March 1970.

52. **Bendixsen, C. L., Lohse, G. E., and Hales, M. P.,** The Third Processing Campaign in the Waste Calcining Facility, Report IN-1474, May 1971.

53. **Wielang, J. A., Lohse, G. E., and Hales, M. P.,** The Fourth Processing Campaign in the Waste Calcining Facility, FY 1971, Report ICP-1004, March 1972.

54. **Wielang, J. A. and Freeby, W. A.,** The Fifth Processing Campaign in the Waste Calcining Facility, FY-1972, Report ICP-1021, June 1973.

55. **Childs, K. F., Donovan, R. I., and Swenson, M. C.,** Ninth Processing Campaign in the Waste Calcining Facility, Report ENICO-1100, April 1982.

56. **Donovan, R. I. and O'Brien, B. H.,** Idaho National Engineering Laboratory, unpublished results.

57. **Swenson, M. C.,** Idaho National Engineering Laboratory, unpublished results.

58. **Bjorklund, W. J. and McElroy, J. L.,** An operating radioactive liquid-fed ceramic melter system, in *Proc. Symp. Waste Management '86*, 1986, 435–440.

59. **Baxter, R. G.,** Design and construction of the defense waste processing facility project at the Savannah River Plant, in *Proc. Symp. Waste Management '86*, 1986, 449–454.

60. **Barnes, S. M., Chapman, C. C., Petkus, T. F., Murawski, T. F., and Pope, J. M.,** Startup and initial experimental results for the West Valley vitrification demonstration project, in *Proc. Symp. Waste Management '86*, 1986, 441.

61. **Leigh, I. W.,** International Nuclear Fuel Cycle Fact Book, Report PNL-3594 Rev. 12, 1992, GE-23.

62. **Leigh, I. W.,** International Nuclear Fuel Cycle Fact Book, Report PNL-3594 Rev. 12, 1992, FR-16.

63. **Merlin, S,** The La Hague vitrification facilities, in *Proc. Symp. Waste Management '86*, 1986, 421.

64. **Kessler, J. L. and Randall, C. T.,** Performance of a large-scale melter and off-gas system utilizing simulated SRP DWPF waste, in *Proc. Waste Management '84*, Tucson, AZ, March 1984, 279.

65. **Randall, C. T. and Sabatino, D. M.,** Off-gas system for the Savannah River plant defense waste processing facility, in *SPECTRUM '86, Proc. Am. Nucl. Soc. Int. Topical Meeting on Waste Management and Decontamination and Decommissioning*, Niagara Falls, New York, September 14–18, 1986, CONF-860905, September 1986, Vol. 1, 1013.

66. **Roth, G.,** Integrated off-gas treatment facilities in vitrification plants, in *Treatment of Gaseous Effluents at Nuclear Facilities*, Goossens, W. R. A., Eichholz, G. G., and Tedder, D., Eds., Harwood Academic Publishers, Chur, Switzerland, 1991, 509–515.

Section IV

Current Separation Technologies in the World
(Research, Development, and Practice)

Today, many countries use nuclear power reactors to support national energy demands. From reprocessing of the spent nuclear fuel, nuclear wastes are produced worldwide. The national policy regarding nuclear waste treatment and disposal varies from country to country. Consequently, different techniques are used or being developed for nuclear waste management by different countries. The major nuclear energy producers in the world can be roughly divided into four regions: North America, Western Europe and the United Kingdom, Russia and Eastern Europe, and Asia. This section provides the readers with up-to-date information regarding the separation technologies for nuclear waste management that are either used or being developed by the countries in these regions. The Asian chapter was written by three chemists representing Japan, China, and Korea. The Russian and East European chapter was co-authored by two scientists from the Russian Academy of Sciences. The Western European chapter was prepared by two scientists from British Nuclear Fuels plc. The North American chapter was written by two chemists from Pacific Northwest Laboratory and from Florida State University. These four chapters contributed by scientists from six countries give us a global view of various approaches to nuclear waste problems undertaken by nuclear energy producers in the world.

Chien M. Wai

Chapter 14

North America

Linfeng Rao and Gregory R. Choppin

I. INTRODUCTION: HISTORICAL REVIEW

Large-scale radiochemical separation technology was developed in the United States during World War II, as part of the Manhattan Project to develop nuclear weapons for military applications. The purpose of these chemical processes was to separate plutonium from uranium and from fission products in the irradiated uranium. The separation principles underlying those processes are still valid. A number of these processes, in some cases with significant modifications, now play important roles in the management of nuclear wastes.

This chapter briefly reviews separation techniques either in use or proposed for defense and power reactor wastes, and for partitioning/transmutation (P/T) processes. In the review of the new technologies, some evaluation of their state of development and possible application is given.

A. SEPARATION TECHNOLOGIES IN THE MANHATTAN PROJECT

The first large-scale radiochemical separation process was based on bismuth phosphate precipitation,[1] in which $BiPO_4$ was used as a carrier for the insoluble phosphates of Pu^{3+} and Pu^{4+}. Because hexavalent uranium and hexavalent plutonium do not form insoluble phosphates, separation of plutonium from uranium was realized by selectively reducing plutonium to the tetra- and/or trivalent states prior to the precipitation. Further purification of plutonium was accomplished by alternately reducing and oxidizing plutonium between successive precipitations. This process was used successfully for the first isolation of hundreds of kilograms of plutonium at the Hanford Site.

The precipitation technique was replaced by solvent extraction, which was more suitable for large-scale, continuous remote operations in which both uranium and plutonium could be isolated in a pure state from the fission products. However, the underlying principle of reduction and oxidation employed in the bismuth phosphate process has been retained in subsequent solvent extraction processes. The first solvent adopted was methyl isobutyl ketone[2] (MIBK or hexone) at Hanford. Hexone forms adduct compounds with coordinatively unsaturated actinide nitrates in such extraction reactions as

$$Pu^{4+}_{(aq)} + 4NO^-_{3(aq)} + 2S_{(org)} \rightarrow Pu(NO_3)_4S_{2(org)}$$

where S = hexone. The tri- and hexavalent actinides form corresponding adduct compounds such as $An(NO3)_3S_3$ and $AnO_2(NO_3)_2S_2$ (where An = actinides), respectively. These neutral hexone compounds of the actinides are soluble to different extents in organic solvents like kerosene, and in hexone itself. The process using hexone is referred to as the REDOX process.

The REDOX process has the disadvantage that hexone tends to decompose slowly in contact with the strong nitric acid used for the aqueous phase. Many other solvent extraction systems have been developed,[3] among which the PUREX process gained the most popularity and remains in use today. The organic extractant used in the PUREX process is tributyl phosphate (TBP), which has the advantages of being more stable, less flammable, and giving better separation than hexone. As important as these advantages is the fact that TBP allows use of nitric acid as the salting agent. Most of the other extractant systems, including hexone, require high concentrations of salts such as aluminum or magnesium nitrates as salting agents for the aqueous phase. Such concentrated salt solutions have the disadvantage that, after use, they are contaminated with radioactivity and contribute to the magnitude of the eventual waste management problems.

Several nonaqueous processes have been developed for separation purposes in the nuclear fuel cycle. For example, the uranium hexafluoride production processes, based on the volatility of UF_6, were used on a large scale to produce the feed material for the enrichment of uranium-235 by gaseous diffusion. Because plutonium also forms a volatile hexafluoride, PuF_6, separation of uranium and plutonium from bulk impurities or fission products by volatilization has been demonstrated as feasible. Molten-salt extraction and molten-metal purification are examples of other nonaqueous processes that have been extensively studied. The former is of interest for reprocessing of the molten-salt reactor fuels and has the advantage of having higher radiation resistance than aqueous systems. The latter has been tested for metallic fuel breeder reactors.

B. SEPARATION TECHNOLOGIES IN NUCLEAR WASTE MANAGEMENT

Research and development in separation technology has been stimulated in recent years by the need to initiate treatment leading to the ultimate disposal of nuclear wastes in a safe and cost-effective manner. While the majority of these radioactive wastes have been generated in the nuclear weapons program (defense wastes), the spent fuel from the operation of nuclear power plants must also be prepared for disposal in a geological repository.

The defense wastes were generated mainly in the reprocessing of irradiated uranium to produce plutonium. The radioactive components are fission products, transuranic elements and their decay products, and nuclides resulting from the neutron activation of structural materials such as cladding. Presently, such wastes are stored in various U.S. Department of Energy (DOE) facilities throughout the country. The basic philosophy is to separate the longer-lived and/or more toxic radioactive species into a relatively small volume for disposal. This fraction is termed the high-level waste or HLW. A much larger volume of less radioactive wastes (low-level wastes, LLW) would also be processed and packaged for disposal in a surface or near-surface repository. This approach is expected to be cost-effective due to the significant reduction in the HLW volume, for which the handling and final storage is extremely costly. This final storage must ensure safety to future generations by "permanent" retention of the wastes.

There are two scenarios for the treatment of the spent fuel from power reactors (presently, light water reactors, LWR): the once-through mode and the reprocessing mode.[4] In the once-through mode, the fuel elements are put into permanent disposal after a cooling period, without any chemical treatment. In the reprocessing mode, chemical separations, basically similar to those for the defense wastes, are required prior to permanent disposal. Present U.S. policy favors the once-through mode. But if nuclear power expands in the future, the economics of recycling unburnt U and product Pu may cause reprocessing of spent fuel to become the accepted approach. If so, substantial research would be necessary to develop improved separation processes, because spent fuel from power reactors has characteristics different from the uranium irradiated for plutonium production, and reactors of advanced designs may further increase these differences. During the period when new technologies would be under development, the existing separation processes could be modified or tailored to meet the requirements of the spent fuel from the new reactors.

As an alternative to direct reprocessing followed by disposal of the resultant nuclear waste, a concept of partitioning/transmutation (P/T) has been under consideration. In the P/T approaches, further treatment separates the long half-lived transuranic elements and fission products, and destroys them by fissioning or by transmutation to stable nuclides. Separation processes would be integral to the success of this approach should it be chosen for reasons of economy or long-term safety (as an alternative to HLW disposal in a geologic repository).

II. AQUEOUS PROCESSES

Aqueous processes have played important roles in the separations of irradiated uranium and are likely to continue, over the near future, as the principal separation processes for the treatment of nuclear wastes. The separation of radioactive elements in these processes is based on the differences in such chemical properties of the dissolved species as the reduction potentials, the complexation strength with various ligands, the affinity to ion-exchange resins or inorganic ion exchangers, and the transport behavior through membranes or in an electric field. Modifications of these processes are necessary, however, to treat wastes of different compositions.

A. SOLVENT EXTRACTION PROCESSES
1. PUREX Process

Due to several advantages over other solvent extraction processes, PUREX remains the generally accepted industrial process for separation of uranium, plutonium, and neptunium from fission products. In the PUREX treatment, uranium and plutonium are co-extracted (as tetravalent cations) into the organic phase from an aqueous nitric acid solution (6 to 8 M) by tributyl phosphate (TBP) while the fission products remain in an aqueous phase. The plutonium is subsequently separated from the uranium by selectively reducing it to Pu^{3+}, which is stripped into the aqueous phase. Depending on the operating conditions, neptunium may remain in the uranium streams and can be separated from uranium by adjusting extraction conditions in subsequent steps. This process could be easily adapted to treat the defense wastes by concentrating the actinide elements into fractions of relatively small volume for storage or for destruction by fissioning in reactors. However, depending on the composition of the wastes, more separation steps (solvent extraction or ion exchange) might have to be added to the PUREX process to obtain better separation of certain elements in the wastes. For example, it is possible to selectively remove long-lived (e.g., ^{99}Tc) or highly radioactive (e.g., ^{137}Cs) fission products from the waste streams using ion-exchange processes.[5,6] Another example is the SREX process,[7] discussed below, in which a crown ether is used to isolate strontium from the PUREX waste streams.

The separation of plutonium and uranium from fission products in the PUREX process is not completely quantitative. The overall losses of plutonium and uranium into the fission products waste streams may range from a few tenths of one percent to several percent. As a result, processes have been proposed that use PUREX for bulk plutonium and uranium separation, and then treat the resultant waste streams with more advanced separations processes using extractants with higher affinity for transuranic elements. One of these processes being studied and developed in the United States is the TRUEX process.

2. TRUEX Process

This solvent extraction process, developed largely at Argonne National Laboratory (ANL),[8] is designed to separate transuranic elements from acidic high-level waste solutions. At present, it seems to be the most promising separation technology as a subsequent step to PUREX for this purpose. A bidentate organophosphorus compound, (octyl(phenyl)-N,N-dibutylcarbamoylmethylphosphine oxide (CMPO), is the extractant. Dissolved in hydrocarbon solvents (alkanes), CMPO can extract transuranic elements from acidic solutions almost quantitatively and selectively. TRUEX treatment of the waste streams from a PUREX separation of plutonium and uranium can reduce the concentrations of residual U and Pu in the wastes by factors of 100 to 1000.

The possibility of formation of a third phase, rich in actinides, is a concern in the TRUEX process. Research is under way to avoid such third-phase formation by various means; for example, addition of TBP shows promise because it not only prevents the third-phase formation, but also enhances the extraction of the transuranics. Elevation of the process temperature or use of different solvents (e.g., aromatic or chlorinated hydrocarbons) may also be useful in reducing or eliminating this third-phase problem. However, further research is needed to ensure that the solution to third-phase formation does not generate new problems. An example would be the use of chlorinated hydrocarbon solvents, which are likely to be unacceptable due to their health and safety hazards.

Since CMPO is a much stronger binding extractant than TBP, complete stripping of actinides from the organic phase of the TRUEX solvent might be difficult. Among the stripping processes being investigated is the use of thermally unstable alkyl diphosphonic acids (TUCS reagents).[9] These compounds form strong complexes with actinides in acidic solutions and decompose relatively easily at elevated temperatures. However, their potential use as stripping agents into strongly acidic solutions of the residual transuranics is uncertain, because a decomposition product, phosphate, is an undesirable component in the waste stream — where it may interfere in the final waste vitrification process.

It is not certain at present whether the TRUEX process would be applicable to the separation of trivalent lanthanides from trivalent transuranic elements.

Extensive laboratory research and development has been conducted on TRUEX, and this process is now ready for pilot-plant testing. A problem with TRUEX is the rather high cost of synthesis of the CMPO reagent, and research in the synthesis method is needed. Diphenyldibutylcarbomoylmethylphosphine oxide has a higher solubility than CMPO, but has been less well studied and has many of the same usage problems.

3. SREX Process

This process, also developed at ANL,[7] is designed for separation of strontium, one of the major sources of radioactivity in the HLW of the defense wastes. A highly specific crown ether extractant for strontium is the principal factor in the SREX process. It can be used to isolate strontium from the waste streams of the PUREX process. Significant research on the SREX process has been done at the laboratory level and this process is ready for pilot-plant tests.

One concern about the SREX process is the cost of the reagent. Like the CMPO in the TRUEX process, the crown ether has to be manufactured on an industrial scale for the large-scale processing of nuclear wastes. The present synthetic process makes SREX much more expensive than classic commercial extractants such as TBP. This economic aspect is aggravated by loss of the extractant in the process due to its chemical and radiolytic degradation, its solubility in the aqueous phase, and emulsification. Consequently, the extent of regeneration of the extractant is an important criterion that might determine the applicability of the SREX process for the large-scale treatment of nuclear wastes.

4. Talspeak Process

This process, developed at Oak Ridge National Laboratory (ORNL),[10] separates lanthanides from trivalent actinides in moderately acidic solutions (pH 2.5 to 3.0). Two versions of the Talspeak process have been proposed for use in the partitioning/transmutation technologies for treating nuclear wastes. In the "normal" version, the trivalent lanthanides are extracted by di(2-ethylhexyl) phosphoric acid (HDEHP) from an aqueous solution of lactic and diethylenetriaminepentaacetic (DTPA) acids at pH 2.5 to 3.0, while the trivalent actinides remain in the aqueous phase. In the "reversed" version,[11] both trivalent lanthanides and actinides are extracted into an organic phase containing HDEHP in the first step, followed by stripping of the trivalent actinides into an aqueous phase containing lactic and DTPA acids. Subsequently, the trivalent lanthanides in the organic phase can be recovered by stripping with 6 M nitric acid.

A disadvantage of the Talspeak process is the formation of radiolytic degradation products of the organic complexing agents and pH buffering agents. Process control may become difficult as a result of the buildup of these degradation products. Therefore, advanced laboratory studies and pilot-plant tests at high radiation levels are required to evaluate the applicability of the Talspeak process in the treatment of HLW.

5. Tramex Process

This process utilizes the extractability and selectivity of liquid cation exchangers, such as trialkylamines or tetraalkylammonium salts, to separate tri- and tetravalent actinides from lanthanides and most other fission products.[12] The separation is usually accomplished using aqueous solutions with high chloride concentrations — which has several disadvantages. For example, because the wastes are usually in nitrate solution initially, added processing steps are required to convert from nitrate to chloride prior to extraction. Furthermore, the concentrated chloride is a corrosive media, so special processing materials would be required.

The partitioning/transmutation approach (to be discussed below) requires the separation of actinides as the first step, and the Tramex process could be used for that purpose. However, it would be much more attractive if the process could be modified to allow extraction from nitrate solutions.

6. Other Extractants

The solvent extraction processes discussed in the previous sections are either already well developed and applied at industrial levels (e.g, PUREX), or well studied at laboratory levels and ready for pilot-plant tests (e.g, TRUEX). To develop highly specific processes for the isolation of a particular radioactive element from nuclear wastes, new extractants are being synthesized and studied in many laboratories. These include stereospecific complexants such as crown ethers and siderophores, bidentate extractants such as diphosphine dioxides and diamides, "soft donor" complexants with active S or N atoms, and cobalt dicarbollide extractant. The diphosphine dioxides and the diamides have good extraction and radiation-resistant qualities, but, like CMPO, do not separate the trivalent lanthanides and actinides. The diamides have the advantage that they do not add phosphate to the waste to be vitrified.

Studies on stereospecific extractants could be of great value for isolation of radioactive elements such as cesium, strontium, technetium, and plutonium. Specific chelating ligands for plutonium have been developed and tested in the laboratory,[13] but, thus far, have not been proven applicable to the isolation of plutonium from nuclear waste streams. However, as discussed previously for CMPO in the TRUEX process and the crown ether in the SREX process, the use of such specific extractants depends on the ease and the economy of their large-scale production and the efficiency of regeneration for recycling.

Studies at Los Alamos National Laboratory (LANL) have demonstrated good separations of trivalent actinides from lanthanides with thio-derivatives of beta-diketones as the extractants.[14] However, these soft donor extractants may not be useful except to separate the heavier trivalent actinides from lanthanide fission products; simpler and more efficient processes already exist for separation of the lighter actinides (thorium through plutonium), based on the redox behavior of these actinides.

The study of cobalt dicarbollide (*bis*-dicarbollycobaltate), first initiated by Czech and Russian scientists, has demonstrated very high selectivity and efficiency for extraction of Cs^+ and Sr^{2+} from 2 to 3 M nitric acid solutions.[15,16] In addition, good chemical and radiation stability has been reported for the hexachloro- and hexabromo-derivatives of dicarbollides. As a result, there is interest within the U.S. DOE in conducting studies of dicarbollide extraction for different radioactive elements and/or under various conditions. A hindrance to the use of the dicarbollides is their employment in present systems of aromatic and halogenated solvents, which impose health and safety hazards. Also, if the extraction must be conducted in acidic solutions, many DOE wastes (now in strongly alkaline solutions) would have to be acidified (in 0.3 to 3 M nitric acid) before extraction, resulting in a large increase in the total waste volume. These problems make it unlikely that the dicarbollide extraction system would be employed for processing highly alkaline wastes (e.g., those at Hanford).

B. ION EXCHANGE AND ADSORPTION PROCESSES

Ion-exchange processes have been successful in the past for the separations of individual rare-earth and transuranic elements. More recently, highly selective organic ion-exchange resins have been developed and show potential in removal of particular radionuclides. Besides, research is underway to synthesize and study various inorganic ion exchangers, which have high capacity and stability. In this section, a brief review is given on the ion-exchange materials that have shown promise in separations of particular radionuclides.

1. Organic Ion-Exchange Resins

A number of interesting ion exchange resins have been developed in a joint research project between ANL and the University of Tennessee, and some of these resins are now commercially available. A very promising resin uses a substituted diphosphonic acid and has high efficiency in retaining various heavy-metal cations from liquid wastes. For example, it is able to remove actinide elements from strongly acidic HNO_3 solutions. Another type of resin uses a crown ether as the functional group and is specific in removing strontium from solutions. At present, these resins are used principally by environmental laboratories for analytical purposes, but they would seem to have promise in acidic reprocessing schemes.

2. Inorganic Ion Exchangers

There is interest in the use of inorganic ion exchangers to treat nuclear wastes, due to the good radiation and chemical stability and the high capacity of these materials. Titanium phosphate, zeolites, silicotitanates, and some clays are among the inorganic materials being studied. Some of these inorganic materials exhibit specificity in the removal of certain cations from the wastes. Their properties have also created interest in their potential as the final waste form for geological storage.

A highly selective cesium ion sieve made by topotactic leaching of a naturally occurring phlogopite mica was demonstrated to be successful in removal of ^{137}Cs from contaminated environments.[17] Similarly, a synthetic material, sodium fluorophlogopite mica ($Na_4Mg_6Al_4Si_4O_{20}F_4$), selectively removes ^{90}Sr from solution and immobilizes it at room temperature.[18]

In the mid-1970s, hydrous titanate ion exchangers were developed at Sandia National Laboratory for immobilizing high-level radioactive waste in ceramic form. These materials are in large-scale use at the Savannah River Site in South Carolina. Based on the research on hydrous titanates, a new class of materials called silicotitanates has been developed recently in a collaboration between Sandia National Laboratory and Texas A & M University.[6] It is claimed that these silicotitanates can remove cesium from radioactive waste solutions in the presence of very high sodium concentrations. Such specificity would indicate a potential use for these materials in treating tank wastes containing ^{137}Cs in concentrated salt solutions, such as those at the Hanford Site. It might be feasible to further process the silicotitanates, with their sorbed radioactive elements, as a method for direct permanent disposal.

At present, further research and development is needed on these inorganic ion-exchange systems before their use in treating nuclear wastes can be fully evaluated.

C. MEMBRANE PROCESSES

The defense wastes stored at the various DOE sites in the United States contain not only radioactive materials, but, usually, high concentrations of salts (e.g., sodium nitrate and nitrite), as well as other nonradioactive chemical reagents. The presence of these nonradioactive ingredients is a complicating factor in treatment of the nuclear wastes. Membrane processes such as dialysis, ultrafiltration, and facilitated transport have been proposed as techniques to separate the bulk salt from the radioactive components.

There are commercially available electrolysis membranes that selectively extract sodium chloride from seawater. Application of an electric field across these membranes results in the passage of monovalent cations or anions through the membranes, but not multivalent ions. As an example of their application, the bulk salt components (sodium nitrate or sodium nitrite) of the Hanford tank wastes could be moved through the membrane, leaving the higher-valence radioactive species in a dilute solution. This would allow a significant reduction in the volume of the radioactive wastes.

The membranes must have high resistance to chemical and radiolytic destruction. Laboratory research is under way to determine the properties of the available membranes and to develop others with the necessary stability.

The "facilitated transport" processes use modified membranes that contain chemical reagents capable of combining with particular species in the solution. These reagents perform a function similar to the extracting agents in solvent extraction processes. Because solvent extraction is a better-developed technique and can accomplish similar objectives, there would not seem to be a strong incentive to develop such facilitated transport processes except, possibly, as an alternative to solvent extraction systems in which phase separation is difficult (e.g., due to the formation of a third phase).

III. NONAQUEOUS PROCESSES

Nonaqueous processes, based on the difference in such properties as volatility of various compounds or redox thermodynamics of elements in molten-salt media, have been used over many years in uranium isotope separation, electrorefining of plutonium metal, and production of metallic fuel for advanced nuclear reactors. There is interest in conducting research of nonaqueous processes as separation technologies for treatment of nuclear wastes.

A. VOLATILITY PROCESSES

Uranium hexafluoride has been used for 50 years in the gaseous diffusion process for uranium isotopic enrichment.[19] Volatility techniques with fluorides have also been used to purify plutonium in isotope separation plants[20] and were studied for use in fuel processing in the molten-salt reactor project at ORNL. The separation of uranium and plutonium from fission products is limited in these processes by the fact that volatile fluorides are formed by several fission products, and, in particular, by iodine and tellurium. Tellurium fluoride (Te_2F_{10}), technetium fluoride (TcF_6), and iodine fluoride, (IF_5), follow the chemistry of UF_6 to some extent, and can contaminate the uranium and plutonium streams in the volatility processes. Both tellurium and iodine in higher oxidation states also form fluorides (TeF_6 and IF_7, respectively), with much higher volatility than UF_6 and PuF_6. As a result, iodine and tellurium can be separated from uranium and plutonium by distillation after oxidation. However, the decontamination of technetium remains a difficult task in the fluoride volatility process because its fluorides diffuse with the UF_6 and PuF_6 streams.

In an early plutonium purification process, PuO_2 was fluorinated to PuF_6 with fluorine at high temperatures (400°C) and in special equipment (e.g., a fluidized-bed reactor).[20] Recently, a more powerful fluorinating agent (O_2F_2, "foof") was developed at LANL and pilot-plant tests demonstrated that it could fluorinate PuO_2 to PuF_6 at ambient temperatures in simpler reactors.

The construction materials for fluoride volatility processes must be highly resistant to fluoride corrosion, because uranium and plutonium hexafluorides are themselves powerful fluorinating agents. Furthermore, the radiolytic decomposition of PuF_6 to PuF_4 and F_2 in the process is another concern, because PuF_4 is less volatile and may deposit in the system. However, additional fluorinating with agents such as ClF_3 should eliminate this problem. Fluoride volatility plants for treating spent fuel on a large scale have not been built in the United States, although they are technically feasible.

There are other separation processes based on the volatility of different compounds. These processes may have promise for separating particular elements from certain types of wastes, but they are not as well developed as the fluoride volatility process. For example, the volatility of $ZrCl_4$ could be used to remove the zirconium cladding on spent fuel elements, in a process similar to the Zirflex process developed at ORNL. The volatility of beta-diketone complexes of trivalent actinides may have some promise but needs more research.[21] In a proposed scheme for transmutation of technetium to the nonradioactive ruthenium, the product Ru is converted to RuO_4 by ozonolysis and separated from the remaining Tc_2O_9, using the higher volatility of RuO_4.[22] Much more research is needed to evaluate these volatility processes for practical use in full-scale separation systems.

B. PYROCHEMICAL PROCESSES

Molten-salt and molten-metal systems are among the pyroprocesses under extensive investigation as possible technologies to treat spent fuel from the Integral Fast Reactor (IFR), molten-salt breeder reactors, and LWRs. These pyroprocesses have the advantage of insensitivity to radiation effects — in contrast to aqueous processes for which radiolysis can be a serious problem, causing degradation of the organic extractants and changing the aqueous-phase chemistry through the radiolysis of water.

Based on research conducted at ORNL on a molten-salt process to treat fuels for the molten-salt breeder reactor,[23] a $LiCl/BeCl_2$ molten-salt system has been proposed in which the separation of transuranic elements would be accomplished by extraction into liquid Bi in the accelerator transmutation system at LANL (see next section).

In IFR fuel processing, ANL has demonstrated a pyroprocessing scheme at the laboratory level.[24] This process uses electrochemical reduction/oxidation at elevated temperatures to separate elements based on the difference in their thermodynamic properties in the molten-salt media. In this process, actinide elements in the spent fuel are dissolved in a molten cadmium electrochemically to form alloys with cadmium. Based on the difference in dissolution potentials between uranium and transuranic elements, uranium is subsequently removed from the molten cadmium by controlling the electrical potentials, and deposited on a solid cathode. Then the cathode is replaced, and the voltage is increased to strip the plutonium from the molten cadmium and deposit it on the cathode. Good separation factors for transuranic elements as a group have been reported from the laboratory experiments and modeling calculations. However, the separation factors for pairs of individual transuranic elements are too small for effective isolation of the pure elements.

Pyroprocessing of LWR fuels is considered by the U.S. DOE as the separation technique of choice for transmutation in the Advanced Liquid Metal Reactor (ALMR) system. ANL has proposed a tentative scheme for pyroprocessing of LWR oxide fuels in which either a modified salt transport process or a lithium process would be used.[25] More research has been performed on the salt transport process, but the lithium process may have the advantage of operating at lower temperatures, resulting in reduced equipment costs. The parameters of these processes are not defined yet, because the research and development of LWR pyroprocessing has not been conducted on a large scale or on fully irradiated fuel.

IV. SEPARATIONS FOR PARTITIONING/TRANSMUTATION

Chemical separations are an indispensable part of the proposals for a transmutation approach to the treatment of nuclear wastes. There are several transmutation concepts, including transmutation in LWRs, in ALMR, and by accelerator irradiation.

The different transmutation concepts impose different requirements on the associated separation techniques. For example, in the transmutation scheme for actinide elements in LWRs, while the PUREX process has many advantages, it does not fully meet the requirements for actinide separation and recovery. Additional separation processes would be required (using TRUEX, Talspeak, Tramex, etc.). If the transmutation is to be performed in ALMRs or with accelerators, either aqueous or pyrochemical processes could be used as separation techniques. However, research and development on the IFR at ANL indicates that pyroprocessing might be more efficient and require smaller equipment than the aqueous processes. Nevertheless, before using pyroprocessing in any transmutation system, further substantial research and development is necessary.

A transmutation scheme has been proposed by LANL to convert ^{99}Tc into stable ruthenium by thermal neutron capture.[22] This scheme requires the separation of Tc from the nuclear waste before transmutation, and/or separation of Ru from Tc after transmutation. Among the techniques proposed for use as steps in these separation schemes are solvent extraction of Tc with liquid amine anion exchangers, ion-exchange separation of cationic Ru species,[5] volatilization separation of RuO_4,[26] and magnetic separation of TcO_4^- from reduced, paramagnetic forms of Ru.[27] All these methods are presently being studied in the laboratory.

The various transmutation schemes are not attractive for treatment and disposal of the Hanford tank wastes. Much more extensive and elaborate separation operations on the tank wastes would be needed to produce actinide fractions for targets in transmutation systems. The total amount of uranium and transuranic elements in the Hanford wastes is not large enough to justify the extra expense of such extensive separation followed by transmutation, when compared to the cost of simpler treatment prior to geologic disposal.

V. SUMMARY

A number of separations processes have been studied at laboratory levels or developed at both laboratory and pilot-plant levels. Each process has its advantages and limitations in possible application to the treatment of nuclear wastes. Only the PUREX process has been used in spent fuel reprocessing. In order to achieve the goal of safe and ultimate disposal of nuclear wastes, additional research and development, through pilot-plant demonstration, is required to evaluate the technical feasibility, cost-effectiveness, and operational safety of these separations technologies.

REFERENCES

1. **Choppin, G. R. and Rydberg, J.,** *Nuclear Chemistry: Theory and Applications*, Pergamon Press, New York, 1980, chap. 20.
2. **Hill, O. F. and Cooper, V. R.,** Scale-Up Problems in the Plutonium Separations Program, *Ind. Eng. Chem.*, 50, 599, 1958.
3. **Culler, F. L.,** Reprocessing of Reactor Fuel and Blanket Materials by Solvent Extraction, in *Progress in Nuclear Energy, Series III, Process Chemistry*, Vol. I, Bruce, F. R., Fletcher, I. M., Hyman, H. H. and Katz, J. J., Eds., McGraw-Hill Book Co. Inc., New York, 1956, chap. 5.2.
4. **Choppin, G. R.,** Chemical Issues in Nuclear Waste Disposal, *Sci. Teacher*, 48, 2, 1981.
5. **Rawlins, J. A. and Bager, H. R.,** CURE: Clean Use of Reactor Energy, Report WHC-EP-0268, Westinghouse Hanford Co., 1990.
6. **Bauer, R.,** Cesium Cut from Radioactive Waste, *C&E News*, July 13, 26, 1992.
7. **Horwitz, E. P., Dietz, M. L. and Fisher, D. E.,** Extraction of Strontium from Nitric Acid Solutions Using Dicyclohexano-18-Crown-6 and Its Derivatives, *Solv. Extr. Ion Exch.*, 8, 557, 1990.
8. **Schulz, W. W. and Horwitz, E. P.,** The TRUEX Process and the Management of Liquid TRUEX Waste, *Sep. Sci. Technol.*, 23, 1191, 1988.
9. **Nash, K. L.,** Actinide Phosphonate Complexes in Aqueous Solutions, Abstracts of Actinides-93 International Conference, Santa Fe, New Mexico, September 19–24, 1993.

10. **Weaver B. and Kappelmann, F. A.,** Talspeak: A New Method of Separating Americium and Curium from Lanthanides by Extraction from an Aqueous Solution of Aminopolyacetic Acid Complex with a Monoacidic Phosphate or Phosphonate, Report ORNL-3559, Oak Ridge National Laboratory, August 1964.

11. **Persson, G. E., Svantesson, S., Wingefors, S. and Liljenzin, J. O.,** Hot Test of a Talspeak Procedure for Separation of Actinides and Lanthanides Using Recirculating DTPA-Lactic Acid Solution, *Solv. Extr. Ion Exch.*, 2, 89, 1984.

12. **Lloyd, M. H.,** An Anion Exchange Process for Americium-Curium Recovery from Plutonium Process Waste, *Nucl. Sci. Eng.*, 17, 452, 1963.

13. **Kappel, M. J., Nitsche, H. and Raymond, K. N.,** Specific Sequestering Agents for the Actinides. 11. Complexation of Plutonium and Americium by Catecholate Ligands, *Inorg. Chem.*, 24, 605, 1985.

14. **Ensor, D. D., Jarvinen, G. D. and Smith, B. F.,** The Use of Soft Donor Ligands, 4-Benzoyl-2,4-dihydro-5-methyl-2-phenyl-3H-pyrazol-3-thione and 4,7-Diphenyl-1,10-phenanthroline for Improved Separation of Trivalent Americium and Europium. *Solv. Extr. Ion Exch.*, 6, 439, 1988.

15. **Rais, J., Tachimori, S. and Selucky, P.,** Synergetic Extraction in Systems with Dicarbollide and Bidentate Phosphonate, *Sep. Sci. Technol.*, 29, 261, 1994.

16. **Esimantovskii, V. M., Galkin, L. N., Lazarev, R. I., Lyubtsev, V. N., Romanovskii, V. N. and Shishkin, D. N.,** Technological Tests of HAW Partitioning with the Use of Chlorinated Cobalt Dicarbolyde (CHCODIC); Management of Secondary Wastes, in *Proc. Symp. on Waste Management*, Tucson, March 1992, p. 801.

17. **Komarneni, S. and Roy, R.,** A Cesium-Selective Ion Sieve Made by Topotactic Leaching of Phlogopite Mica, *Science*, 239, 1286, 1988.

18. **Paulus, W. J., Komarneni, S. and Roy, R.,** Bulk Synthesis and Selective Exchange of Strontium Ions in $Na_4Mg_6Al_4Si_4O_{20}F_4$ Mica, *Nature*, 357, 571, 1992.

19. **Choppin, G. R. and Rydberg, J.,** *Nuclear Chemistry: Theory and Applications*, Pergamon Press, New York, 1980, chap. 13.

20. **Hyman, H. H., Vogel, R. C. and Katz, J. J.,** Fundamental Chemistry of Uranium Hexafluoride Distillation Processes for the Decontamination of Irradiated Reactor Fuels, in *Progress in Nuclear Energy, Series III, Process Chemistry*, Vol. I, Bruce, F. R., Fletcher, I. M., Hyman, H. H. and Katz, J. J., Eds., McGraw-Hill Book Co. Inc., New York, 1956, chap. 6.1.

21. **Steinberg, M., Powell, J. R. and Takahashi, H.,** APEX Nuclear Fuel Cycle for Production of Light Water Reactor Fuel and Elimination of Radioactive Waste, *Nucl. Tech.*, 58, 437, 1982.

22. **Dewey, H. J., Jarvinen, G. D., Marsh, S. F., Marsh, N. C., Schroeder, N. C., Smith, B. F., Villareal R., Walker, R. B., Yarbro, S. L. and Yates, M. A.,** Status of Development of Actinide Blanket Processing Flowsheets for Accelerator Transmutation of Nuclear Waste, Report LA-UR-93-2944, Los Alamos National Laboratory, 1993.

23. **Rosenthal, M. W., Haubenreich, P. N. and Briggs, R. B.,** The development Status of Molten-Salt Breeder Reactors, Report ORNL-4812, Oak Ridge National Laboratory, 1972.

24. **Steunenberg, R. K., Pierce, R. D. and Burris, L.,** Pyrometallurgical and Pyrochemical Fuel Processing Methods, in *Progress in Nuclear Energy, Series III, Process Chemistry*, Vol. IV, Pergamon Press, Oxford, 1969, chap. 6.2.

25. **Mcpheeters, C. C. and Pierce, R. D.,** Nuclear Waste from Pyrochemical Processing of LWR Spent Fuel for Actinide Recycle, Report ANL-IFR-165, Argonne National Laboratory, 1992.

26. **Walker, R. B.,** Flowsheet Report for Baseline Actinide Blanket Processing for Accelerator Transmutation of Waste, Report LA-UR92-1241, Los Alamos National Laboratory, 1992.

27. **Abney, K. D., Schroeder, N. C., Kinkead, S. A. and Attrep, M., Jr.,** Separation of Technitium from Ruthenium after the Accelerator Transmutation of Technitium, Report LA-UR92-39, Los Alamos National Laboratory, 1991.

Chapter 15

Europe and United Kingdom

Robert G. G. Holmes and Harry Eccles

I. INTRODUCTION

The separation processes and technologies employed in the management of wastes from the uranium fuel cycle are very dependent on factors such as the waste types (liquid, solid, or gaseous), its radiological classification, the origin of the waste (mining, fuel fabrication, power generation, reprocessing, decontamination, and decommissioning), and the national policy regarding waste treatment and disposal. The geographical area covered by this discussion, mainland Europe and the United Kingdom, (Figure 1), represents a number of nations, all of which have their own plans and policies. It does not include those nations in the former Eastern Bloc; these are covered in Chapter 16.

Within mainland Europe and the United Kingdom there are a number of countries that are actively involved in some aspect of the nuclear industry (Figure 1). Many countries use nuclear power reactors to support national energy demands. Mining for uranium, although a historically important activity, is much reduced in mainland Europe due to the current price of uranium. Isotopic enrichment of uranium for use in fuel is carried out by several countries — France, United Kingdom, Germany, and Holland — with fuel fabrication also carried out by France, Spain, United Kingdom, Italy, Belgium, Germany, and Sweden.[1] It is perhaps in spent reactor fuel management that Europe and United Kingdom show the greatest diversity with policies ranging from direct disposal, e.g., Sweden,[2] to a commitment to a closed fuel cycle, e.g., France and United Kingdom;[3] yet other nations are utilizing existing European reprocessing capabilities while exploring long-term strategies that will ultimately influence the choice of separative process and waste management. Within Europe, principally the United Kingdom and France, there are segregation processes and wastes associated with defence activities; these will not be specifically discussed within this presentation, although the separation technologies used tend to be similar to those deployed in the commercial nuclear industry. The major stages of the nuclear fuel cycle are shown in Figure 2.

Separative technologies are employed in Europe to address the following:

- To recover radionuclides for recycle or reuse, for example, for returning plutonium to the fuel cycle as reactor fuel or mixed oxide (MOX) fuel[4]
- A reduction in the levels of nuclide concentration in waste streams so they can be discharged to the environment (this principally applies to liquid and gaseous or aerial effluent)
- The preparation of the waste for disposal (this applies to solid waste or its precursors)
- To remove radionuclides for destruction or incineration by transmutation without power generation; this concept is, however, currently in its infancy in Europe[5]

All reprocessing within mainland Europe and the United Kingdom is currently achieved using a plutonium and uranium extraction process, PUREX. This topic is the subject of a separate publication in this series.[6] This presentation limits itself to separative technologies

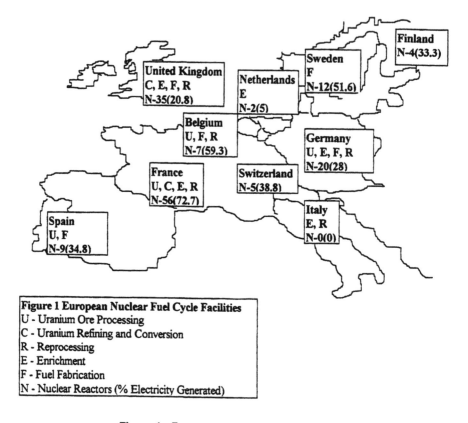

Figure 1 European Nuclear Fuel Cycle Facilities
U - Uranium Ore Processing
C - Uranium Refining and Conversion
R - Reprocessing
E - Enrichment
F - Fuel Fabrication
N - Nuclear Reactors (% Electricity Generated)

Figure 1 European nuclear fuel cycle facilities.

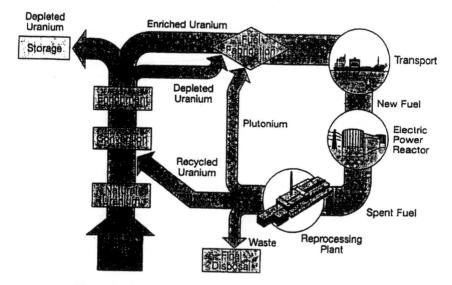

Figure 2 Illustration of the key stages of the nuclear fuel cycle.

deployed against wastes, but not including reprocessing of irradiated nuclear fuel or wastes generated from the reprocessing of fast reactor fuels. Similarly, wastes generated from the enrichment of ^{235}U are not addressed, as their impact on the nuclear fuel waste management, per se, is negligible. The categories of waste considered will embrace solid, liquid, and gaseous waste, although the greatest diversity of technologies is deployed against the category of liquid waste. It is also worthy of note that the separative technologies deployed depend on the categorization of waste. European categorization is in line with IAEA recommendations,[7] but the categories (for example, solid high level, intermediate-level waste, and low level waste in the United Kingdom) are not consistent in detail within mainland Europe, and are certainly not consistent with the U.S. definitions of waste categories.[8]

Accepting the range of polices, strategies, and detailed waste definitions operated by the European nuclear industries, they are, however, single-minded in the principles that underpin the use of separation in waste management. These principles are

- To ensure that all aspects of the nuclear industry are both safe and environmentally acceptable
- To employ cost-effective processes in the treatment of waste

The presentation concentrates on those technologies that are already deployed in mainland Europe and the United Kingdom, but also discusses those processes that have been developed to the point that they could be deployed. In addition, the presentation draws attention to techniques that are currently the subject of research if it appears they may offer particular advantages. Whilst many techniques are utilized to prepare waste for disposal or discharge, the discussion is limited to those processes that could or do achieve some partition or segregation. It does not cover technologies that have been extensively covered elsewhere, which, although preparing waste for disposal or discharge, do not strictly involve segregation or partition, e.g., encapsulation[9] and organic waste destruction.[10]

II. SOLID WASTES

A. INTRODUCTION

Separation processes are an integral part of waste management strategy for solid wastes and are motivated by the following:

- To remove materials that are potentially hazardous or detrimental to a disposal philosophy
- To recover potentially valuable components of the waste to recycle or reuse material or items
- To decategorize the waste or to allow the waste to be consigned to a lower cost category of waste
- To minimize waste volumes

In view of the relative complexity of most separative techniques, and minimal advantages to segregation of materials as compared to disposal, possibly the least effort and lowest technology is deployed against solid waste. The possible exception to this is in the area of recovery of fissile materials, particularly plutonium from waste from the back end of the fuel cycle, where considerable effort has already been expended in segregating plutonium from other nuclides (plutonium-contaminated material [PCM], which has some equivalence to U.S. TRU waste). A major impediment in deploying any separative process remains, however, the potential for generation of secondary wastes that in turn require treatment and the attendant exposure dose accrued from the separative process.

There are a number of types of solid wastes that are currently receiving or might receive treatment to achieve separation. These wastes include mine tailings, process wastes from nuclear fuel cycle activities, and decommissioning wastes.

B. MINE TAILINGS

By virtue of their bulk and the comparatively low levels of activity associated with the waste, the policy for such tailings has tended to be containment or impoundment prior to disposal or permanent impoundment.[11] Separative processes that have been deployed include segregation of sludges by simple lagoon techniques.[12] One of the problems associated with mine tailings is the enhanced acidity and subsequent leaching of tailings due to the oxidation of sulfides by bacteria such as *Thiobacillus ferro oxidans*, following the general equation (1). It is in this area where research effort is currently being directed,[13] and technologies may be generated to effectively separate radionuclides from mine tailings by leaching.

$$4FeS_2 + 2H_2O + 15O_2 \rightarrow 2Fe_2(SO_4)_3 + 2H_2SO_4 \qquad (1)$$

C. PROCESS WASTES

Process wastes, particularly from reprocessing irradiated nuclear fuel, fall into three main categories:

- High-level waste, which is either spent fuel for which a clear decision has been made not to carry out separation, or solidified high-level liquid waste
- Intermediate-level or medium active waste that is either fuel element debris from which the main radionuclides have already been removed, or solids generated in a separative process that do not require regeneration (e.g., spent ion exchange resins, filter materials, or precipitates)
- Technological 'trash' that falls into the lower waste categories (low-level waste)

A range of technologies has been applied or investigated for treatment of these three waste streams, with the initial emphasis on segregation to allow more effective further separation.

1. Segregation

The most prudent technique that is used in waste treatment per se is segregation, and this will have a significant impact on the selected separation process. This process involves the removal of items that may be troublesome in further treatment of the waste or are prescribed from the disposal route. Typical examples are gas cylinders and free liquid that are excluded from low-level waste disposal.[14] This process is achieved either by visual sorting at the point of origin or nonintrusive interrogation techniques[15] deployed against containerized waste, such as X-ray interrogation followed by physically sorting and removing the item from the waste.

2. Waste Washing and Recovery

In the fuel cycle, particularly after the separation of uranium from plutonium, a body of contaminated waste is produced that has been in contact with plutonium. This

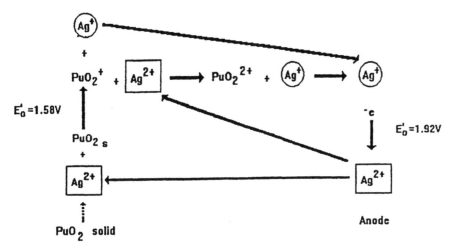

Figure 3 Scheme of silver(I)/silver(II) electrooxidation technique for the dissolution of plutonium dioxide.[19]

plutonium-contaminated material (PCM) has been the subject of a considerable fund of development work initially driven by the defence industry to separate and recover the plutonium contaminant. The current position in the United Kingdom and France is to minimize the generation of PCM and dispose of this waste rather than rely on a downstream process. Nonetheless, a large number of processes have been explored to separate resident plutonium contaminant from PCM. In general, the processes involve separation of the nuclide residues either by dissolution or destruction of the substrate, for example, the mixed-acid process,[6,16] and the microbial degradation of organic ion-exchange resins,[17] or low-level waste.[18]

These processes are usually developed as a volume reduction process, with recovery being of secondary importance. Many, therefore, cannot be considered separative processes unless combined with a subsequent leaching process, such as leaching of incinerator ash, for which processes such as the silver(II) oxidative process can be used.[19] This technology has been developed in the United Kingdom at AEA Dounreay,[20] and in France[16,19,21] (see Figure 3), to recover PuO_2 for recycle. The process and a number of other dissolution techniques have been presented in a sister publication to this text.[16] The process is, however, only a precursor for a conventional recovery or treatment process.

A further category of material that may be amenable to recovery is metallurgical slags. This material shares many of the features of, for example, incinerator ash, and like incinerator ash, the separation of plutonium from the waste is preceded by dissolution followed by conventional separation. This process has also been described in a previous publication in this series,[18] where the experience of the French has been detailed.

D. DECOMMISSIONING AND DECONTAMINATION

The nuclear industry generates a substantial quantity of solid wastes that are generally only contaminated at their surface. Typical of these waste types are decommissioning wastes and redundant items of equipment. There is a strong financial incentive to use separative methods in these materials. This includes:

- High cost of disposal options for higher waste categories
- Recycle of scrap material to the metallurgical industry
- Reuse of decontaminated items

A range of techniques are utilized against these wastes, which are largely process equipment and metal or concrete construction wastes.
Decontamination falls into two main categories:

- The removal of the contaminated surface by physical methods
- The removal of the contaminated surface by chemical decontamination

Contamination is held on a surface by a range of mechanisms, including chemical and physical bonding, cracks and surface irregularities, etc. Physical methods either remove the contaminant by a washing process or by removal of the entire contaminated surface. Methods that have been used include strippable coatings that simply remove loose contamination after their application[22] and washing with water or Arklone systems,[23] with either jetting[24] or ultrasonic enhancement of the decontamination process.[25] These techniques can either be employed with dip tanks or deployed by spray systems. In addition, the aqueous systems can be enhanced with suitable surfactants.

More aggressive methods involve abrading the contaminated surface by scarification,[26] high-energy jetting,[27] sand blasting,[28] or removal of surfaces with solid CO_2.[29] Although most of these methods have been tested, there remains the problem of the volumes generated and the subsequent treatment of the secondary waste. Relatively few can be regarded as strictly separative techniques. These methods find practical application in decontaminating transport flasks in the United Kingdom[30] and in France.[31]

The use of chemical decontaminants is both well known and well documented. In this method the chemical system can be used for:

- Removing the contaminated surface by techniques such as electrochemical dissolution[32] or fluoride dissolution of steels or concretes[33]
- Dissolving or suspending the contaminant, using an oxidant and complexant;[34] a wide range of citrate and other chelating complexants have been employed[35]
- Acid dissolution, e.g., using nitric acid or hydrofluoroboric acid

A particular success in the decontamination field is the decommissioning of the Capenhurst Gaseous Diffusion plant (United Kingdom) to return some 10,000 t of aluminium, 1700 t of steel, and 300 t of other metals to the domestic scrap market by chemical decontamination of the material.[36,37]

A further component in the return of metals to the scrap metal market is metal melting. Scrap-melting devices have been installed in Sweden.[38] The separation is achieved in these furnaces by removal of ^{137}Cs in slag while fusing ^{60}Co in the metal billet itself.[39] This process addresses one of the main concerns relating to recycle of metal by proposing to utilize the recovered metal as waste containers.

Several of the solid waste treatment routes (e.g., leaching), as a result of decontamination, have the drawback that they in turn produce liquid effluents that have to be treated (see Section III.F).

III. LIQUID WASTES

A. INTRODUCTION

The composition and volume of liquid wastes arising from the various stages of the uranium nuclear fuel cycle are diverse and varied. The radioactivity of these liquid wastes varies from a few $GBq \cdot m^{-3}$ due to naturally occurring radionuclides such as uranium and thorium (from uranium ore processing) to about 50 $TBq \cdot m^{-3}$ due to fission products such as ruthenium and zirconium (from fuel reprocessing). Similarly, the nature and concentration of nonradioactive components will vary from small/negligible as in reactor waste liquors, to appreciable as in the aqueous raffinate that contains numerous heavy metals discarded from the uranium ore concentrate purification process. These parameters in conjunction with some physical properties such as density, viscosity, surface tension, turbidity, and emulsifying ability, along with quantity of arisings, will greatly influence the selection of the separation technology. The selected technology will also be capable of accommodating other variables that will arise due to perturbations in day-to-day operations. In the future, separation technologies will also be selected that are not only robust, flexible, simple, reliable, cost-effective, etc., but also generate recyclable products.

B. URANIUM REFINING

Uranium concentrate milled from ore is converted to UF_6 for enrichment using combinations of treatments including nitric acid dissolution, solvent extraction, calcining, hydrofluorination, and fluorination. Liquid wastes essentially arise from:

- Purification of the uranium ore concentrate
- Scrubbing of off-gas from denitrification of uranyl nitrate; reduction of UO_3 to UO_2; conversion of UO_2 to UF_4, and UF_4 to UF_6

These spent alkaline scrubber liquors generally contain insignificant amounts of natural uranium, and therefore do not require treatment. Treatment may, however, be required for other reasons such as the presence of appreciable amounts of nitrate/nitrite or fluoride ions. The latter can be removed by simply treating with lime and removing the precipitated calcium fluoride prior to recycling the regenerated alkaline scrubber liquor. At present, nitrate/nitrite scrubber liquors are cotreated with other alkaline liquid wastes, as described later.

The aqueous raffinate discarded from the purification of uranium concentrate (by solvent extraction using the solvent tributyl phosphate in odorless kerosene) represents the major liquid waste in uranium refining. Approximately 5 m^3 of aqueous raffinate per tonne uranium processed is discarded, depending on flow sheet conditions. This acidic liquid waste is rich in alkali and alkaline-earth metals, aluminium, iron, and natural thorium (^{232}Th), with lesser quantities of natural uranium, ^{226}Ra, ^{230}Th, and their daughter products, present as nitrates or nitrato complexes.

The current treatment technology involves initially the selective removal of the radionuclides such as radium by coprecipitation with barium as the corresponding sulfate. The resultant liquor is neutralized with a suitable alkali, thus precipitating heavy metals including aluminium, uranium, and thorium as their hydroxides. At this stage the spent nitrate/nitrite scrubber liquor would be combined with the comparatively pure alkaline nitrate liquor. This combined liquor has subsequently been used for agricultural purposes.

Unfortunately, this treatment route has found limited and restricted application and, therefore, alternative processes have been evaluated. These alternatives have included:

1. The selective removal of radionuclides, in particular thorium isotopes using a solvent extraction route employing tri-octyl phosphine oxide (TOPO)
2. The use of a chelate ion-exchange resin, Duolite ES 467, which contains aminophosphonic acid groups that are highly selective for thorium ions
3. Precipitation of thorium from the uranium ore concentrate raffinate using fluoride ions
4. Biosorption of uranium and thorium using fungal biomass[40]

Although all of the above techniques were capable of handling, as received, raffinate from the uranium ore concentrate refinery, the TOPO solvent extraction route proved to be more acceptable and was proven at pilot-plant scale.[41]

A major constraint to the removal of thorium isotopes and trace quantities of other radionuclides such as uranium and radium from the uranium ore concentrate (UOC) raffinate is the comparative high acidity, typically 1.0 to 1.5 M in nitric acid. Removal or a reduction in this acidity would allow more established techniques to be used, such as:

1. Conventional sulfonic acid ion exchangers[42]
2. Removal by complexation–ultrafiltration using water soluble macromolecules[43]
3. Solvent extraction using amines[44]

All of the above would achieve the prime objective of separating the radionuclides from the alkali, alkaline-earth metals, aluminium, and iron and allow aqueous effluent discharge.

The recovery of nitric acid and nitrate associated with metal ions in the UOC raffinate has been recently demonstrated,[45] using a conventional solvent extraction flow sheet but with a novel bulk liquid-membrane contactor. The resultant deacidified raffinate containing the thorium, uranium, and radium nuclides would now be more amenable to such techniques as described above.

C. FUEL FABRICATION

Fuel fabrication facilities include a variety of operations including conversion of UF_6 to UO_2, sintering and finishing fuel pellets, components and tubing preparation, and final fuel assembly. Unlike the head-end UOC refining processes, fuel fabrication employs essentially dry mechanical routes, and, therefore, the number and volume of liquid wastes are comparatively small. Furthermore, as the ^{235}U content is now enriched, up to 5%, the majority of the waste streams are categorized as residues having an intrinsic value and hence are reworked. Consequently, liquid wastes will contain negligible amounts of uranium, but again may contain nitrate and ammonium ions, whilst spent scrubber liquors will be rich in fluoride. To ensure that liquid wastes contain negligible amounts of uranium, treatment with ammonia or other alkali is practised to produce the corresponding diuranate, which is recycled. This simple but effective separation technique has not been challenged to date. If, however, alternatives are required, some of the processes previously described would be applicable.

Although little or no technology development has been specifically targeted at liquid wastes generated from fuel fabrication, process development has resulted in some liquid wastes being eliminated and others significantly reduced. Improvements in contactors and control systems, whilst adapting alternative filtration equipment, have led to waste minimization. The development of the integrated dry route (IDR) for conversion of enriched uranium hexafluoride to UO_2 powder, replacing the wet ammonium diuranate

Table 1 Liquid Waste Arising from the Various Reactor Types

	Reactor type			
	PWR	**BWR**	**HWR**	**GCR**
Average volume of liquid waste (m³) arising per year for a 1000-MW(e) plant	10,000	20,000	17,000	15,000

route,[46] has radically influenced the quantity of liquid wastes now being discarded by fuel fabricators.

D. REACTOR SYSTEMS

According to the different types of reactors now operating commercially in mainland Europe and the U.K., pressurized water reactor (PWR), boiling water reactor (BWR), hot water reactor (HWR), gas cooled reactor (GCR), and magnox reactor, different waste streams arise. These streams are different both in activity content and the amounts of liquids generated.

Water-cooled and moderated reactors generate more liquid waste than the gas-cooled plants. Gas-cooled reactors produce only a few liquid waste streams and these include fuel-cooling pond water and laundry drain. In all cases liquid waste will also arise from decontamination operations. Typical volumes of liquid wastes arising per year for the various reactors are presented in Table 1. A large part of the soluble radioactivity appears in the primary coolant of water-cooled reactors or the fuel storage pond. Minor amounts of these liquids can leak or be transferred and as a result radioactivity appears in various other streams of the reactor plant.

Three basic treatment methods for liquid wastes predominate, namely filtration, ion exchange, and evaporation. Many other techniques are also used, e.g., centrifugation, ultrafiltration, and flocculation, but these are generally used for specific applications where the more common techniques are not available or in conjunction with other techniques to improve efficiencies.

Although evaporation achieves the highest decontamination factors of all treatment processes, it is insensitive to the chemical composition of effluents and is relatively expensive. Flocculation and chemical precipitation, and particularly simple filtration are less expensive but also less effective and lead to relatively large amounts of residues, especially filter aids. Ion exchange is generally applied in the treatment of comparatively clean water such as from primary circuits, fuel element storage ponds, etc. and it is well suited for this purpose. It is possible, however, that substitution of evaporation for ion exchange could lead to a decrease in volume of the final waste products and facilitate and improve the conditioning of these products.

It can be stated that the treatment of radioactive liquid effluents from reactor facilities has reached a high degree of maturity. As most treatment processes generally produce a secondary waste, e.g., a solid residue, alternative techniques involving membrane filter, electromagnetic filter, centrifugal clarifier, and reverse osmosis have recently been evaluated. These processes are, however, still under development or at an early stage of application.[47] Other approaches have concentrated on improvements to ion-exchange resins either by substitution with other materials such as inorganic adsorbents or using modified organic materials. The use of grafted cellulose has been evaluated and this appears to offer several advantages over other systems.[48]

E. FUEL REPROCESSING

Normally, the reprocessing operations are carried out according to the classical PUREX process, which involves separation of fission products, uranium, and plutonium using a solvent extraction technique involving diluted tri-*n*-butyl phosphate (TBP).[6]

Besides high-level wastes, large amounts of medium-active or intermediate-level wastes are generated in each step of the fuel reprocessing operations from initial receipt of the fuel and storage in ponds, until the Pu and U extractions and purifications. Obviously, the characteristics and amounts of such wastes are very variable according to the type of fuel reprocessed. Nevertheless, as a common feature, some typical waste streams are inevitably produced such as:

1. Cooling pond liquid wastes — The interim storage of spent fuel is normally of 5 to 10 years duration in water ponds. With magnox fuels the period is much shorter due to the interaction of water with magnox resulting in corrosion of cladding. The quality of the water in fuel ponds is maintained by recirculation through ion-exchange columns, and the regenerant liquors from such columns constitute the liquid waste. The activity in the pond water will also be controlled to some extent by this ion-exchange process, and, therefore, the spent regenerant liquors themselves will require some form of treatment. Such liquors would be combined with other similar liquid wastes for treatment, e.g., evaporation followed by encapsulation.

    Inorganic ion-exchange materials such as zeolites have replaced some of the organic ion-exchange resins so that the regeneration stage can be omitted and the spent zeolite directly encapsulated into a concrete matrix.

2. High level waste — In the primary stages of the solvent extraction circuit, fission products are separated from uranium and plutonium. The aqueous raffinate from this separation process is highly radioactive, and, apart from evaporation to reduce the volumetric bulk and storage in specifically designed vessels, the concentrated highly radioactive and acidic waste receives little or no treatment. In the last 5 years, however, this liquid waste has been further treated using a combination of denitrification and vitrification techniques to produce a stable, glassified solid waste in which the various radionuclides are encapsulated.

3. Off-gas scrubber solutions — In order to prevent undue release of NO_x and active gases into the environment during fuel dissolution and thermal denitration of uranyl nitrate, caustic scrubber solutions are generally used for the off-gas treatment. These spent scrubber liquors are usually evaporated and ultimately sent for encapsulation along with other intermediate-level wastes.

4. Solvent alkaline wastes — This waste generated during the TBP cleanup operations for recycling purposes is mainly composed of a mixture of NaOH and Na_2CO_3 in which TBP radiolytic degradation products are also present. As with other alkaline liquid wastes, evaporation followed by encapsulation in a concrete matrix is the preferred treatment route.

5. Second and third uranium and plutonium cycle wastes — From the extensive purification cycles for Pu and U, always using the same solvent extraction technique, some aqueous alpha-contaminated raffinates arise. Such wastes currently are evaporated and the concentrated liquor combined with others prior to encapsulation.

The PUREX process is among the most proven industrial processes to date. There are, however, several steps/stages of this process that need further refinement in order to minimize the production of radioactive wastes and to meet the challenges of higher burn-up fuels of the future. The mainland European nuclear industry has not been lethargic to these challenges and during the last two decades significant R & D programs have focused attention on three or four major areas.

Probably the most scrutinized waste of the nuclear fuel cycle arises from the highly active cycle, namely the HA raffinate.[49] The current treatment and disposal options are largely influenced by this liquid waste being of mixed character, i.e., containing both α– and β/γ–emitting radionuclides; if, for example, long-lived emitters were absent, then treatment and disposal options could be significantly different. This speculation has led workers in France[50] and Italy[51] to pursue alternative extractants to TBP. Extractants such as N,N-dialkylamide have been evaluated, but for several reasons have not been adopted.

This apparent lack of success probably initiated an alternative approach involving the removal of specific radionuclides from the HA raffinate. Removal techniques again involved the use of highly selective, near-specific liquid–liquid extractants as previously mentioned, plus others,[52] and inorganic ion exchangers. Materials examined have been phosphotungstate,[53] hydrous titanium oxide, manganese dioxide, titanium phosphate, polyantimonic acid, and hexacyanoferrate.[54] The latter five were incapable of adsorbing either Pu, Np, or Am from a 4 M nitric acid simulated waste, however. The treatment of alkaline solvent wash waste, although not as demanding as for the HA raffinate, nevertheless requires a multistage operation involving evaporation, interim storage, treatment, and encapsulation. Removal of the radionuclides directly from this waste would be both cost-effective and environmentally prudent. Researchers in France,[55] amongst others, have been evaluating and comparing the following:

- Solvent extraction using tributylacetohydroxamic acid as the extractant
- A combination of inorganic ion-exchanger materials including active charcoal, titanium oxide, iron oxide, and zeolite
- A combination of chemical precipitation with ultrafiltration

The workers concluded that each of these new treatment processes for the decontamination of this particular waste was successful in achieving an alpha DF of at least 160. Their respective performances, however, differed considerably in terms of selectivity and volume reduction.

Although the majority of the TBP–diluent phase is washed and recycled, a small, but nonetheless troublesome quantity needs disposal, but first requires treatment. As this is an organic hydrophobic waste, techniques hitherto mentioned have proved either unsuitable or inappropriate. Processes have been developed that employ simple chemical pretreatments to remove radionuclides prior to incineration[56] or direct pyrolysis[57] involving a suspension of TBP–diluent in lime ($Ca(OH)_2$). Similarly, the chemical composition of ILWs may also be different, in particular, the nitric acid content. Consequently, additional/alternative separative technologies to those already described elsewhere in this chapter can be employed. One such technology, liquid membranes, has received particular attention by several mainland European research groups working independently or in collaboration.[58,59] The inherent characteristics of this technology have allowed newer, possibly more expensive, but highly selective extractants to be considered. Some of the more notable studies have involved the use of ionizable crown ethers and functionalized calixarenes either in supported liquid membranes or polymer-based membranes. One of the targets for such materials was the removal of cesium from high-salinity liquid waste, although the research program per se involved other metals such as actinides and strontium.[58] As with conventional solvent extraction, the chemical composition of the solvent strongly influences the selectivity, affinity, etc. for particular metal species. The influence of the membrane solvent on strontium transport from reprocessing solutions has been studied.[59] The workers concluded that n-hexylbenzene, when added to dicyclohexyl-18-crown-6 as extractant supported on Celgard 2500, improved Sr permeability.

Liquid-membrane technology is greatly influenced by interfacial interactions/actions, which in turn influence mass transfer. Not surprisingly, therefore, mass transfer and other related parameters have been studied by several groups.[60] Although many of these studies were not targeted at the nuclear fuel cycle, the information gathered is nonetheless pertinent to a sound understanding of membrane technology. Simple separative technologies such as chemical precipitation and filtration have been further developed and married. This has resulted in a technology that is capable of dealing with and separating minute quantities of actinides. The key to the success lies in the addition of an adsorbent to the liquid waste to be treated prior to an ultrafiltration stage. The adsorbents that have been evaluated range from nickel hexacyanoferrate for cesium removal to ferric hydroxide for actinides coupled with cross-flow filtration using an ultrafiltration membrane.[61]

In recent years other technology combinations have been examined, such as electrokinetic dewatering and electrochemical ion exchange. These two electrical processes have been developed to bench-top scale for the treatment of active liquid wastes. Electrokinetic dewatering is an attractive low-pressure process for the final stage of slurry concentration prior to immobilization in cement. The electric field-induced solid–liquid separation process at a microporous membrane has enabled waste treatment flocs to be concentrated from 5 to 40% solids — with high DFs. In electrochemical ion exchange, adsorption and elution at both organic and inorganic exchangers are controlled by an external voltage. While DFs of approx. 2000 are achievable with up to 75% loadings of the exchange medium on adsorption, polarity-reversed elution into water can generate a concentration product (0.25 m^3) compatible with vitrification.[62]

The commercial use of microorganisms for effluent treatment, as in the active sludge process, has been in existence for nearly two centuries. More recently, however, indigenous microorganisms for the removal/recovery of actinides and toxic heavy metals has been extensively studied worldwide. The mechanistic uptake of metals by microorganisms can be categorized into two groups, namely bioaccumulation and biosorption. The former is exemplified by the removal of actinides using Citrobacter sp. The organism is capable of generating inorganic phosphate from an organic phosphate such as glycerol 2-phosphate, thus precipitating actinides as their corresponding phosphate.[63] On the other hand, bacteria, fungi, and algae are capable of biosorbing actinides, and several genera have been studied for uranium removal.[64] The biosorption of uranium is fast occurring at or in the cell wall and, generally, because nonmetabolic processes are involved, the metal does not enter into the inner cell, unlike bioaccumulation, which is generally dependent on metabolic processes.

Biosorption/bioaccumulation processes lend themselves to the treatment of radioactive liquors that are of low-level waste category and only slightly acidic, i.e., pH value about 2 or greater, but can operate up to pH values of about 10.[40]

F. DECOMMISSIONING AND DECONTAMINATION

Waste steams vary widely in activity and chemical composition. They often, however, result from the mixture of HNO_3 and NaOH and may contain a large variety of complexing agents. When and wherever possible, decontamination liquid wastes are usually mixed as far as they are compatible for subsequent treatment. Some of the wastes produced are problematic to handle because of high concentrations of chemicals, which in many cases cause difficulties in the solidification processes of the wastes. To avoid this problem, modern methods have been developed in which a very low concentration of chemicals is used. In addition, ion-exchange techniques are used to remove the bulk of the radioactivity in the waste, thus allowing the decontamination liquor to be recycled.

As decommissioning of now redundant nuclear facilities becomes increasingly more important, better decontamination or perhaps novel procedures, are needed not only to minimize waste arisings, but also to reclassify the bulk solid wastes, and in doing so, reduce the cost of decommissioning per se.

The majority of the separation processes previously described have been considered for cleaning up decontamination liquors. Some have found greater favor than others, for example, adsorption using inorganic materials, chemical precipitation in conjunction with ultrafiltration, and membranes. Although these processes offer some potential, the presence of chelating agents such as EDTA, and citrate in some decontamination solutions complicate the separation. Oxidation of the chelating agents prior to metal removal has been evaluated[65] whilst alternative approaches include the simultaneous elimination of heavy metals and chelating agents by means of strong-base anion-exchange resins and electrodeposition. The resins are used in the bicarbonate form and regenerated by means of magnesium carbonate, which is produced from magnesium oxide, water, and carbon dioxide under pressure. Laboratory-scale experiments demonstrate the feasibility of the proposed process. Macroporous exchange resins showed the best performance.[66]

IV. GASEOUS WASTES

A. INTRODUCTION

Gaseous waste, or aerial effluent, is a waste that is common to most processes and activities in the nuclear industry. The treatment and separation techniques employed fall into four major applications:

- Fuel fabrication and dry-powder systems
- Reactor systems
- Fuel reprocessing
- Waste treatment systems

Typically, the volumes of aerial effluent are large and are dictated by the large flow rates required to secure containment.

The separation techniques fall into two main categories. The first is a barrier to particulates and aerosols; the second is to remove volatiles. Elements of both technologies are used in many off-gas treatment processes, although the nature of the removal process, particularly for volatiles, depends on the challenge to the off-gas systems.

B. FUEL FABRICATION AND DRY SYSTEMS

Dry-powder systems, such as fuel fabrication plants, and those that handle oxides of plutonium, have special problems. The off-gas for dry systems are usually fitted with a high efficiency particulate air (HEPA) filter system that provides a barrier to the escape of particles of solid. Glove boxes themselves are protected by HEPA filters, filtered to their local exhaust systems to limit ingress of solids into plant and building ventilation systems. These local systems have relatively low volumetric flows, and hence can utilize some novel systems, for example, high gradient magnetic filters for PuO_2[67] and cyclone systems.[68]

Table 2 Volatile Radionuclides Produced under Normal Operating Conditions by LWR

Element	Isotope mass number
Hydrogen	3
Carbon	14
Argon	37, 41
Krypton	83m, 85m, 85, 87, 88, 89
Xenon	131m, 133m, 133, 135m, 135, 137, 138
Iodine	131, 132, 133, 135

From *Design of Off-Gas and Air Cleaning Systems at Nuclear Power Plants,* IAEA Technical Report No. 274, International Atomic Energy Agency, Vienna, Austria, 1987. With permission.

C. REACTOR SYSTEMS

The challenge to the off-gas system from a reactor[69] is shown in Table 2, while a typical reactor off-gas system[70] is illustrated in Figure 4. One of the main nuclides of concern for removal from the gaseous effluent from reactor systems is iodine. Iodine is generally removed by charcoal, either relying on its natural capacity to absorb elemental and organic iodides,[71] or combined with an additive (e.g., triethylene diamine, or KI).[72] HEPA filters are also generally employed in treatment systems to remove particulates.[73]

D. REPROCESSING PLANT

Off-gas systems for a reprocessing plant are required to deal with a number of different elements,[74] and these are shown in Table 3. The requirements include removal of particulates, either solids or aerosols, removal of radioactive volatiles, and removal of inactive volatiles from the process. The reliance of the PUREX process on aqueous nitric acid systems in mainland Europe and the U.K. results in oxides of nitrogen (NO_x) as the

Table 3 Volatile and Potentially Volatile Radioactive Nuclides in LWR Liquid Wastes

Element	Isotope mass number
Hydrogen	3
Krypton	85
Iodine	129
Selenium	79
Technetium	99
Ruthenium	103, 106
Rhodium	106
Tellurium	123m, 127m
Antimony	124, 125, 126m, 126
Cesium	134, 135,137

From *Control of Semivolatile Radionuclides in Gaseous Effluents at Nuclear Facilities,* IAEA Technical Report No. 220, International Atomic Energy Agency, Vienna, Austria, 1987. With permission.

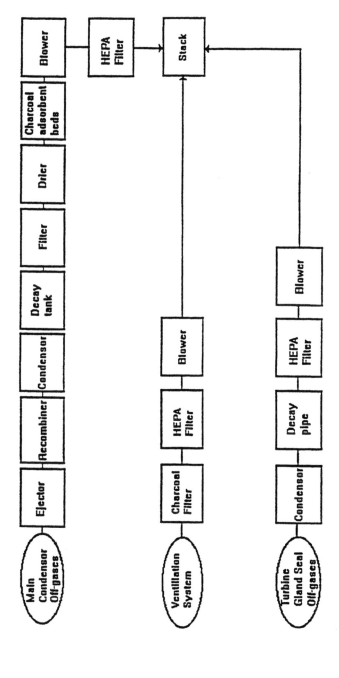

Figure 4 Typical gaseous waste from a BWR. (From *Design of Radioactive Waste Management Systems at Nuclear Power Plants*, IAEA Safety Series No. 79, International Atomic Energy Agency, Vienna, Austria, 1986. With permission.)

266

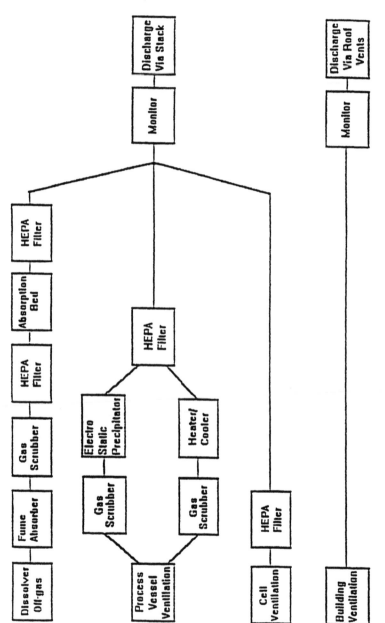

Figure 5   THORP off-gas system.

main volatile species, along with considerable quantities of water in the humid process air streams. The off-gas system(s) for the Thermal Oxide Reprocessing Plant (THORP) is shown in Figure 5, which gives an indication of a reprocessing off-gas system.

1. HEPA Filters and Electrostatic Precipitators
In all reprocessing plants, the off-gas is polished using HEPA filters. The HEPA filter requires that the gas stream be sufficiently dry to preclude condensation in the filter. This is achieved by a dehumidifier and by heating the gas stream prior to filtration. HEPA filter performance is well established, with a typical filter capable of removing 99.9% of particles in the 0.1- to 0.3-μm size range (representing the range of least efficiency).[75]

Electrostatic precipitators (ESPs) are used in reprocessing plants for the removal of particles from gas streams. They work by ionizing the gas stream between a high-voltage electrode and an earthed electrode. This causes the particles to become charged and consequently attracted to the earthed electrode where they are collected. ESPs can be used for collection of particles of <2 μm, at high efficiencies.[76]

2. Scrubbers
Scrubber systems are used to remove volatiles and particulates. Nitric acid-scrubbing systems remove oxides of nitrogen for reuse and to act as a primary scrub for particulates and species such as ruthenium and tritium (as ^3HHO). Typical acid scrubbers are packed columns with countercurrent liquid–gas flow. Caustic scrubbers are used to remove iodine, ^{14}C as CO_2, residual NO_x, and any remaining ruthenium. The spent scrub liquor is treated to enable discharge or disposal. These off-gas systems are well proven on PUREX plants. A number of variations have been investigated, although, as yet, none have been deployed.

One variation was proposed for the German reprocessing plant (now abandoned) and was based on a philosophy for iodine removal using a silver nitrate-impregnated silicic acid preparation, AC 6120.[77] A host of other technologies to remove components of the off-gas have been the subject of studies or pilot-scale tests. These include corona discharge methods to remove iodine,[78] and metal-impregnated zeolites.[79]

3. ^{85}Kr Removal
An area of debate in the nuclear industry is the removal of ^{85}Kr. Although in terms of quantity ^{85}Kr represents a major Bq discharge from reprocessing plants, it represents a low radiological risk based on dilution and the absence of major pathways to man. None of the existing reprocessing plants in mainland Europe utilize a ^{85}Kr removal system, but a significant body of research has been carried out to investigate ^{85}Kr removal. Techniques that have been studied include the ACHAT system based on activated charcoal,[80] see Figure 6, cryogenic absorption in dichlorodifluoromethane,[81] see Figure 7, and cryogenic distillation.[82] All potential processes are currently under review[83] to confirm that they are not sufficiently low cost to justify their inclusion in reprocessing plants.

E. WASTE TREATMENT
Most waste treatment off-gas systems utilize unit processes. The waste treatment process often resorts to high temperatures, as encountered in incineration and vitrification processes. Typical systems are shown in Figure 8 for the AVM in France[84] and PAMELA,[75] a pilot-scale rig in Germany. Special problems in vitrification include normally

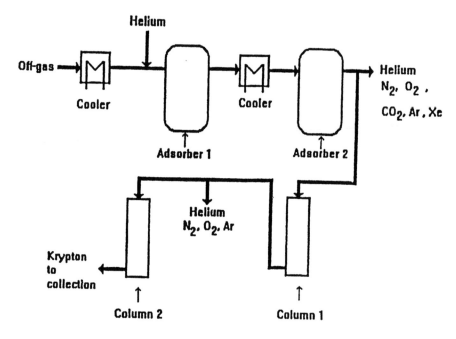

Figure 6 ACHAT krypton gas treatment system.

Table 4 Decontamination Factors Reported
for AVM Process at Different Stages[84]

Nuclide	Campaign number	Calciner and melter	First scrubber	Condensor	Acid recovery column
^{106}Ru	1	3.2	2.3	255	650
	2	5.7	2.8	306	6440
^{137}Cs	1	9.6	8.4	420	24600
	2	16.5	5.26	404	7060
^{144}Ce	1	17.6	3.1	5130	72
	2	26.2	2.3	4240	307
^{90}Sr	1	20.3	24.7	6000	480
	2	33.8	28.0	3105	475

low-volatility species such as ^{106}Ru, ^{137}Cs, ^{144}Ce, and ^{90}Sr. Some DFs for AVM are shown in Table 4. BNFL in the United Kingdom operates a vitrification process, substantially based on the AVM process, in which the BNFL off-gas system utilizes an electrostatic precipitation system, which is not used on either the AVM or the PAMELA systems. Other high-temperature systems, such as incinerators, face problems of high-temperature gas flows with high particulate loadings, coupled with corrosive flue gases. The off-gas system depends very much on the design and service of the incinerator itself; typical systems include an incinerator developed by NUKEM in Germany[85] and a cyclone device for low-level waste.[86] The unique problem associated with incineration being displayed in the KfK system is the removal of particulates from a hot air stream prior to quenching, which is carried out using two ceramic filters.[87] The system has been operating since 1989.

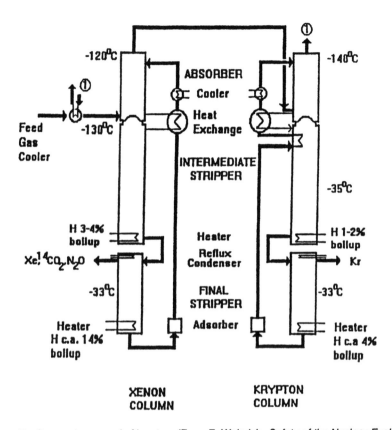

Figure 7 Cryogenic removal of krypton. (From F. Weinrich. *Safety of the Nuclear Fuel Cycle,* Eds. K. Ebert and R. V. Ammon, VCH Publishers, Cambridge, 189, 1989. With permission.)

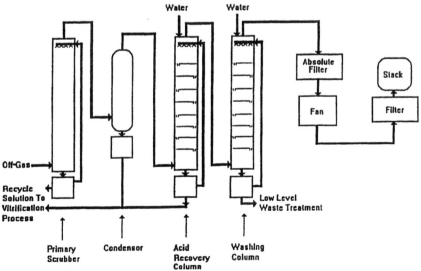

Figure 8 AVM off-gas treatment system.[84]

Other waste treatment processes, e.g., aqueous liquid effluent systems, use off-gas cleanup technologies that are found in reprocessing plants.

V. SUMMARY

One of the major commitments of the nuclear industry is the effective management of its various and diverse wastes. This requires wastes to be treated and/or disposed of within the strict confines of national and international regulations and directives, whilst ensuring that the most cost-effective processes are employed. Separation processes have played a major role in achieving these objectives and will continue to do so. Although these criteria will continue to be of paramount importance, it is anticipated that quantitative targets will be revised generally, becoming increasingly more stringent. This will require more efficient and cleaner technologies either incorporated into mainstream processes or, in the interim, as an end-of-pipe solution.

The need for more efficient/cleaner technologies will in turn influence many scientific and engineering disciplines as smart materials, intensive equipment, proactive control systems, and better modeling packages will be required. The development of new technologies will require a greater integration of multidisciplinary teams at all stages of development and demonstration whilst finally requiring more enlightened production and commercial managers.

These new and novel processes will be based on married technologies such as electrochemical membranes, with many replicating separations achieved by natural biological systems. As in the past, the nuclear industry in the future will lead in the development of more effective, efficient, and environmentally friendly waste management separation technologies.

REFERENCES

1. *Nuclear Engineering International*, World Nuclear Industry Handbook, Reed Business Publications, 1994.
2. Evaluation of Spent Fuel as a Final Waste Form, IAEA Technical Report Series No. 320, IAEA,Vienna, 1991.
3. *IAEA Yearbook 1993*, IAEA, Vienna, 1993.
4. Plutonium Fuel — An Assessment, Organisation for Economic Co-Operation and Development, Paris, 1989.
5. **M. Salvatores and C. Prunier**. *Endeavor*, 17(3), 116, 1993.
6. **W. Schulz, L. Burger and J. Navratil**. *CRC: Science and Technology of Tributyl Phosphate*, Vol. III, CRC Press, Boca Raton, FL, 1990.
7. *Radioactive Waste Management: An IAEA Source Book*, IAEA, Vienna, 1992.
8. **R. Holmes and N. Griffin**. International Perspectives on LLW Processing, Recycle and Disposal, International Low-Level Waste Conference, Monterey, U.S.A., in press.
9. Treatment of Spent Ion Exchange Resins for Storage and Disposal, IAEA Technical Report Series No. 254, IAEA,Vienna, 1985.
10. **J. Williams, J.E. Kitchin and W.H. Burton**. Removal of Organics from Radioactive Waste, Vol. II, Literature Review, DOE Report No. DOE/RW/89/059, 1989.
11. Current Practices and Options for Confinement of Uranium Mill Tailings, IAEA Tech. Report No. 209, IAEA, Vienna, 1981.
12. Safe Management of Wastes from the Mining and Milling Of Uranium and Thorium Ores, IAEA Safety Standards, No. 85, IAEA,Vienna, 1987.

13. R. Cripps. The Recovery of Metals by Microbial Leaching, *Biotechnol. Lett.*, 2(5), 225–230, 1980.
14. Operational Experience in Shallow Ground Disposal of Radioactive Wastes, IAEA Technical Report Series No. 253, IAEA, Vienna, 1985.
15. T.H. Green, D.L. Smith, K.E. Burgoyne, D.J. Maxwell, G.H. Norris, D.M. Billington, R.G. Pipe, J.E. Smith and C.M. Inman. Sampling Methods and Non-Destructive Examination Techniques for Large Radioactive Waste Packages, EUR-13884, Commission of the European Community, 1992.
16. J. Bourges and M. Lecomte. *Science and Technology of Tributyl Phosphate*, Vol. III, Eds., W. Schulz, L. Burger and J. Navratil, CRC Press, Boca Raton, FL, 147, 1990.
17. E. Tusa. IVO's Resin-Eating Bacteria Make Light Work of Waste Treatment, *Nucl. Eng. Int.*, 37(451), 39, 1992.
18. E. Tusa. Microbial Treatment of Radioactive Waste at the Loviisa NPP, Waste Management '89, 15th International Waste Mangement Sympopsium Conference, Tucson, 485, 1989.
19. M. Lecomte, J. Bourges and C. Madic. Applications of the Plutonium Dioxide Oxidising Dissolution Process, Recod '87, Nuclear Fuel Reprocessing and Waste Management, Paris, I, 441, 1987.
20. D.F. Steele. Electrochemistry and Waste Disposal, *Chem. Br.*, October, 915, 1991.
21. J.Bourges, C.Madic, G.Koehly and M.Lecomte. Dissolution Du Bioxyde De Plutonium En Milieu Nitrique Par L'Argent(II) Electrogenre, *J. Less-Common Metals*, 122, 303, 1986.
22. M.J. Saunders and M.G. A. Pengelly. Large Scale Use of Strippable Coatings in Preventative, Tie-down and Decontamination Applications, Department of Environment, Report No. DOE-RW-88.025, 1985.
23. S.C. Gaudie, J.D. Wilkins and A.D. Turner. Evaluation of Arklone for the Decontamination of Noncombustible Plutonium Contaminated Materials, United Kingdom, AEA Report No. AERE-G-3603, 1986.
24. R.D. Bond. Decontamination by High Pressure Water Jetting — A Power Reactor Application, United Kingdom, AEA Report No. AEEW-M-2253, 1985.
25. G. Naud, J.P. Gauchon and M. Lattaud. Ultrasonic Washing and Electrodecontamination as Means to Change Waste Classification, SPECTRUM 92, Boise, 805, 1992.
26. Solving Decontaminable Flooring Problems, *Nucl. Eng. Int. (United Kingdom)*, 34(421), 37, 1989.
27. N.A. Johnson and T. Johnson. Recent Developments in High Pressure Water Technology, Decommissioning and Demolition 1992, Manchester, 90, 1992.
28. F. J. Sandalls. Removal of Radiocaesium from Urban Surfaces Contaminated as the Result of a Nuclear Accident, United Kingdom, AEA Report No. AERE-R-12355, 1987.
29. CO_2-Pellet Cleaning Process Cuts Costs, *Nucl. Eng. Int. (United Kingdom)*, 37(451), 53, 1992.
30. B.G. Williams. Decontamination of Flasks for the Transport of CEGB fuel, IAEA TECDOC-556, 27, 1990.
31. P. L. Chometon. Operating Experiences in the Field of Decontamination of GCR and LWR Spent Fuel Transport Casks and Unloading Facilities, IAEA-TECDOC-556, 37, 1990.
32. F. Bregani and A. Garofalo. Decontamination for Decommissioning Purposes, *Energ. Nucl.*, 8, 117, 1991.
33. J. Hanulik. Process for Decontaminating Radioactively Contaminated Metal or Cement-containing Materials, U.S. Patent No. 5,008,044, 1991.
34. C. Musikas and J. Livet. Dissolving Solution for Metal Compounds Containing Dibutyl Phosphate Ions and Decontamination Process Using this Solution, United Kingdom Patent Application GB 2228491A, 1990.
35. M.J. Peach and R.L. Skelton. Chemical Decontamination for Decommissioning, *Proc. Inst. Mech. Eng.*, C400/8, Decommissioning of Radioactive Facilities, 151, 1988.
36. G. Goodall and B. Gillespie. Decontamination of Surfaces, United Kingdom Patent Application GB 21991329 B, 1989.
37. S. Baxter. Decommissioning to *de minimis* at Capenhurst, *Nucl. Eng. Int.*, 38(457), 23, 1992.
38. V.S. Hedlund and C. Bergman. Melting Radioactive Scrap at Studsvik, IAEA-SM-303/116P, 1988.

272

39. M. Seidler and M. Sappok. Melting Radioactive Scrap, *Nucl. Eng. Int.*, 32(392), 43, 1987.
40. G.M. Gadd and C. White. Removal of Thorium from Simulated Acid Process Streams by Fungal Biomass. Potential for Thorium Desorption and Reuse of Biomass and Desorbent, *J. Chem. Technol. Biotechnol.*, 55, 39, 1992.
41. H. Eccles and A. Rushton. Separative Technologies for the Removal of Thorium from an Acidic Waste Liquor, *Sep. Sci. Technol.*, 28, 59, 1993.
42. M.R.S. Grade. The Behaviour of Thorium Nitrate with Amberlite IR120, *Rev. Port. Quim. Lisbon*, 6, 123, 1964.
43. B. Chaufer and A. Deratami. Removal of Metal Ions by Complexation and Ultrafiltration Using Water Soluble Macromolecules, *Nucl. Chem. Waste Manage.*, 8, 175, 1988.
44. D.J. Carswell and J.J. Lawrance. Solvent Extraction with Amines 1. The System Th-HNO$_3$-Trioctylamine, *J. Inorg. Nucl. Chem.*, 11, 69,1959.
45. M. Cox, H. Eccles, S. Ivakhno and A. Rogatinski. Recovery and Concentration of Nitric Acid Using a Bulk Liquid Membrane Contactor, *Proc. ISEC'93*, Eds. D.H. Logsdail and M.J. Slater, Vol. 2, Elsevier Applied Science, New York, 1993.
46. L. Arcuri, L. Pietrelli, and C. Rizzello. Experience with a Uranyl Nitrate/Uranium Dioxide Conversion Pilot Plant, Report No. ENEA-RT-COMB-84-9, ENEA, 1984.
47. H. Krause. Management of Radioactive Wastes from Nuclear Power Plants, IAEA, CN-43/80, Radioactive Waste Management, Seattle, WA, 3, 1983.
48. R. Sandeaux. Decontamination of Nuclear Plant Fluids With Grafted Celluloses, *Industrie Min.-Mines et Carrieres Les Techniques*, 73, 44, 1991.
49. R. Thompson. A Review of Techniques for Partitioning Highly Active Waste, AERE-R 11966, 1986.
50. C. Musikas. Potentialilty of Monorganophosphorus Extractants in Chemical Separations of Actinides, *Sep. Sci. Tech.*, 23, 1211, 1988.
51. G.M. Gasparini and G. Grossi. Long Chain Distributed Aliphatic Amides as Extracting Agents in Industrial Applications of Solvent Extraction, *Solv. Extr. Ion Exch.*, 4, 1233, 1986.
52. C. Musikas. New Extractants for the Nuclear Hydro-Metallurgy, *Proc. ISEC'90*, Ed. T. Sekine, Elsevier Science Publishers, New York, 297, 1990.
53. J.M. Adnet. Extraction Selective des Actinides des Effluent de Haute. Etude des Possibilities Offertes par les Proprietes Redox des Actinides. CEA-R-5615, 1992.
54. J.E. Cross and E.W. Hooper. The Application of Inorganic Ion Exchangers to the Decontamination of Radioactive Liquid Effluents, *Ion Exchange for Industry*, Ed., M. Streat, Ellis Hoorwood, Chichester, England, 457, 1988.
55. L. Cecille, I.W. Cumming, J.F. Dozol and G. Gasparini. New Alternative Processes for the Treatment of Alkaline Solvent Wash Waste, SPECTRUM '86, 179, 1986.
56. H. Eschrich. Studies on the Treatment of Organic Wastes Part V. The EUROWATT Process, Rep ETR-287, Eurochemic Mainland European Company for the Chemical Processing of Irradiated Fuels, N701, 1980.
57. A. Chrubasik, G. Kemmler and L. Scherbaum. Process for the Treatment of Spent TBP/Kerosene (PUREX) Solvents, SFEN-RECOD 87/006s/JG/da, 773, 1987.
58. Community Research and Development Programme on Radioactive Waste Management and Storage. Shared Cost Action 1990–1994, 153, 1990.
59. J.F. Dozol, J. Casa and A.M. Sastre. Influence of Membrane Solvent on Strontium Transport from Reprocessing Concentrate Solutions Through Flat Sheet Supported Liquid Membranes, *Sep. Sci. Technol.*, 28, 2007, 1993.
60. A.M. Urtiaqa and J.A. Irabien. Internal Mass Transfer in Hollow Fibre Supported Membranes, *A. I. Chem. E.*, 39(3), 521, 1993.
61. W. Heafield and M. Howden. The Future Treatment of Liquid Effluents at Sellafield, *Nucl. Eng. Int.*, 149, 1988.
62. A.D. Turner, W.R. Bowen, N.J. Bridger, A.R. Junkinson and D.R. Cox. Electrical Processes for the Treatment of Liquid Wastes, SPECTRUM 86, 199, 1986.

63. **E.J. Plummer and L.E. Macaskie.** Actinide and Lanthanum Toxicity Towards a *Citrobacter sp.* Uptake of Lanthanum and a Strategy for the Biological Treatment of Liquid Wastes Containing Plutonium, *Bull. Environ. Contam. Toxicol.*, 44, 173, 1990.

64. **M.P. Pons and M.C. Fuste.** Uranium Uptake by Immobilised Cells of *Pseudomonas* Strain EPS 5028, *Appl. Micorbiol. Biotechnol.*, 39, 661, 1993.

65. **K.-J. Muller, T. Bolch and K. Merz.** Festbettelektrolyse zum Abbau bzw. zur Ruckgewinnung von organnischen Komplexbildnern im Abwasser. *Galvanotechnik*, 79, 172, 1988.

66. **W.H. Holl.** Simultaneous Ellimination of Heavey Metals and Chelating Agents from Waste Waters, *Recent Developments In Ion Exchange*, Eds., P.A. Williams and M.J. Hudson, Elsevier Applied Science, New York, 161, 1990.

67. **J.H.P. Watson and C.H. Boorman.** A Permanently Magnetised High Gradient Magnetic Filter for Glove-Box Cleaning and Increasing HEPA Filter Life, Proc. 21st DOE/NRC Nuclear Air Cleaning Conference, San Diego 1990, Ed. M.W. First, 2, 762, 1991.

68. **P. Stallard, P. Scowen, P.W. Oates and P. Meddings.** Development of a Multitube Axial Flow Cyclone Separator System for Use in Nuclear Gas Cleaning Systems, Proc. 21st DOE/NRC Nuclear Air Cleaning Conference, Boston, 1988.

69. Design of Off-Gas and Air Cleaning Systems at Nuclear Power Plants. IAEA Technical Report No. 274, IAEA, Vienna, Austria, 1987.

70. Design of Radioactive Waste Management Systems at Nuclear Power Plants, IAEA Safety Series No. 79, IAEA, Vienna, Austria, 1986.

71. **M. G. Evans and J.J. Hillary.** Recent Studies on the Performance of Impregnated Charcoals for Trapping Methyl Iodide, After Exposure to Moist Air, Gaseous Effluent Treatment in Nuclear Installations, Proc. Mainland European Conference, Luxembourg, 475, 1985.

72. **J.J. Hillary and L. R. Taylor.** The Trapping of Methyl Iodide in CO_2 on KI-Impregnated Charcoal at Elevated Temperatures and Pressures, Gaseous Effluent Treatment in Nuclear Installations, Proc. of Mainland European Conference, Luxembourg, 499, 1985.

73. **R. Skeldon, S. Taylor, C. Fern and M. Stead.** High Efficiciency Particulate Air Filter Technology From 1980 to 1985 in the Central Electricity Generation Board, Gaseous Effluent Treatment in Nuclear Installations, Proc. Mainland European Conference, Luxembourg, 163, 1985.

74. Control of Semivolatile Radionuclides in Gaseous Effluents at Nuclear Facilities, IAEA Technical Report 220, IAEA, Vienna, Austria, 1982.

75. Design and Operation of Off-Gas Cleaning Systems at High Level Liquid Waste Conditioning Facilities, IAEA Technical Report No. 291, IAEA, Vienna, Austria, 1988.

76. **R.K. Sinnott.** *Chemical Engineering, Vol. 6, An Introduction to Chemical Engineering Design*, Pergamon Press, Oxford, 1991.

77. **J. M. Maurel and D. Vigla.** Behaviour of a Sorbent Material (AC6120) for Iodine Removal in the Presence of Nitrogen Dioxide, Proc. 19th DOE/NRC Nuclear Air Cleaning Conference, Seattle, 1986, Ed. M.W. First, II, 730, 1987.

78. **N.S. Holt, A.L. Goldsmith and I. Denniss.** The Development of Corona Discharge for Iodine Removal From Nuclear Fuel Reprocessing Plant Off-Gases, Nuclear Air Cleaning and Treatment Conference 1992, Denver, 102, 1993.

79. **M. Jacquemin and H. Schuettelkopf.** [129]I Trapping in Irradiated Fuel Reprocessing Plants, NEA Specialist Meeting on [129]I, Paris, OECD, 84, 1977.

80. **H. Ringel, B. Brodda, T. Burbach and R. Printz.** Development of a Process for Adsorptive Separation of Kr-85 from the Off-Gas of Nuclear Facilities, Proc. 21st DOE/NRC Nuclear Air Cleaning Conference, San Diego 1990, Ed. M.W. First, I, 299, 1991.

81. **F. Weirich.** Krypton Removal from the Dissolver Off-gas with the Solvent R-12, In: *Safety of the Nuclear Fuel Cycle*, Eds. K. Ebert and R.V. Ammon, VCH Publishers, Cambridge, 189, 1989.

82. **L.P. Geens, W.R.A. Goosens and J. Marien.** Krypton Recovery From Reprocessing Off-Gases by Cryogenic Distillation. Gaseous Effluent Treatment in Nuclear Installations, Proc. Mainland European Conference, Luxembourg, 219, 1985.

83. **S. Judd**. A Review of the Separation and Immobilisation of Krypton Arising from Nuclear Fuel Reprocessing Plant, HMIP, DOE Report No. DOE/HMIP/RR/92/019, 1992.

84. **J. Coste, A. Jouan, C. Papault and C. Porteau**. Vitrification of High Level Waste Solutions at Marcoule — France, Mainland European Nuclear Conference 79, Hamburg, *Trans. Am. Chem. Soc.*, 31, 528, 1979.

85. **A. Chrubasik**. Development of an Incineration System for Radioactive Waste, Radioactive Waste Management 2, London, I, 181, 1989.

86. **J.L. Matteman, A.H. Eenink and A.J. Gcutjes**. The KEMA Incinerator for Radioactive Waste, SPECTRUM 1986, I, 534, 1986.

87. **H. Leibold, R. Mai and J. G. Wilhelm**. Steps Towards the Minimization of Particulate Emissions from a Low-level Waste Incineration Facility, Proc. 21st DOE/NRC Nuclear Air Cleaning Conference, San Diego, Ed. M.W. First, I, 510, 1991.

Chapter 16

Russia and Eastern Europe

Boris F. Myasoedov and Igor A. Lebedev

I. INTRODUCTION

The development of safe methods for disposal of highly radioactive wastes (HRW) is one of the most important factors for successful development of nuclear energetics and other industries that exploit radionuclides. Radioactive wastes arise in many processes that are connected with the reprocessing of materials containing radionuclides (the mining and reprocessing of uranium ores, uranium enrichment, the generation of spent nuclear fuel, and the isolation and purification of separate radionuclides). The method of the closed nuclear cycle employed in Russia and other East European countries[1] is to reprocess HRW before their disposal so that long-lived radionuclides (LRN), which are the most hazardous components of these wastes, may be isolated more completely. LRN are transuranium isotopes (Np, Pu, Am, and Cm) as well as some fission products (isotopes of Cs, Sr, lanthanides, etc.). HRW solutions usually contain large concentrations of a nitric acid, nitrates of aluminum, zirconium, and other elements. The content of LRN in them vary over a wide range. At present the principal methods of LRN removal from HRW are chemical ones, namely extraction, flotation, and sorption. Also, principally new solutions to this problem are considered, for example, transmutation, i.e., transformation of LRN into short-lived or stable isotopes by irradiation in a special reactor or an accelerator.[1,2] An important problem is also to choose a method for solidifying HRW prior to their disposal. According to Russian research, vitrification is the best method of radioactive waste solidification that provides the minimum migration of radionuclides into the environment. At present the apparatus for large-scale vitrification of the nuclear industry wastes that are free of LRN is being developed in Russia.

II. METHODS OF RADIONUCLIDE ISOLATION FROM WASTES

A. EXTRACTION

The radioactive wastes remaining after spent fuel reprocessing are usually water solutions containing large concentrations of strong acids (often nitric acid) and salts. Different extractants can be used for deep purification of such HRW from LRN. Recent investigations show that bidentate neutral organophosphorus compounds (BNOC) are the most effective extractants for this purpose. Two types of BNOC have been studied: tetraalkyl(aryl)alkylene(arylene) diphosphine dioxides (DO):[3]

$$R'R''PO-X-PO(R''')_2$$

and dialkyl(diaryl)(dialkylcarbamoylmethyl) phosphine oxides (CMPO, or carbamoyls):[4,5]

$$R'R''PO-X-CO-N(R''')_2$$

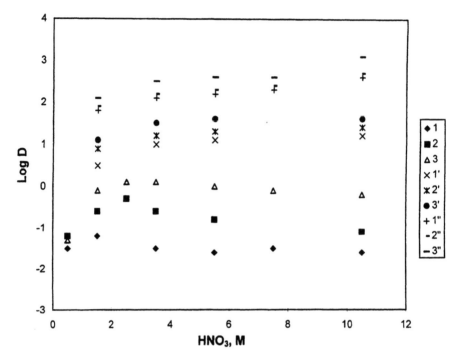

Figure 1 Extraction of Am(III) (1 to 3), U(VI) (1′ to 3′) and Pu(IV) (1″ to 3″) by 0.05 M dichloroethane solutions of CMPO from HNO$_3$ solutions.[5] 1, 1′, 1″: R′ = R″ = phenyl, R‴ = ethyl; 2, 2′, 2″: R′= phenyl, R″ = butyl, R‴ = ethyl; 3, 3′, 3″: R′ = R″ = butyl, R‴ = ethyl.

where R′, R″, R‴, and X are alkyl or aryl groups.

The wide possibilities of varying structures of these compounds enable researchers to change both extractive power and selectivity of these reagents together with their compatibility with solvents. Thus, the radical nature of the phosphorus atom sufficiently influences the extraction capacity of the reagent. As seen from Figure 1 the aryl-substituted reagents have the highest extraction capacity with respect to Am(III) as well as to U(VI) and Pu(IV).

The change of the substituent at phosphorus atoms influences not only reagent extraction capacity, but also its solubility in both phases. Thus, tolyl-substituted reagents are much more compatible with a wide range of diluents than phenyl-substituted ones. Aryl-substituted reagents are less soluble in water and in HNO$_3$ than alkyl-substituted ones. The nature of substituents at nitrogen atoms in CMPO mainly influences their solubility in organic and water phases, i.e., butyl-substituted reagents are much more compatible with diluents than ethyl-substituted and more soluble in HNO$_3$ than octyl-substituted. The change of bridge structure between functional groups influences reagent extraction capacity. Thus, the reagents with methylene bridges between functional groups have the highest extraction capacity. The increase of linear bridge length results in a decrease of reagent extraction capacity.

When studying possibilities of isolating transplutonium elements (TPE) from various media, one usually sees an unusually high efficiency of extraction from perchloric acid media by aryl-substituted CMPO.[6] The efficiency of extraction is so high that addition of very small amounts of perchloric acid to solutions of other acids or complexing agents

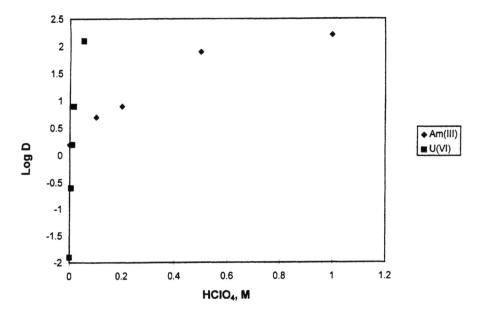

Figure 2 Effect of HClO$_4$ on extraction of Am(III) from 3 M HNO$_3$ and U(VI) from 6 M H$_3$PO$_4$ by CMPO in dichloroethane.[6] For Am(III): 0.05 M solution, R$'$ = R$''$ = phenyl, R$'''$ = ethyl; for U(VI): 0.1 M solution, R$'$ = R$''$ = tolyl, R$'''$ = ethyl.

results in a formidable increase in the efficiencies of extraction of TPE, lanthanides, U(VI), and Pu(IV) and makes it possible to concentrate these elements from various solutions. Figure 2 represents the effect of perchloric acid on Am(III) extraction from HNO$_3$ and U(VI) extraction from H$_3$PO$_4$ solutions by CMPO. It is shown that Am extraction is sufficiently increased. The addition of small amounts of HClO$_4$ allows the extraction of U(VI) from phosphoric acid quantitatively.

Together with many advantages, DO and CMPO have substantial drawbacks that cannot be overcome even by changing their structures; e.g., they are not practically dissolved in aliphatic diluents. One means to eliminate these drawbacks is to add modifiers such as tributyl phosphate (TBP) to solutions of bidentate reagents. These are compatible with many organic solvents. Addition of TBP not only results in eliminating the third phase, but also provides a nonadditive increase in distribution coefficients (K$_d$) of TPE (synergistic effect). The addition of TBP allowed application of the aliphatic hydrocarbons widely used in radiochemical industries as solvents for CMPO.[7,8] A mixture of 0.2 M CMPO (R$'$ = phenyl, R$''$ = octyl, and R$'''$ = i-butyl) and 1.2 M TBP in dodecane was used for Pu and Am extraction from 7.5 M HNO$_3$ (96% of Pu and Am were extracted at one step).[9]

Some alternatives may also be found from the study of principally new diluent classes such as fluoroethers and fluoroderivatives of some other compounds. These diluents have good compatibilities with reagents and their complexes, high densities, and high boiling points (>1300°C). They are poorly soluble in water and have high chemical and radiation stabilities.[10]

One of the most important requirements of reagents for TPE extraction is radiation stability. It was shown that the extraction properties of phenyl-substituted DO and CMPO do not change even under high gamma- and alpha-radiation.[11] BNOC solutions of a

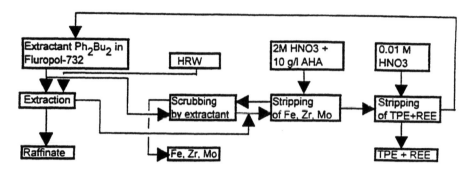

Figure 3 Scheme of isolation of TPE and lanthanides from HRW by 0.1 M solution of diphenyl(dibutylcarbamoylmethyl) phosphine oxide in Fluoropol-732.[31]

solvating diluent (Fluoropol-732) were suggested for isolating actinides from strongly acidic solutions of HRW. Such systems have the advantage of the possible use of readily available and efficient reagents such as carbamoyls and the avoidance of the addition of solvating TBP. The extraction of TPE and accompanied elements from the model acidic HRW solutions (2.5 M HNO$_3$) by CMPO (Ph$_2$Bu$_2$, i.e., $R' = R'' =$ phenyl, $R''' =$ butyl) in Fluoropol-732 was studied. It was shown that lanthanides and Zr(IV) are extracted as effectively as TPE; Mo(VI) and Fe(III) are extracted to a lesser extent than TPE, while alkaline and alkaline-earth metals and Cr, Co, and Ni are not extracted at all. Mo(VI), Fe(III), and Zr(IV) can be separated from TPE and lanthanides by multiple stripping of the organic phase by acetohydroxamic acid (AHA).

The technological scheme of actinide and lanthanide isolation using an 18-step set of centrifuge extractors with Ph$_2$Bu$_2$ in Fluoropol-732 and with AHA as stripping reagent has been studied (Figure 3). With a model HRW solution containing more than 13 g/l of lanthanides and actinides and up to 5 M HNO$_3$, a possibility has been shown to isolate more than 99.5% lanthanides and actinides as well as to purify them of Fe, Zr, and Mo (coefficient of purification >50) and to transfer them to a weakly acidic extract with concentrating by four to six times. This scheme has been experimentally tested at the Mayak Plant (Chelyabinsk).

The stripping solution for TPE and lanthanides from the extract contains as low as 0.02 to 0.04 M HNO$_3$. This makes it possible to choose any other water system for group separation of these elements. However, tests have shown that the big loss of extractant due to its high water solubility is the main disadvantage of this extraction system. To eliminate this disadvantage, the use of BNOC with isoamyl and n-octyl substituents at the nitrogen atom has been suggested. Water solubility of these extractants is three to five times lower than that of Ph$_2$Bu$_2$ while extractivity is practically the same. Therefore, solutions of diphenyl (diisoamyl or dioctyl)carbamoylmethyl phosphine oxides in Fluoropol-732 diluent may be used to isolate actinides from radioactive wastes of various compositions.

Chlorinated cobalt dicarbollide (CCD) is a very effective extractant for Cs, Sr, and tervalent ions from acidic wastes.[2,12] Extraction of Cs by CCD is most effective, K_d(Cs) = 11.5, for extraction by 0.3 M CCD from 5 M HNO$_3$ solutions.[12] Extraction of bi- and tervalent ions increases considerably if polyethylene glycol is added to this extractant. Thus, 0.037 M CCD in nitrobenzene containing 1% polyethylene glycol extracts Sr and lanthanides from 0.3 M HCl with K_d >10.13. The scheme of LRN removal from waste solutions by CCD extraction was worked out previously.[1] This scheme allows one to extract more than 99% of the Cs, Sr, TPE, and lanthanides and to obtain separate fractions of these elements. The scheme is now being tested at the Mayak Plant (Chelyabinsk) in order to develop the method to fractionally isolate radionuclides from discharging solutions.

A liquid–liquid extraction system based on water-soluble polymers such as polyethylene glycol has been suggested for extraction of actinides from various salt solutions, including HRW. These systems are of interest from a practical point of view because they contain no organic solvents (which are usually volatile, explosive, and toxic), making it possible to use well-known water-soluble reagents. Extraction of actinides from sulfate, carbonate, phosphate, rhodanide, and nitrate solutions in the presence of different complexing agents has been investigated.[14] Conditions have been found for quantitative group extractions of the TPE and for separation of these elements from U, Th, and lanthanides.

TPE and many other elements can be extracted from alkaline solutions in the presence of complex-forming reagents by extractants of various classes: quaternary ammonium bases, amines, alkylpyrocatechols (DOP), alkyl derivatives of amine alcohols (AA), and β-diketones.[16] Some of these extractants, especially DOP and α-oxy-5-alkylbenzyl-diethanolamine are characterized by a high extraction capacity in relation to TPE. With their help it is possible to isolate effectively these elements from the alkaline and carbonate solutions with alkali concentration up to 5 to 6 M (Figure 4). The following extractant can be used for Cs extraction from alkaline radioactive wastes with high salt concentration:[15]

$$(CH_3)_3C-C_6H_2CH_3(OH)-CH_2-(OH)C_6H_2CH_3-C(CH_3)_3$$

A 1 M solution of this extractant in nitrobenzene extracts Cs from LiOH solution (pH = 12.77) with a K_d = 30.

B. FLOTATION

In addition to extraction, ion flotation is successfully applied for actinide and lanthanide isolation from acidic aqueous solutions.[17-19] When using this method, a surface-active substance (SAS) is added to the solution that forms nonsoluble products with the extracted ions. These products are carried away from the solution by an air flow. Laurylphosphoric acid (LPA), $CH_3(CH_2)_{11}OPO(OH)_2$, can be used as the SAS. For the ratio LPA:metal = 4, almost 100% of the Th(IV) and Pu(IV) is isolated by LPA from nitric acid solutions (up to 5 M HNO$_3$).[17] Precipitates that are formed by LPA with ions of inactive elements (Al, Bi, and Fe) were used as a special kind of sorbents to purify solutions containing microconcentrations of radionuclides.[18] For example, as much as 96% of the Eu^{3+} (TPE imitator) is extracted from 0.1 M HNO$_3$ by aluminium laurylphosphate. Unfortunately, the extraction of trivalent ions by this reagent significantly decreases when the acidity of solution increases. Other SAS — $(C_6H_5)_2POCH_2PO(C_6H_5)_2$ (PPO) and $(C_6H_5)_2POCH_2PO(C_6H_4CH_3)_2$ (TPO) — are used for TPE and lanthanide isolation from more acidic solutions.[19] For example, 76% of the Am and 91% of the Pu is extracted by TPO from 3.1 M HNO$_3$ for the ratio SAS:metal = 5.2.

Undoubtedly, the flotation methods are useful for TPE and lanthanide isolation from different kinds of water solutions, which can contain both considerable concentrations and microconcentrations of these elements.

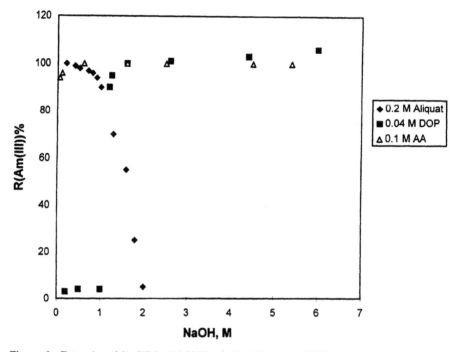

Figure 4 Extraction of Am(III) by 0.2 M Aliquat-336.OH, 0.04 M DOP in toluene, and 0.1 M AA in hexane in the presence of 0.025 M tartaric acid as a function of the NaOH concentration.[16]

C. SORPTION

Sorption methods that use columns with ion exchangers or other sorbents for purifying liquid radioactive wastes are less popular in Russia and other East European countries than extraction methods. Nevertheless, these methods in special cases are used to separate some radionuclides. Thus, the application of zeolites (natural zeolite-4A, zeolite-13X, and synthetic mordenite) to isolate cesium from the overflow solutions of radiochemical industries has been investigated.[20] It was found that mordenite is the best sorbent. After the solution passes through a mordenite column the content of cesium is 3000 times less; also, mordenite shows the best selectivity with respect to cesium.

III. METHODS FOR SOLIDIFYING WASTES

The waste disposal is the completion stage of nuclear waste reprocessing. This stage must guarantee the most complete isolation of radionuclides from the biosphere. Because the half-lives of TPE isotopes are sufficiently long, this isolation must last many thousand years. Solidification, i.e., the conversion of radioactive wastes into solid materials, followed by their disposal in deep mines, is considered to be the safest method of storing wastes in Russia and East European countries. The comparative tests of different solidification methods, i.e., cementation (Portland cement), bituminization, and vitrification (borosilicate glass), showed that vitrification was the safest method.[21] The potential hazard of the methods correlate (as for ^{137}Cs) as 3000:200:1, respectively. The rate of ^{137}Cs washout from cement is 3×10^{-5} to 4×10^{-4} g/(cm^2·d)[22] (depending on the cement composition), and from glass it is 10^{-8} to 10^{-7} g/(cm^2·d).[23,24] For example, in the case of glass that contained 10% Na_2O, 40% SiO_2, and up to 50% of other metal oxides, the rates of ^{137}Cs, ^{90}Sr, and ^{239}Pu washout were $(1$ to $8) \times 10^{-7}$, $(5$ to $10) \times 10^{-7}$, and $(3$ to $10) \times 10^{-8}$ g/(cm^2·d), respectively.

Investigations showed that phosphate and borosilicate glass had the most resistance as regards washing out of radionuclides. The rate of TPE washout from such glass containing more than 10000 Ci/l alpha-active radionuclides was 10^{-9} to 10^{-7} g/(cm^2·d).[25] It was found that such factors as the total content of radionuclides, the uniformity of their distribution in glass, and the glass homogeneity influence the rate of washout the most. Special investigations showed that the rate of washout practically remains constant if the size of impurity particles in glass is not more than 20 μm.[26] ^{239}Pu can be uniformly distributed in glass if the total content of Pu is not more than 3% and the glass remains melted under 1000°C not more than 8 h.[27] The rate of radionuclide washout from glass significantly increases if the total content of their oxides in glass exceeds 40 to 50% (because of depolymerization of glass silicon–oxygen framework),[24] or if the total number of alpha decays exceeds 10^{19}/cm^3 (because of the destruction of the material structure).[25]

The possibility of using borosilicate and aluminum phosphate glass to dispose of wastes that contain sulfates was investigated.[28,29] It was determined that aluminum phosphate glass could contain more sulfate and the rate of washout from it was less.[28] On the other hand, sulphates as a dispersed phase may be introduced (up to 22%) into borosilicate glass, and the rate of ^{137}Cs washout in this case is 10^{-6} to 10^{-5} g/(cm^2·d).[29] The long storage (more than 10 years) of vitrified wastes that contain up to 10^6 Bq/kg radionuclides (^{137}Cs, ^{90}Sr, etc.) on the open-earth platform does not practically affect the rate of washout.[30]

REFERENCES

1. **Romanovskii, V. N., Galkin, B. Ya., Lazarev, L. N., Lyubtzev, R. I., Rogozin, Yu. M.**, New approaches to reprocessing in Russia, in *Proc. Int. Conf. Technology Exposition on Future Nuclear Systems: Emerging Fuel Cycles and Waste Disposal Options (GLOBAL-93)*, American Nuclear Society, Washington, D.C., 1993, 71.
2. **Nikiforov, A. S., Rozen, A. M.**, Accumulation of actinides and some problems of radiochemistry, *Radiokhimiya*, 33(5), 1, 1991.
3. **Chmutova, M. K., Kochetkova, N. E., Myasoedov, B. F.**, Polydentate neutral organophosphorus compounds as extractants of transplutonium elements, *J. Inorg. Nucl. Chem.*, 42, 897, 1980.
4. **Chmutova, M. K., Kochetkova, N. E., Koiro, O. E., Myasoedov, B. F., Medved, T. Ya., Nesterova, N. P., Kabachnik, M. I.**, Extraction of transplutonium elements with diphenyl (alkyl)(dialkylcarbamoylmethyl) phosphine oxides, *J. Radioanal. Chem.*, 80, 63, 1983.
5. **Myasoedov, B. F., Chmutova, M. K., Kochetkova, N. E., Koiro, O. E., Pribylova, G. A., Nesterova, N. P., Medved., T. Ya., Kabachnik, M. I.**, Effect of the structure of dialkyl (aryl)(dialkylcarbamoylmethyl) phosphine oxides on their extraction capacity and selectivity, *Solv. Extr. Ion Exch.*, 4, 61, 1986.
6. **Chmutova, M. K., Litvina, M. N., Nesterova, N. P., Myasoedov, B. F., Kabachnik, M. I.**, Extraction of Am(III), Eu(III) and U(VI) from perchloric acid and mixtures of acids by dialkyl(aryl)(diethylcarbamoylmethyl) phosphine oxides, *Solv. Extr. Ion Exch.*, 10, 439, 1992.
7. **Chmutova, M. K., Pribylova, G. A., Nesterova, N. P., Myasoedov, B. F., Kabachnik, M. I.**, The extraction of americium(III) from nitric acid with the mixtures of neutral phosphorus organic reagents, *Radiokhimiya*, 31(3), 73, 1989.
8. **Pribylova, G. A., Chmutova, M. K., Nesterova, N. P., Myasoedov, B. F., Kabachnik, M. I.**, Extraction of americium(III) by solutions of diaryl(dialkylcarbamoylmethyl) phosphine oxides in aliphatic diluents, *Radiokhimiya*, 33(4), 70, 1991.

9. **Mathur, J. N., Murali, M. S., Natarajan, P. R., Badheka, L. P., Banerji, A., Michael, K. M., Kapoor, S. C., Dhumwad, R. K.,** Tail-end purification of Am from Pu loading effluents using a mixture of octyl(phenyl)-*N*,*N*-diisobutylcarbamoulmethylphosphine oxide and tri-*N*-butylphosphate, *J. Radioanal. Nucl. Chem.* (Letters), 165, 219, 1992.

10. **Pribylova, G. A., Chmutova, M. K., Babain, V. A., Shadrin, A. Yu.,** Extraction of Am(III) by dialkyl(diaryl) dialkylcarbamoylmethylphosphine oxides solutions in mixed and fluorocontaining solvents, in *Actinides-89,* Nauka, Moscow, 1989, 311.

11. **Kochetkova, N. E., Chmutova, M. K., Lebedev, I. A., Myasoedov, B. F.,** Investigation of radiation stability of tetraphenylmethylene diphosphine dioxide, *Radiokhimiya,* 23(3), 420, 1981.

12. **Korolev, V. V., Afonin, M. A., Kopyrin, A. A., Proyaev, V. V., Romanovskii, V. V., Viznyi, A. N., Romanovskii, V. N., Lazarev, L. N.,** Separation of cesium, strontium, cerium(III) and europium from nitric acid solutions by extractant based on chlorinated cobalt dicarbollide, *Radiokhimiya,* 32(3), 97, 1990.

13. **Selucky, P., Plesek, J., Rais, J., Kyrs, M., Kadlecova, L.,** Extraction of fission products into nitrobenzene with dicobalt tris-dicarbollide and ethyleneoxy substituted cobalt bis-dicarbollide, *J. Radioanal. Nucl. Chem. (Articles),* 149, 131, 1991.

14. **Molochnikova, N. P., Shkinev, V. M., Myasoedov, B. F.,** Two-phase aqueous systems based on poly(ethylene glycol) for extraction separation of actinides in various media, *Solv. Extr. Ion Exch.,* 10, 697, 1992.

15. **Gulis, G., Mikulai, V.,** Recovery of cesium from intermediate level liquid radioactive waste, *J. Radioanal. Nucl. Chem. (Articles),* 150, 255, 1991.

16. **Karalova, Z. K., Myasoedov, B. F., Bukina, T. I., Lavrinovich, E. A.,** Extraction and separation of actinides and lanthanides from alkaline and carbonate solutions, *Solv. Extr. Ion Exch.,* 6, 1109, 1988.

17. **Mezhov, E. A., Samatov, A. V., Troyanovskii, L. V.,** Flotation isolation of actinoids and lanthanoids from nitric acid solutions, *Radiokhimiya,* 31(5), 45, 1989.

18. **Mezhov, E. A., Samatov, A. V., Troyanovskii, L. V.,** Flotation withdrawal of polivalent nuclides using sorbents based on laurylphosphoric acid, *Radiokhimiya,* 34(6), 132, 1992.

19. **Mezhov, E. A., Samatov, A. V., Troyanovskii, L. V.,** Isolation of actinides and lanthanides by flotation from nitric acid solutions in the form of the complexes with diphosphineoxides, *Radiokhimiya,* 33(2), 77, 1991.

20. **Bronic, J., Subotic, B.,** Removal of cesium radioisotopes from solutions using granulated zeolites, *J. Radioanal. Nucl. Chem. (Articles),* 152, 359, 1991.

21. **Barinov, A. S., Ozhovan, M. I., Sobolev, I. A., Ozhovan, N. V.,** Potential hazard of solidified radioactive wastes, *Radiokhimiya,* 32(4), 127, 1990.

22. **Plecas, I., Peric, A., Drlyaca, J., Kostadinovic, A.,** Leaching behavior of 137Cs in cement, *J. Radioanal. Nucl. Chem.* (Letters), 154, 309, 1991.

23. **Lifanov, F. A., Stefanovskii, S. V.,** Silicate glasses and vitroceramics for immobilization of radiactive ash arising from incineration of organic wastes, *Radiokhimiya,* 32(3), 166, 1990.

24. **Stefanovskii, S. V., Ivanov, I. A., Gulin, A. N., Lifanov, F. A.,** Incorporation of sodium-containing radioactive wastes into the loam-based glass, *Radiokhimiya,* 35(3), 106, 1993.

25. **Nikiforov, A. S., Polyakov, A. S., Kulichenko, V. V., Matyunin, Yu. I.,** Problem of localization of transplutonium elements, *Radiokhimiya,* 32(2), 43, 1990.

26. **Karlina, O. K., Ozhovan, M. I., Popov, M. V.,** Comparative analysis of water resistance of glass composites and homogeneous glass matrixes for immobilizing radioactive wastes, *Radiokhimiya,* 35(3), 120, 1993.

27. **Matyunin, Yu. I., Astakhov, M. N., Savin, M. F.,** The investigation of immobilization of alpha-radionuclides in phosphate glassy compositions, *Radiokhimiya,* 32(3), 154, 1990.

28. **Ivanov, I. A., Gulin, A. N., Stefanovskiy, S. V.,** Diffusion of sodium cations and water stability of glasses for immobilization of middle-active wastes, *Radiokhimiya,* 33(5), 122, 1991.

29. **Ozhovan, M. I., Karlina, O. K.,** Preparation and properties of the glass composition materials for the solidification of radioactive wastes, *Radiokhimiya,* 34(2), 143, 1992.

30. **Ozhovan, M. I., Batyukhnova, O. G., Barinov, A. S., Devyatkova, L. I., Semenov, K. N., Shcherbakova, T. D., Manyukova, N. N., Titov, V. N.,** The testing of vitrified radioactive wastes in the open site conditions, *Radiokhimiya,* 34(4), 100, 1992.

31. **Myasoedov, B. F., Chmutova, M. K., Smirnov, I. V., Shadrin, A. Yu.,** Isolation of actinides and separation from rare earth elements and selected fission products by bidentate neutral organophosphorus compounds, in *Proc. Int. Conf. and Technology Exposition on Future Nuclear Systems: Emerging Fuel Cycles and Waste Disposal Options (GLOBAL-93),* American Nuclear Society, Washington, D.C., 1993, 581.

Chapter 17

Asia: Japan, China, and Korea

Zenko Yoshida, Yadong Li, and Joon-Hyung Kim

I. JAPAN

Forty-two nuclear power plants are in commercial operation and 11 plants are currently under construction in Japan, as of July 1993.[1] The spent fuels of the power reactors are subjected to the reprocessing to recover plutonium and unfissioned uranium. The Tokai reprocessing plant based on the PUREX process has been operated by Power Reactor and Nuclear Fuel Development Corporation of Japan (PNC) since 1981. A new commercial reprocessing plant at Rokkasho-mura is scheduled to begin operation in 1999. Reprocessing capability of the Tokai plant is reported to be about 0.7 ton/d and that of Rokkasho-plant will be 800 ton/year. From the reprocessing of the spent fuel, high-level liquid waste (HLLW) containing actinide elements as well as fission product elements (FP) is created. The accumulated amount of HLLW that had been generated from Tokai plant by March 1993 was 516 m³. The current Japanese policy calls for solidifying HLLW into a stable form such as vitrified waste, storing the waste for 30 to 50 years, and finally disposing the waste in deep geological formations.

One of the strategies for the efficient and safe management of HLLW is based on the partitioning–transmutation concept that can minimize the contents of long-lived radionuclides in the wastes resulting in the long-term reduction in repository risk. The targets of the partitioning are transuranium elements (TRU), heat-generating ^{90}Sr + ^{137}Cs, and Tc + noble metals. In this decade, a management of Np in HLLW has become more important from the viewpoint of environmental safety. This is based on ICRP Publication 30, which introduced a new concept of annual limit on intake (ALI) and pointed out that ^{237}Np should be assessed as one hundred or more times hazardous than the result on the basis of maximum permissible concentration (MPC). ^{99}Tc has also been regarded to be long-term hazardous, and additionally, to possibly have beneficial uses as a catalyst, corrosion inhibitor, etc.

Under such circumstances, much research has been done on the partitioning of HLLW. Wet-chemical partitioning methods such as three- or four-group partitioning of HLLW have been studied in the Japan Atomic Energy Research Institute (JAERI) and PNC, and pyrometallurgical methods to recover TRU have been studied in the Central Research Institute of Electric Power Industry (CRIEPI). Several processes were proposed by these organizations. Many related fundamental works have been conducted in universities, institutes, and companies.

In 1988 the Japanese goverment submitted "A Proposal to Exchange Scientific and Technological Information Concerning Options Making Extra Gains of Actinides and Fission Products Generated in Nuclear Fuel Cycle (OMEGA)" under the OECD/NEA international cooperation program. Presently, the related research works are extensively performed as organized in the OMEGA project.

Table 1 Distribution Ratios of Ions in the Extraction with 0.5 M DIDPA + 0.1 M TBP+ n-Dodecane

Ions	[HNO$_3$] 0.5 M	4 M	Ions	[HNO$_3$] 0.5 M	4 M
Am(III)	7.4	0.05	Fe(III)	>10^3	>10^3
La(III)	3.6	0.02	Ru	about 0.01	about 0.01
Nd(III)	7.8	0.07	Rh	<10^{-2}	<10^{-2}
Gd(III)	38	0.2	Pd	about 0.1	<10^{-2}
U(VI)	>10^3	>10^3	Cs	<10^{-2}	<10^{-2}
Np(VI)	>10^3	>10^3	Sr	<10^{-2}	<10^{-2}
Np(IV)	>10^3	740	Zr	>10^3	>10^3
Pu(IV)	>10^3	>10^3	Mo	>10^3	500

A. THREE- OR FOUR-GROUP PARTITIONING OF HLLW

Kubota et al.[2-6] of JAERI have promoted continual and phased research for the development of partitioning of elements in HLLW since 1973. They proposed a few partitioning processes combining several wet-chemical separation reactions.

One of the main characteristics of the processes proposed is an employment of the novel extractant, diisodecyl phosphoric acid (DIDPA), which is the product of Daihachi Chemical Industry Co., Ltd., Japan.

The DIDPA extraction is advantageous for the partitioning of the TRU group. Am(III) and Cm(III) can be extracted from a solution of fairly low acidity such as 0.5 M HNO$_3$ and both tetra- and hexavalent U, Np, and Pu ions and even Np(V) can be extracted. Distribution ratios of selected ions between 0.5 or 4 M HNO$_3$ and n-dodecane of 0.5 M DIDPA + 0.1 M TBP are summarized in Table 1. Trivalent ions can be separated from tetravalent ions by appropriate back-extraction procedures. DIDPA is more tolerant of radiation than TBP and the decomposition products of DIDPA, if present, hardly affect the extraction behavior of the elements. DIDPA also exhibits a pronounced applicability to the mutual separation of trivalent Am, Cm, and lanthanides.

The partitioning process fractionates elements in HLLW into three groups, i.e., TRU, Sr+Cs, and the other elements groups, as shown in Figure 1(a). The process consists of the solvent extraction of TRU with TBP and DIDPA and separation of Sr and Cs with an inorganic ion exchanger. Am(III) and Cm(III) are separated from coextracted lanthanides with pressurized cation-exchanger columns. The feasibility of the process was demonstrated with a mixer-settler of seven stages using an actual HLLW and it was confirmed that >98% Pu was extracted with 30% TBP, and >99.99% Am and Cm were extracted with 0.5 M DIDPA + 0.1 M TBP. At the TBP extraction step, 86% of Np remained in the aqueous phase, suggesting that most of Np was pentavalent in the HLLW feed solution, and at the successive DIDPA extraction step >90% of Np was extracted and could not be stripped with 4 M HNO$_3$.

In 1984 a new plan was introduced to develop a more advanced partitioning process that would fractionate elements into four groups, i.e., TRU, Tc + noble metals, Sr + Cs, and the other elements groups. Here, the study was focused mainly on the separation of Np and Tc. The process proposed is shown in Figure 1(b).

DIDPA extraction or oxalate precipitation was recommended for the partitioning of the TRU group. As shown in Table 1, tri-, tetra-, and hexavalent TRU elements can be extracted with DIDPA. It is noteworthy that even Np(V) is extracted by DIDPA, although

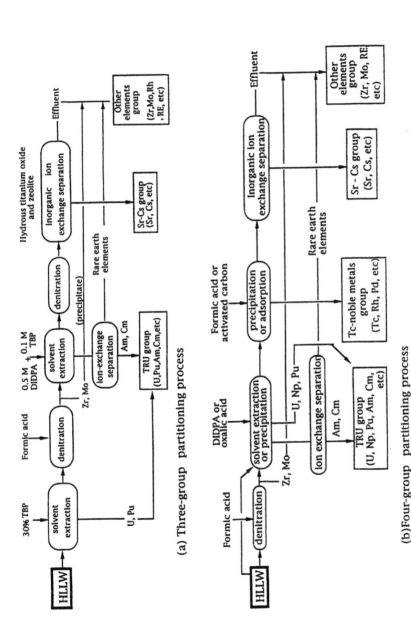

Figure 1 Processes for three- and four-group partitioning of HLLW.

the extraction rate is rather slow from 0.5 M HNO$_3$, from which Am(III) and Cm(III) are extracted more efficiently. It was found that an addition of hydrogen peroxide enhanced the rate of the extraction of Np(V). The oxidation state of Np extracted with DIDPA from 0.5 M HNO$_3$ solution with hydrogen peroxide was IV in the organic phase, while the initial oxidation state in the aqueous solution was V. Rapid reduction of Np(V) to Np(IV) occurred in the presence both of hydrogen peroxide and DIDPA. When the concentration of HNO$_3$ was 4 M, the Np extracted was tetravalent, and the reduction of Np(V) even in the aqueous phase was observed. Experimental results of the continuous extraction using mixer-settler with seven stages showed that 93% of Np(V) in 0.5 M HNO$_3$ feed solution (45°C) was recovered. As a result, DIDPA can extract simultaneously U, Np, Pu, Am, and Cm from HLLW if the acidity is previously adjusted to 0.5 M. The back-extraction of Am + Cm, Np(IV) + Pu(IV), and U(VI) can be achieved with 3 M HNO$_3$, 1 M oxalic acid, and 5 M phosphoric acid, respectively.

An alternative proposal involves the recovery of TRU as oxalate precipitates. Precipitation behavior of Np ions of various oxidation states was investigated with various compositions of simulated HLLW solutions. When the crystalline oxalic acid (0.5 M) was added to the simulated HLLW solution containing Nd, Sr, and Cs together with Np(IV), quantitative recovery of Np(IV) was attained. If the initial oxidation state of Np in the solution was VI or V, about 50 or 35% of the Np was precipitated. It was found that the precipitation efficiency of Np(V) increased by the addition of the other components to the simulated HLLW, though the precipitation mechanism has not yet been fully elucidated. For example, when oxalic acid (1.0 M) was added to the simulated HLLW of 14 components (Nd, Sr, Cs, Fe, Cr, Ni, Na, Mo, Ba, Rb, Zr, Te, Rh, and Pd) and kept at 25°C for longer than 2 h, >99% of the Np(V) could be precipitated. In order to minimize an operation time for the recovery of Np the oxidation state of Np should be adjusted to IV. The addition of ascorbic acid was effective for the quantitative precipitation of Np within 5 min.

For the separation of Tc, a precipitation or adsorption method is recommended. When the HLLW containing Tc was denitrated with formic acid under reflux condition for 3 h, >95% of the Tc precipitated at a pH higher than 2.0. If the concentrations of Rh and Pd in the solution were less than $5 \times 10^{-3} M$, Tc did not precipitate, which indicates that Tc coprecipitates with Rh and Pd. Tc can be leached selectively from the precipitate with a 30% hydrogen peroxide solution.

An alternative method based on the adsorption onto activated carbon was developed for the recovery of Tc from the solution containing Rh and Pd of less than $5 \times 10^{-3} M$ to which the coprecipitation technique is not available. Distribution coefficients of 300 to 400 (ml/g) were obtained when a 0.5 to 2 M HNO$_3$ solution containing trace amounts of Tc was mixed with a fresh activated carbon and shaken at 25°C for 1 h. When increasing the concentration of HNO$_3$ above 2 M, the distribution coefficient decreased significantly. The adsorption capacity of the adsorbent was determined to be 0.60 meq. Tc per gram of the fresh activated carbon. If the fresh activated carbon had been treated by 4 M HNO$_3$ before the adsorption experiment, both the distribution coefficient and the adsorption capacity decreased. Tc adsorbed onto the activated carbon can be quantitatively eluted with potassium thiocyanide solution of pH higher than 4.

Sr and Cs can be removed from solution by passing through a column packed with a mixture of hydrous titanium oxide and zeolite. The decontamination factors of these elements were more than 10^5, which was confirmed by the tests with actual HLLW. Employment of such inorganic exchangers as above is promising, because the exchanger

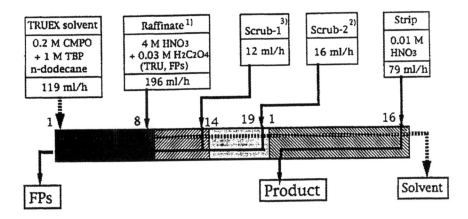

Figure 2 Flow sheet conditions for countercurrent experiments for the partitioning of HLLW with CMPO + TBP + *n*-dodecane solvent. 1) Highly active raffinate from FBR spent fuel-reprocessing experiment; 2) $0.3\ M\ HNO_3 + 0.1\ M\ H_2C_2O_4$ (for run 1), $0.3\ M\ HNO_3$ (for run 2), 3) $7.7\ M\ HNO_3 + 0.03\ M\ H_2C_2O_4$ (for run 2).

containing Sr and Cs can be converted by calcination to mineral-like compounds suitable for their immobilization, and the total volume of the secondary wastes can be minimized.

A feasibility study of the four-groups partitioning process will be performed with an actual HLLW at the recently built Nuclear Fuel Cycle Safety Engineering Research Facility (NUCEF) of JAERI.

B. CMPO EXTRACTION FOR THE PARTITIONING OF HLLW

Basic studies[7] to separate actinides from HLLW have been performed in PNC, as a part of efforts to widen the options for future waste management and to increase the fuel utilization efficiency, because the minor actinides such as Am, Cm, and Np are being recognized as fuels in the future FBR cycle. They employ octyl(phenyl)-*N*,*N*-diisobutylcarbamoylmethyl-phosphine oxide (CMPO), which is used in the TRUEX process originally developed by Horwitz et. al.[8] for the extraction of actinides from various solutions in combination with the PUREX solvent.[7,9] Investigations are mainly focused on confirming the fundamental distribution data of major components in the HLLW, and verifying the compatibility of the TRUEX solvent to the actinides partitioning process of HLLW. Countercurrent flow sheet tests using real highly active raffinate originated from FBR spent fuel-reprocessing experiments were performed by bench-scale mixer-settlers with 19 stages for extraction-scrubbing and 16 stages for stripping (Figure 2). Oxalic acid could effectively lower the distribution ratio of such troublesome elements as Zr and Mo, and oxalic acid was added to the scrubbing solution as well as the feed solution. In flow sheet test run 2, a dual scrubbing procedure was employed for improving the decontamination of Ru and for avoiding formation of an oxalate precipitate in the scrubbing and stripping banks.

The extraction behaviors of most of components such as actinides and lanthanides were in good agreement with those predicted from the distribution ratios determined by the batchwise extraction studies, except for a peculiar behavior of Ru, which might be present as species of complicated chemical forms in the raffinate. A series of countercurrent tests showed more than 10^3 decontamination factors for major actinide elements.

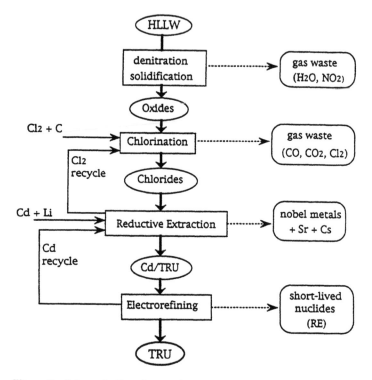

Figure 3 Schematic flow diagram for pyrometallurgical partitioning of HLLW.

Lanthanides and some Ru were coextracted in the actinides stream, therefore, further investigations are required for the removal of them. In successive countercurrent runs after 1992, an effort for the selective stripping of U/Np/Pu/Am has been continued.

Many other studies on the CMPO extraction have been carried out at several laboratories in connection with the partitioning of HLLW; e.g., the extraction of Tc, Np, in particular Np(V), Zr, and Mo; the effect of oxalic acid on the extraction behavior of the ions; and the phenomena of the third-phase formation in the CMPO extraction system.

C. PYROMETALLURGICAL PARTITIONING OF TRU FROM HLLW

The partitioning of TRU from HLLW by pyrometallurgical methods has been studied and a process was proposed by CRIEPI.[10-12] The proposal involves successive transmutations of TRU in a metallic fuel FBR. The process as shown in Figure 3 consists of (1) a denitration of HLLW to prepare oxides, (2) chlorination of oxides, (3) reductive extraction of TRU from molten chlorides with liquid Cd–Li, and (4) electrorefining of TRU. It was claimed that the pyrometallurgical process is advantageous to minimize the amount of the secondary radioactive waste and to make the facilities compact, though the purity of the TRU product is generally lower than that obtained by the wet-chemical processes.

Microwave heating was selected as the preferred technique for the denitration of HLLW, based on a few advantages such as a lower amount of secondary waste and operating safety. Chlorination using chlorine gas together with carbon as the reductant was recommended, taking advantage of reuse of chlorine gas after processing of the salt wastes. It was confirmed that all of Mo and part of Zr, Fe, and U are volatilized at 650 to 850°C at the chlorination step; therefore, these elements could be recovered by means of appropriate techniques.

TRU are separated from molten KCl + LiCl by extraction with liquid Cd–Li, where Li serves as the reducing agent and Cd as the solvent for recovery of the metals reduced. In order to evaluate the purity of TRU recovered, it is necessary to obtain the distribution coefficients between the salt and Cd phases and the related thermodynamic data such as activity coefficients for TRU and lanthanide elements in both phases. The distribution coefficients of lanthanide elements were determined and the following conclusions were reached. More than 99% of the lanthanides are extracted into liquid Cd–Li. The distribution coefficients of lanthanides can be expressed as $D_{Ln} = D_{Li^3} K_{Ln'}$, where $K_{Ln'}$ is the apparent equilibrium constant, which corresponds to the extraction of lanthanide(III) in the salt phase. The extractabilities of Ce, Pr, and Nd are nearly equal, and that of Y is the least, which can be explained satisfactorily on the basis of such thermodynamic data of the elements as the activity coefficients of chlorides and metals in salt and Cd phases. The calculation using the thermodynamic data is useful for the prediction of separation of lanthanides.

D. MISCELLANEOUS

The coprecipitation method is powerful for the separation of metal ions in the solution because of its simplicity, rapidity, and high reliability. Kimura et al.[13,14] studied the coprecipitating behaviors of actinides and FP with such precipitates as barium sulfate or bismuth phosphate, and proposed methods to recover actinides by combination with the oxidation-state control in the presence of oxidizing or reducing reagents. These techniques can be applied to the removal of actinides from nuclear wastes, in particular, from the sulfate- or phosphate-containing liquid wastes.

Kimura et al.[15,16] developed the method for the separation of trivalent actinides and lanthanides and FP by extraction chromatography with DHDECMP- and TBP-loaded Amberlite XAD-4 resins. On the basis of the fundamental results such as the preparation of adsorbent, the determination of column capacity, and adsorption/desorption behavior of ions, they concluded that extraction chromatography is promising for recovery of actinides and lanthanides from acidic waste solutions.

Extraction chromatography with DHDECMP-loaded resin has been studied also by Takeshita, et al.[17]

Toshiba Co. Ltd. has performed the preliminary studies to develop a process for the recovery of TRU in HLLW, which involves (1) the separation of TRU and lanthanides by the oxalate-precipitation technique, and then (2) the separation of TRU from lanthanides based on the electrodeposition in a KCl–LiCl eutectic salt.

Yoshida et al.[18] proposed a new concept of an electrochemical process for spent fuel reprocessing, SREEP, which consists of (1) spent fuel dissolution by electrolytic oxidation; (2) recovery of platinum-group metals by electrodeposition; (3) successive TBP-extraction of U, Np, and Pu after the electrolytic preparation of U(VI), Np(VI), and Pu(IV); (4) recovery of Cs and Sr by electrolytic ion-transfer extraction; and (5) separation of transplutonium and lanthanide elements by electrodeposition in the molten salt. The SREEP concept is applicable to the partitioning of HLLW. The electrochemical method is advantageous, because precise control of the electrolytic potential may enhance the selectivity in the separation, the amount of the secondary radioactive waste can be minimized, and electrochemical operation can be remotely controlled by automatic devices.

Motojima[19] proposed a method for the removal of Ru from PUREX process as well as from HLLW. It was found that the RuO_4 formed by the addition of Ce(IV), is extractable into n-paraffin oil, and the RuO_4 extracted is readily reduced by solvent yielding a RuO_2 suspension. The RuO_2 can be filtrated through ordinary filter paper made of cellulose fiber. Hydrogen peroxide is effective for decomposition of such stable Ru complexes as nitrosylruthenium nitro or nitrato complexes into species that can be oxidized to RuO_4. The author afterward proposed an advanced method involving the regeneration of Ce(IV) by the electrolytic oxidation.

Researchers at PNC have been studying the recovery of noble metals from HLLW as well as from the insoluble residue generated at the spent fuel dissolution step of the PUREX process. The main processes are based on (1) the volatilization of RuO_4 generated by oxidation, (2) extraction of platinum-group metals into lead metal, and (3) the electrodeposition of noble metals at the solid electrode.

Procedures for the recovery of Cs and Sr from HLLW have been developed. The process proposed by Akiba et al. consists of (1) the denitration of HLLW, (2) Cs adsorption onto ferrierite column followed by the elution by 5 M NH_4Cl or NH_4NO_3 solution, and (3) Sr adsorption onto an A-type zeolite column followed by the elution with 0.05 M EDTA solution.

II. CHINA

Although there are only two nuclear power stations in China, fairly large amounts of nuclear wastes have been generated from the reactors for military research, isotopes production, and from nuclear fuel production, spent fuel reprocessing, the application of isotopes and nuclear techniques, as well as mining and milling. According to the program of the nuclear industry, more than 20 nuclear facilities will be decommissioned and the capacity of nuclear power will reach 10,000 MW by 2015. Thus, more nuclear wastes will be generated.

China has decided to reprocess the spent fuels from nuclear reactors. Therefore, much attention is paid to the separation of the fuel-reprocessing wastes. The research and development of separation technologies for recovering nuclides and noble metals from HLLW have been ongoing since the nuclear industry was established in 1950s. Since the beginning of 1980s, much more separation technologies have been studied, especially for partitioning of HLLW. The objective is to separate the long-lived alpha-emitting nuclides, some of them being reused, others being treated specially as high-level or TRU wastes. The residual solution after the separation can be treated as an intermediate-level waste; thus, the volume of HLLW would be significantly reduced. If the transmutation technique were practical in the future, the long-lived nuclides separated would be turned into short-lived radionuclides. Extensive progress has been made in the separation technology for the liquid wastes, though little work has been done for the separation of solid and gaseous wastes.

A. TRPO EXTRACTION FOR THE PARTITIONING OF HLLW

The extraction of actinide and lanthanide elements from HLLW by trialkyl(C_6–C_8) phosphine oxide (TRPO) has been studied at the Institute of Nuclear Energy and Technology of Tsinghua University and Beijing Institute of Nuclear Engineering since 1982. TRPO is synthesized in China and specially used to separate actinides and lanthanides from HLLW. It was determined that 30 vol% TRPO–kerosene was the optimum organic phase. Figure 4 shows the distribution ratios of constituents in HLLW between HNO_3 and

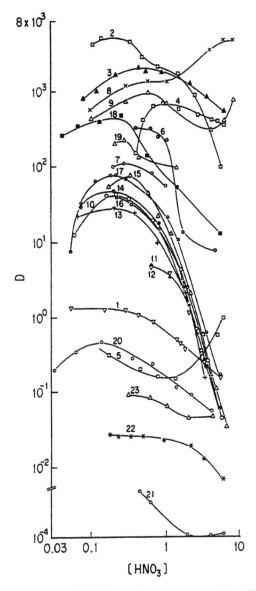

Figure 4 Distribution ratios of HLLW constituents between 30 vol% TRPO–kerosene and HNO₃ of different concentrations (25°C). 1, HNO₃; 2, Th⁴⁺; 3, UO₂²⁺; 4, Np⁴⁺; 5, NpO₂⁺; 6, NpO₂⁺; 7, Pu³⁺; 8, Pu⁴⁺; 9, PuO₂²⁺; 10, Am³⁺; 11,Am³⁺; 12, Cm³⁺; 13, Ce³⁺; 14, Nd³⁺; 15, Pm³⁺; 16, Sm³⁺; 17, Eu³⁺; 18, Tc⁷⁺; 19, Zr⁴⁺; 20, Ru⁴⁺; 21, Cs⁺; 22, Sr²⁺; 23, Fe³⁺. Numbers 11 and 12 were simulated liquid waste: RE, 0.05 *M*; Zr, 0.02 *M*; Fe, 0.035 *M*. (From Jiao, R., Wang, S., Fan, S., Liu, B., Zhu, Y., *J. Nucl. Radiochem.*, 7, 65, 1985. With permission.)

30 vol% TRPO–kerosene.[20] More than 99% of U(VI), Np(IV), Np(VI), Pu(IV), and Pu(VI) in 0.2 to 1 *M* HNO₃ can be extracted by one-stage extraction. Also, >95% of Pu(III), Am(III), and lanthanides can be extracted, while much less FP such as Sr, Cs, and Ru are extracted. This makes it feasible to separate actinides and lanthanides in HLLW

from FP effectively. The stripping behavior of actinides loaded in 30% (volume) TRPO–kerosene was studied.[21] Americium and lanthanides were stripped with 5 to 6 M HNO$_3$, Np and Pu with 0.5 M oxalic acid, and U with sodium carbonate. The cross contaminations between the three fractions were very little.

Based on the extraction and stripping results, a flow sheet for the removal of actinides from HLLW by TRPO extraction was proposed recently, as shown in Figure 5(a), and verified by multistage countercurrent cascade extraction experiment with synthetic HLLW.[22] The oxidation state of Np is adjusted to IV by electrolytic reduction in order to improve its removal efficiency. When HNO$_3$ concentration of the feed solution was 1 M, more than 99.9% of the Am, Pu, Np, and U could be recovered from HLLW after a few stages of extraction. The actinides stripped are divided into three groups, Am + lanthanides, Np + Pu, and U. The FP fall into three categories: lanthanides, Pd, Zr, and Mo coextracted with actinides; Tc, Fe, Cr, Ag, and Ru slightly extracted; and Sr, Cs, Ba, Cd, Ni, and Rh scarcely extracted. These results demonstrate that the TRPO process can well meet the demands to remove actinides from HLLW. A hot pilot test is expected to be performed.

TRPO is cheap and available in China. This, along with its high extraction efficiency, makes the process promising. The TRPO process, however, has some disadvantages, i.e., it works only at a relatively low HNO$_3$ concentration (0.1 to 1 M), and therefore, HLLW from the reprocessing has to be denitrated before the extraction. Moreover, lanthanides cannot be separated from transplutonium elements.

B. DHDECMP EXTRACTION PROCESS

Since the early 1980s the extraction of actinides in HLLW by dihexyl-N,N-diethyl carbamyl methylene phosphonate (DHDECMP) has been studied at the China Institute of Atomic Energy.[23-25] On the basis of the extraction behavior of actinides, lanthanides, and FP, the ideal composition of the extraction phase was selected to be 30 vol% DHDECMP–diethyl benzene(DEB), which extracted all actinides and lanthanides from 2 to 6 M HNO$_3$ directly with high efficiency.

Based on the primary studies, a DHDECMP extraction process for the separation of actinides from HLLW was proposed as shown in Figure 5(b). The process has been verified through single-stage and multistage cascade extraction experiments with a simulated HLLW of 3 M HNO$_3$.[23] The cascade experiments showed that in the six extraction stages, the recoveries of U, Np, Pu, Am, and Gd were 99.95, 99.40, 99.95, 99.99, and 99.70%, respectively. With six stripping stages for the recovery of Pu, Am, and Gd, the stripping efficiencies were 96.58, >99.65, and >99.70%, respectively. Then with six stripping stages for Np and U, the efficiencies were 99.95 and 98.02%, respectively. The above results demonstrate that a DHDECMP process is feasible for direct use with acidic HLLW, requiring neither prior denitration nor neutralization. The disadvantages are the fairly high price of the reagent and the inability to separate lanthanides from actinides.

A combined DHDECMP–HDEHP process is being developed at the China Institute of Atomic Energy. This work comprises two tasks: (1) recovery of actinides and lanthanides from HLLW of high acidity, i.e., 2 to 6 M HNO$_3$, by DHDECMP–DEB, and (2) further separation of lanthanides from actinides in less acidic solution, i.e., 0.2 to 0.5 M HNO$_3$, by HDEHP. It can be seen from the results obtained that actinides can be recovered with an efficiency higher than 99.9%, while more than 95% of lanthanides remain in the aqueous phase.[26]

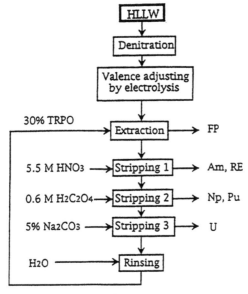

(a) TRPO extraction process

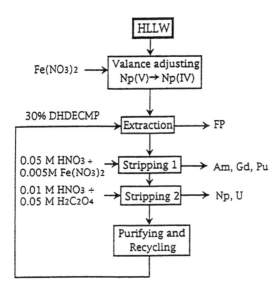

(b) DHDECMP extraction process

Figure 5 Schematic flow sheet of TRPO extraction (a) and DHDECMP extraction (b) processes for the removal of actinides and lanthanides from HLLW. (Figure 5(a) from Song, C., Xu, J., Zhu, Y., *J. Nucl. Radiochem.*, 14, 193, 1992. With permission.)

C. CROWN ETHER EXTRACTION

China started the research on the extraction of actinides by crown ethers, for elemental as well as isotopic separations, in 1980. A series of such crown ethers as dibenzo crown ethers (DB-18C6, -24C8) and dicyclohexyl crown ethers (DCH-18C6, -21C7, -24C8, -27C9, -30C10) were studied for separating U and Pu from HCl and HNO3 media. The DCH-crown ethers diluted in 1,2-dichloroethane extracted U(IV) effectively from 8 to 8.5 M HCl and U(VI) from about 7 M HCl solution.[27,28] The extraction efficiency of U(VI) from HNO_3 solution was three orders of magnitude lower than that from HCl. The extraction mechanism, complex formation, and thermodynamics were studied for U.[29-31] In this connection, the isotopic separation of $^{140}Ce/^{142}Ce$ and $^{235}U/^{238}U$ by DCH-18C6 and -24C8 were studied,[32,33] and appreciable separation factors were obtained.

Recently, the study on the extraction of Sr by crown ethers was started, for the further treatment of HLLW from which the long-lived alpha-emitting nuclides have been removed. This work is being conducted at the China Institute of Atomic Energy and Institute of Nuclear Energy and Technology of Tsinghua University. Recent results were presented in the 4th National Conference on Nuclear and Radiochemistry, China, 1993.[34] Yang and Han[34] studied the extraction of Sr in HNO_3 and found that DCH-18C6 in 1,1,2,2-tetrachloroethane is promising for the extraction of Sr in 1.4 to 2.0 M HNO_3. With single-stage back-extraction with water, the stripping ratio of Sr was 99%. He et al.[34] investigated the extraction of Sr in HLLW with DCH-18C6 in octanol and found that this system showed fairly large selectivity for Sr, with only a small part of K and Mo being coextracted and other cations like Na, Cs, Ni, Fe, Cr, Al, etc. being hardly extracted. It was found that the ions Al^{3+} and Fe^{3+} have a salting-out effect on the extraction. A cascade experiment demonstrated that the removal of Sr from the synthetic waste was 99% with four-stage extraction and four-stage stripping processes.

D. MISCELLANEOUS

For a method to remove Cs, related research was started recently. The extraction of Cs^+ with $[(\pi\text{-}B_9C_2H_{11})_2Co]^-$ anion[34] and ion exchange with titanium ferrocyanide and zinc ferrocyanide[34] are being studied. Besides the main research and development mentioned above, a number of related studies has been performed for the separation or recovery of nuclides from liquid nuclear wastes. For instance, the absorption, ion exchange, liquid-membrane methods, solvent extraction, and synergistic extraction with binary or ternary systems have been studied.

III. KOREA

The dramatic economic growth of the Republic of Korea over the past three decades has been accompanied by rapid increase in electricity demand, with nuclear power having assumed a key role in meeting this demand. The country's first nuclear power plant, Kori-1, went into commercial operation in 1978. Today Korea has nine nuclear power plants (eight PWRs and one CANDU) in operation with a generating capacity of 7616 MWe. In addition, two PWRs are under construction, three units (two PWRs and one CANDU) are under design, and five CANDUs and eight PWRs are planned with total capacity of about 23,000 MWe in 2006.

The cumulative wastes from nuclear power operation amounted to about 42,585 drums by the end of 1993 and will reach up to approximately 170,000 drums by 2000. The capacity of storage facilities in nuclear power plants will reach the limit in the near future. Under these circumstances, a centralized repository for low-level radioactive waste

(LLW) is scheduled to be constructed by the end of 1995. The cumulative amount of spent fuels reached 2168 MTU at the end of 1993. Spent fuel assemblies are currently stored in pools at reactor sites.

In July 1988, the Korea Atomic Energy Commission (AEC) set two main goals for radioactive waste management: (1) a repository for LLW will be constructed by December 1995, and (2) an away-from-reactor interim storage facility for spent fuels will be constructed by December 1997. In December 1988, the AEC approved the plans that (1) the first repository for LLW will be a rock cavern type with a capacity of 25,000 drums and that (2) the first wet-type interim storage facility with a capacity of 3000 MTU will be built in a coastal area. In mid-1986, a regional survey for a repository site was initiated to identify suitable areas. However, siting processes have not been successful.

The Korea Atomic Energy Research Institute (KAERI) is responsible for all nuclear related R & D and has been designated to carry out the radioactive waste management projects. KAERI is currently collecting radioisotope wastes that will be ultimately disposed of in a repository when available. Current major tasks are to acquire a site for central management of radioactive wastes and to construct a repository for the low- and intermediate-level wastes and spent fuel interim storage facility.

After the first research reactor, TRIGA Mark-II, reached critical condition in 1962, the research and development on the separation of liquid wastes using ion-exchange resins started in Korea. In the 1980s the separation technologies, especially for LLW, were our concerns.

A. NATURAL EVAPORATION

Liquid radioactive waste from research laboratories at KAERI is to be treated at the radioactive waste treatment facility. After volume reduction by evaporation, the condensate, a very low-level liquid waste, as well as laundry wastes were planned to be discharged into the environment after dilution. The zero release concept was introduced because of the absence of important dilution rivers near the KAERI site. Evaporation characteristics and environmental impact assessment of a nonboiling forced draft evaporation system were performed.[35] A full-scale evaporation facility of 1 ton/h is in operation at KAERI. The decontamination factors of the system vary between 10^3 to 10^4.

B. TREATMENT OF LAUNDRY WASTES BY ULTRAFILTRATION AND REVERSE OSMOSIS

Fundamental and pilot-scale experiments were conducted to develop a laundry waste treatment system that is composed of a preconcentration step with a reverse osmosis (RO) unit, a volume reduction step with ultrafiltration (UF), and the final purification step with a RO unit.[36] At the first RO unit, the waste was concentrated over its critical micelle concentration (CMC) on the basis of its surfactant concentration. The performance of the UF process was investigated by adsorption experiments of radionuclides on the micellar surface and the separation of the micelles. The RO/UF combination method is promising for the treatment of radioactive laundry wastes. The operation of the process is characterized by the surfactant concentration in the laundry wastes. The RO process is available for the treatment of low concentration of surfactant in the wastes. The degree of volume reduction in the RO unit is limited by the concentration of surfactant in the wastes. A reasonable surfactant concentration of 2000 mg/l is obtained, which is a little higher than CMC, where good decontamination factors of ^{60}Co and ^{137}Cs are achieved and the permeation rate is not decreased significantly. The wastes that are preconcentrated above the CMC are treated efficiently by a UF membrane. This behavior is directly linked with the micelle formation and the binding degree of the radionuclides to the micelle surface. Under the experimental conditions, the overall volume reduction factor was 250 and the average decontamination factors for ^{60}Co and ^{137}Cs were 110 and 20, respectively.

C. REMOVAL OF ^{60}CO FROM LIQUID RADIOACTIVE WASTE CONTAINING SEA SALT

The removal of ^{60}Co from liquid radioactive waste of high salt concentration by precipitation and filtration was studied.[37] From the experimental results it can be concluded that Co in the waste containing sea salt is effectively removed by the precipitation with added alkali and filtration. The poor filtration property of the precipitate was improved by the use of an appropriate polymer flocculant to aid agglomeration of the suspended particles.

D. SEPARATION OF CS AND SR USING ZEOLITE

The characteristics of mass transfer in an inorganic ion-exchange system where Cs and/or Sr are separated by means of zeolites was investigated.[38,39] Batchwise experiments were carried out using various kinds of zeolites. It was found that the mass transfer rate was controlled mainly by liquid-film diffusion. The mass transfer coefficients in the film were in the range of 10^{-3} to 10^{-4} cm/s, while the apparent diffusivity inside the particles was found to be on the order of 10^{-6} cm^2/s. The best type of zeolite is type AW500 for the recovery of Cs and 13X for Sr in terms of ion-exchange capacity. Experiments on the temperature effect showed that the capacity is not influenced by the temperature in the AW500-Cs, AW300-Cs, and 13X-Sr systems. The distribution coefficients increased with an increase of pH of the solution, and were about 10^3 and 10^3 to 10^4 cm^3/g for the AW500-Cs and 4A-Sr systems, respectively.

REFERENCES

1. *1993 World Directory of Nuclear Utility Management*, Sixth Edition, Payne, J., Ed., American Nuclear Society Inc., Illinois, July 1993.
2. Kubota, M., Nakamura, H., Tachimori, S., Abe, T., Amano, H., Removal of transplutonium elements from high-level waste, Paper IAEA-Sm-246/24, Proc. Int. Symp. Management of Alpha-Contaminated Wastes, Vienna, Austria, 1981.
3. Kubota, M., Yamaguchi, I., Okada, K., Morita, Y., Nakano, K., Nakamura, H., Partitioning of high-level waste as pretreatment in waste management, in *Proc. Scientific Basis of Nuclear Waste Management VII*, Boston, McVay, G. L., Ed., Mat. Res. Soc. Symp. Proc., Vol. 26, Elsevier, New York, 1984, 551.
4. Morita, Y., Kubota, M., Behavior of neptunium in chemical process of partitioning long-lived radionuclides from high-level waste, *J. Nucl. Sci. Tech.*, 22, 658, 1985.
5. Morita, Y., Kubota, M., Extraction of neptunium with di-isodecyl phosphoric acid from nitric acid solution containing hydrogen peroxide, *Solv. Extr. Ion Exch.*, 6, 233, 1988.
6. Kubota, M., Dojiri, S., Yamaguchi, I., Morita, Y., Yamagishi, I., Kobayashi, T., Tani, S., Development of a partitioning method for the management of high-level liquid waste, in *High Level Radioactive Waste and Spent Fuel Management*, Vol. II, Slate, S. C., Kohout, R., Suzuki, A., Eds., The American Society of Mechanical Engineering, New York, 1989, 537.
7. Kawata, T., Ozawa, M., Nemoto, S., Togashi, A., Hayashi, S., Preliminary study on the partitioning of transuranium elements in high level liquid waste, in Proc. Information Exchange Meeting on Actinide and Fission Product Separation and Transmutation, Mito, OECD/NEA, 1991, 186.
8. Horwitz, E. P., Shulz, The TRUEX process: A vital tool for disposal of U.S. defense nuclear waste, in *Proc. New Separation Chemistry for Radioactive Waste and Other Specific Applications*, Rome, Elsevier, New York, 1991, 21.

9. Ozawa, M., Nemoto, S., Togashi, A., Kawata, T., Onishi, K., Partitioning of actinides and fission products in highly active raffinate from purex process by mixer-settlers, *Solv. Extr. Ion Exch.*, 10, 829, 1992.

10. Inoue, T., Sakata, M., Miyashiro, H., Matsumura, T., Sasahara, A., Yoshiki, N., *Nucl. Technol.*, 93, 206, 1991.

11. Sakata, M., Miyashiro, H., Inoue, T., Development of pyrometallurgical partitioning of transuranium elements prior to transmutation in metallic fuel FBR, in Proc. Int. Conf. Fast Reactors and Related Fuel Cycles, Vol. II, Kyoto, Japan, 1991.

12. Sakata, M., Hijikata, T., Inoue, T., Equilibrium distribution of rare earth elements between molten KCl-LiCl eutectic salt and liquid cadmium, *J. Nucl. Mater.*, 185, 56, 1991.

13. Kimura, T., Kobayashi, Y., Akatsu, J., Separation and determination of neptunium, plutonium, and americium using coprecipitation with barium sulfate, *Radiochim. Acta*, 39, 179, 1989.

14. Kimura, T., Sequential separation of neptunium, plutonium, americium, and curium using coprecipitation with bismuth phosphate, *J. Radioanal. Nucl. Chem., Articles*, 139, 307, 1990.

15. Kimura, T., Extraction chromatography in the TBP-HNO$_3$ system, II. Characteristics of the TBP/XAD-4 resin on separation of actinide elements, *J. Radioanal. Nucl. Chem., Articles*, 141, 307, 1990.

16. Kimura, T., Akatsu, J., Extraction chromatography in the DHDECMP-HNO$_3$ system. II. Characteristics of the DHDECMP/XAD-4 resin on separation of trivalent actinide elements, *J. Radioanal. Nucl. Chem., Articles*, 149, 25, 1991.

17. Takeshita, K., Kumagai, M., Takashima, Y., Matsumoto, S., Koga, S., Endo, Y., Development of CMP-impregnated solid extractant for TRU removal from radioactive liquid wastes, *Inst. Chem. Eng. Symp. Ser.*, 119, 95, 1990.

18. Yoshida, Z., Aoyagi, H., Mutoh, H., Takeishi, H., Sasaki, Y., Uno, S., Tachikawa, E., Spent-fuel reprocessing based on electrochemical extraction process (SREEP), in Proc. Actinides-93 Int. Conf., Santa Fe, September 1993, *J. Alloy. Comp.*, 213:214, 453, 1994.

19. Motojima, K., Removal of ruthenium from PUREX process, *J. Nucl. Sci. Tech.*, 26, 358, 1989.

20. Jiao, R., Wang, S., Fan, S., Liu, B., Zhu, Y., Trialkyl(C$_6$-C$_8$) phosphine oxide for the extraction of actinides and lanthanides from high active waste, *J. Nucl. Radiochem.*, 7, 65, 1985.

21. Zhu, Y., Song. C., Recovery of Np, Pu, and Am from highly active waste; trialkyl phosphine oxide extraction, in *Transuranium Elements, A Half Century*, Morss, L. R., Fuger, J., Eds., American Chemical Society Books, Washington, D.C., 1992, chap. 32.

22. Song. C., Xu, J., Zhu, Y., The removal of actinide elements from high level radioactive waste by trialkyl phosphine oxide (TRPO) extraction; the stripping of actinide elements from loaded TRPO organic phase, *J. Nucl. Radiochem.*, 14, 193, 1992.

23. Zhao, H., Ye, Y., Yang, X., Lin, Z., Extraction of Np(IV), Pu(IV), and Am(III) by bidentate organophosphorus extractant, *Inorg. Chim. Acta*, 94, 189, 1984.

24. Zhao, H., Fu, L., Wei, X., Liu, S., Ye, G., Yang, L., Jiang, J., Separation of actinides and lanthanides from nuclear power reactor fuel reprocessing waste by bidentate organophosphorous extractant, *At. Ener. Sci. Tech.*, 24, 66, 1990.

25. Zhao, H., Progress on the separation of actinides and lanthanides from high-level waste, *At. Ener. Sci. Tech.*, 27, 277, 1993.

26. Shu, R., Zhao, H., Hu, J., Huang, H., Study on the separation of actinides and lanthanides recovered from from high-level liquid waste (HLLW) by CMP process, *At. Ener. Sci. Tech.*, 27, 242, 1993.

27. Xu, S., Zhang, W., Gu, Z., Extraction of uranium(VI) with a series of five DCH (dicyclohexyl) crown ethers in HCl medium, *At. Ener. Sci. Tech.*, 20, 425, 1986.

28. Zhang, W., Xu, S., Han, Y., The study of the extraction of uranium(IV) chloride with crown ethers, *At. Ener. Sci. Tech.*, 20, 420, 1986.

29. Wang, W. J., Chen, B., Solvent-extraction complex of uranium(VI) with cis, syn, cis-dicyclohexano-18-crown-6, *Inorg. Chim. Acta*, 117, 81, 1986.

30. Chao, W., Luo, W., Study on the extraction of uranyl chloride with crown ethers, *At. Ener. Sci. Tech.*, 21, 456, 1987.

31. **Wang, W. J., Lin, J., Wang, A.,** Thermodynamics and coordination characteristic of the hydro-nium-uranium(VI)-dicyclohexano-24-crown-8 extraction complex, *Inorg. Chim. Acta,* 149, 151, 1988.

32. **Wen, X., Luo, W., Wang, D.,** Isotope separation of cerium and uranium with crown ethers, *J. Nucl. Radiochem.,* 8, 118, 1986.

33. **Han, Y., Luo, W., Gao, S., Guo, J.,** The isotope separation effect of uranium by extraction with dicyclohexyl-24-crown-8, *J. Nucl. Radiochem.,* 15, 56, 1993.

34. Abstracts of the 4th National Conference on Nuclear and Radiochemistry, Xi'an, China, 1993.

35. **Jung, K.J., Lee, K.W., Yoo, J.W., Kim, J.H., Park, H.H.,** A study on the natural evaporation system for the treatment of the VLAW, in Proc. SPECTRUM '90, 481, 1990.

36. **Lee, K.W., Park, S.C., Park, H.H., Kim, J.H.,** Treatment of laundry wastes by the combination of ultrafiltration and reverse osmosis, in Proc. 1993 Int. Conf. on Nuclear Waste Management and Environmental Remediation, 733, 1993.

37. **Yim, S.P., Kim, Y.M., Park, H.H., Shin, J.I., Kim, J.H.,** Removal of ^{60}Co from the liquid radioactive waste in high salt concentration, in Proc. WM '93, 1673, 1993.

38. **Lee, E.H., Lee, W.K., Yoo, J.H., Park, H.S.** Separation behavior of Cs and Sr on the various zeolites, *J. Korean Ind. Eng. Chem.,* 4, 731, 1993.

39. **Lee, E.H., Yoo, J.H., Park, H.S.,** Ion exchange kinetics for Cs and Sr in batch zeolite system, *J. Korean Ind. Eng. Chem.,* 4, 739, 1993.

INDEX